Long-Time Behavior of Second Order Evolution Equations with Nonlinear Damping

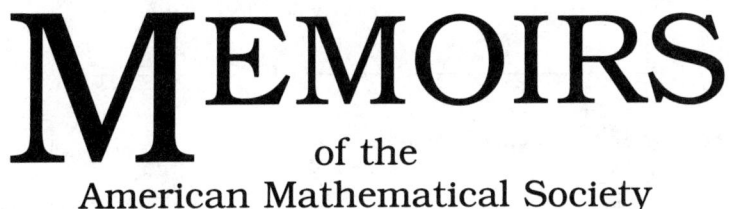
of the
American Mathematical Society

Number 912

Long-Time Behavior of Second Order Evolution Equations with Nonlinear Damping

Igor Chueshov
Irena Lasiecka

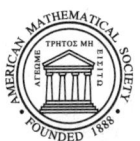

September 2008 • Volume 195 • Number 912 (third of 4 numbers) • ISSN 0065-9266

American Mathematical Society
Providence, Rhode Island

2000 *Mathematics Subject Classification.* Primary 37L30; Secondary 34G20, 47H20.

Library of Congress Cataloging-in-Publication Data

Chueshov, Igor, 1951–
  Long-time behavior of second order evolution equations with nonlinear damping / Igor Chueshov, Irena Lasiecka.
      p. cm. — (Memoirs of the American Mathematical Society, ISSN 0065-9266 ; no. 912)
    "Volume 195, number 912 (third of 4 numbers)."
    Includes bibliographical references and index.
    ISBN 978-0-8218-4187-7 (alk. paper)
    1. Attractors (Mathematics) 2. Evolution equations, Nonlinear. 3. Differentiable dynamical systems. I. Lasiecka, I. (Irena), 1948– II. Title.
  QA614.813.C49   2008
  514′.74—dc22   2008020750

Memoirs of the American Mathematical Society

This journal is devoted entirely to research in pure and applied mathematics.

Subscription information. The 2008 subscription begins with volume 191 and consists of six mailings, each containing one or more numbers. Subscription prices for 2008 are US$675 list, US$540 institutional member. A late charge of 10% of the subscription price will be imposed on orders received from nonmembers after January 1 of the subscription year. Subscribers outside the United States and India must pay a postage surcharge of US$38; subscribers in India must pay a postage surcharge of US$43. Expedited delivery to destinations in North America US$53; elsewhere US$130. Each number may be ordered separately; *please specify number* when ordering an individual number. For prices and titles of recently released numbers, see the New Publications sections of the *Notices of the American Mathematical Society*.

Back number information. For back issues see the *AMS Catalog of Publications*.

Subscriptions and orders should be addressed to the American Mathematical Society, P. O. Box 845904, Boston, MA 02284-5904, USA. *All orders must be accompanied by payment.* Other correspondence should be addressed to 201 Charles Street, Providence, RI 02904-2294, USA.

Copying and reprinting. Individual readers of this publication, and nonprofit libraries acting for them, are permitted to make fair use of the material, such as to copy a chapter for use in teaching or research. Permission is granted to quote brief passages from this publication in reviews, provided the customary acknowledgment of the source is given.

Republication, systematic copying, or multiple reproduction of any material in this publication is permitted only under license from the American Mathematical Society. Requests for such permission should be addressed to the Acquisitions Department, American Mathematical Society, 201 Charles Street, Providence, Rhode Island 02904-2294, USA. Requests can also be made by e-mail to `reprint-permission@ams.org`.

Memoirs of the American Mathematical Society (ISSN 0065-9266) is published bimonthly (each volume consisting usually of more than one number) by the American Mathematical Society at 201 Charles Street, Providence, RI 02904-2294, USA. Periodicals postage paid at Providence, RI. Postmaster: Send address changes to Memoirs, American Mathematical Society, 201 Charles Street, Providence, RI 02904-2294, USA.

© 2008 by the American Mathematical Society. All rights reserved.
Copyright of this publication reverts to the public domain 28 years
after publication. Contact the AMS for copyright status.
This publication is indexed in *Science Citation Index*®, *SciSearch*®, *Research Alert*®,
CompuMath Citation Index®, *Current Contents*®/*Physical, Chemical & Earth Sciences*.
Printed in the United States of America.

∞ The paper used in this book is acid-free and falls within the guidelines
established to ensure permanence and durability.
Visit the AMS home page at `http://www.ams.org/`

10 9 8 7 6 5 4 3 2 1   13 12 11 10 09 08

Contents

Preface   viii

Chapter 1. Introduction   1
   1.1. Description of the problem studied   1
   1.2. The model and basic assumption   4
   1.3. Well-posedness   8

Chapter 2. Abstract results on global attractors   17
   2.1. Criteria for asymptotic smoothness of dynamical systems   18
   2.2. Criteria for finite dimensionality of attractors   22
   2.3. Exponentially attracting positively invariant sets   28
   2.4. Gradient systems   32

Chapter 3. Existence of compact global attractors for evolutions of the second order in time   38
   3.1. Ultimate dissipativity   39
   3.2. Asymptotic smoothness: the main assumption   53
   3.3. Global attractors in subcritical case   56
   3.4. Global attractors in critical case   63

Chapter 4. Properties of global attractors for evolutions of the second order in time   90
   4.1. Finite dimensionality of attractors   90
   4.2. Regularity of elements from attractors   101
   4.3. Rate of stabilization to equilibria   113
   4.4. Determining functionals   120
   4.5. Exponential fractal attractors (inertial sets)   122

Chapter 5. Semilinear wave equation with a nonlinear dissipation   125
   5.1. The model   125
   5.2. Main results   127
   5.3. Proofs   132

Chapter 6. Von Karman evolutions with a nonlinear dissipation   140
   6.1. The model   140
   6.2. Properties of von Karman bracket   141
   6.3. Abstract setting of the model   142
   6.4. Model with rotational forces: $\alpha > 0$ 144
   6.5. Non-rotational case $\alpha = 0$ 152

Chapter 7. Other models from continuum mechanics   158

7.1.	Berger's plate model	158
7.2.	Mindlin-Timoshenko plates and beams	164
7.3.	Kirchhoff limit in Mindlin-Timoshenko plates and beams	167
7.4.	Systems with strong damping	173

Bibliography 179

Index 183

Abstract

We consider abstract nonlinear second order evolution equations with a *nonlinear damping*. Questions related to long time behaviour, existence and structure of global attractors are studied. Particular emphasis is put on dynamics which - in addition to nonlinear dissipation - have noncompact semilinear terms and whose energy may not be necessarily decreasing. For such systems we first develop a general theory at the abstract level. This general theory is then applied to nonlinear wave and plate equations exhibiting the aforementioned characteristics. This way we are able to provide new results pertaining to several open problems in the area of structure and properties of global attractors arising in this class of PDE dynamics.

Received by the editor September, 2004.

1991 *Mathematics Subject Classification.* Primary 37L30; Secondary 34G20, 47H20.

Key words and phrases. global attractors, dimension, structure of attractors, nonlinear dissipation, asymptotic smoothness.

Research of the second author was partially supported by the NSF Grant DMS-0104305 and ARO Grant DAAD19-02-10179.

Preface

Our main aim is the study of long time behavior of second order evolutions with *nonlinear dissipation*. Questions such as global dissipativity of evolution, existence of global attractors and their properties (structure, dimensionality etc.) are of prime concern to this work.

It is known, that nonlinear dissipation in hyperbolic and hyperbolic-like dynamics, as exhibited by second order evolution, has been a source of technical difficulties in the analysis of dynamical systems. There are two fundamental reasons for this. First - the lack of regularity, hyperbolic flows with nonlinear dissipation are not C^1-maps. Second - stability. The nature of instability in hyperbolic dynamics is intrinsically infinite-dimensional. Conservative (energy preserving) dynamics have infinitely many eigenvalues that are unstable. Thus long time behaviour is associated with drastic changes of essential part of the spectrum of the linearized operator (in contrast with parabolic-like dynamics). The dissipation to be effective must be non-compact. This makes the analysis of long time behaviour more delicate and contributes to an array of technical difficulties. For instance, in a canonical case of wave equation with a nonlinear dissipation, while there are several results on the existence of attractors, the study of dimensionality of the attractors seems a much more subtle issue with only few results available in the literature. This motivates our efforts to study attractors arising in the context of canonical hyperbolic and hyperbolic-like dynamics. Our benchmark motivating problems are wave and plate equation with a nonlinear damping and "critical" nonlinear terms. By this we mean that semilinear terms in the equation are not compact with respect to the topology of the phase space. In addition, dynamical systems that are not necessarily subjected to conservative forces (i.e. the energy may not be decreasing) are considered as well. A prototype for this kind of models are von Karman type of evolutions discussed in details in this monograph.

The monograph is written in a rather self-contained manner and contains proofs or well documented references for all the results claimed or used.

Acknowledgement. The work of the second author was done in part while visiting Scuola Normale Superiore, Pisa, Italy. The hospitality and support of the Scuola Normale is acknowledge and greatly appreciated. Particular thanks are extended to Professor G. Da Prato for many stimulating discussions.

Igor Chueshov, Irena Lasiecka

CHAPTER 1

Introduction

1.1. Description of the problem studied

Let $\mathcal{A}$ and M be linear positive, selfadjoint operators densely defined on a Hilbert space $\mathcal{H}$. Let $V \equiv \mathcal{D}(M^{1/2}) \subset \mathcal{H}$ and we shall assume that $\mathcal{D}(\mathcal{A}^{1/2}) \subset V$. We consider the following second-order abstract equation:

(1.1) $$\begin{cases} Mu_{tt}(t) + \mathcal{A}u(t) + k \cdot D(u_t(t)) = F(u(t), u_t(t)), \\ u|_{t=0} = u_0 \in \mathcal{D}(\mathcal{A}^{1/2}),\ u_t|_{t=0} = u_1 \in V = \mathcal{D}(M^{1/2}), \end{cases}$$

where k is a positive parameter, and operators D and F satisfy suitable hypotheses formulated in Assumption 1.1 below. These hypotheses guarantee, in particular, the existence and uniqueness of a flow $S_t(u_0, u_1) \equiv (u(t), u_t(t))$ generated by the dynamics in (1.1). The focus of this monograph is on long time behaviour of the model in (1.1). Issues such as existence and properties of global attractors, exponential attractors and determining functionals are of particular interest. The emphasis is paid to flows generated by equations with *nonlinear dissipation D* and a *non-compact nonlinear term F*. In addition, dynamical systems with energy function that is *not necessarily decreasing* will be considered as well. It is well known, these features lead to rather substantial mathematical difficulties in the analysis of long time behaviour of hyperbolic-like flows (as considered above) -particularly at the level of studying dimensionality and structural properties of the said attractors. General and rich theory developed for dynamical systems [4, 59, 61, 101] is often non applicable to these more special classes of dynamics. Indeed, majority of results in the area of dynamical system theory are geared to parabolic-like flows which display very different (from hyperbolic) properties. While some of these general theories can still be applied to flows generated by hyperbolic equations, more delicate situations of nonlinear damping which lead to flows which are not C^1, can not be treated within that framework. On the other hand, many physically important mathematical models, such as von Karman evolutions of nonlinear elasticity, nonlinear plate and wave equations display all of the above mentioned specifications: nonlinearity of dissipation which is often not quantified at the origin, non-monotone behaviour of energy and the lack of compactness of the nonlinear term in the equation. This motivates our interest in the problem and our desire to construct rather general treatment which is geared toward hyperbolic-like dynamics with nonlinear dissipation and encompasses several models of physical interest.

To accomplish this, we shall first formulate and prove appropriate generalizations of several abstract results in the area of dynamical systems. This will be done with an eye to apply these results to a class of hyperbolic like PDE models we have in mind. The second task is to prove - by PDE methods - that the imposed abstract conditions are indeed satisfied for the class of evolutions in (1.1). Thus, the contribution of this monograph is at two levels: (i) the abstract one within the realm of dynamical system theory and (ii) the PDE part, where verification of needed conditions requires new PDE estimates for hyperbolic-like evolutions. The final goal and task is to verify that specific models of mathematical physics under considerations in this monograph (wave equations, von Karman evolutions etc.) fit the framework described. Ultimately, we will be able to solve and answer several open questions which have been asked in the literature in the context of long time behaviour of specific nonlinear hyperbolic-like evolutions. As far as concrete PDE models are concerned, we have chosen here nonlinear wave equations and nonlinear plate equations as the main (motivating) examples. These two models are not only important from the stand point of applications, but also representative from the theory point of view. They are rich enough to display major difficulties encountered in the analysis of long time behaviour of hyperbolic-like dynamics. On the other hand, we wish to stress that techniques introduced in the book are more general and clearly apply to other models as well.

To orient the reader, we shall outline the content of the monograph and the goals we aim to achieve. Chapter 1 introduces evolutions described by (1.1) and provides preliminary background material related to quantitative properties of solutions under consideration. Chapter 2 describes abstract results pertinent to existence and properties of attractors within the context of general dynamical systems. These abstract results are then applied to the second order evolutions given by (1.1). This is done in Chapter 3, which provides results on existence of global attractors, and in Chapter 4, which describes various structural properties and characteristics of the attractors. More specifically, for evolutions given by (1.1), under appropriate hypotheses to be specified later within the context, Chapters 3 and 4 provide general results on:

- Existence of global attractors.
- Estimates for fractal dimension of attractors.
- Regularity of elements on attractors.
- Uniform convergence rates (e.g., exponential, algebraic etc.) of a single trajectory to an equilibrium and bounded sets to attractors.
- Existence of exponential attractors (inertial sets) and existence o determining functionals.

The final part of the book -Chapters 5, 6 and 7- deal with applications of the general theory. We start with two benchmark motivating models (i) semilinear wave equation with nonlinear damping and local nonlinearity, treated in Chapter 5, and (ii) von Karman system of elasticity with a nonlinear dissipation, treated in Chapter 6. In this latter case, we shall consider two distinct sub- models displaying very different properties: von Karman evolutions with rotational forces, which is characterized by finite speed of propagations, and (ii) von Karman evolutions without rotational inertia, characterized by the infinite speed of propagation. As we shall see, these two models driven by nonlocal nonlinear terms are vastly different and require different treatments. The obtained results are likewise different with physical characteristics of the respective systems coming to play.

In Chapter 7 we consider several other examples from continuum mechanics which can be also covered by our general theory. These examples demonstrate other types (in comparison with wave and Karman equations) of nonlinearities and dampings and include Berger, Mindlin-Timoshenko and Kichhoff models of plates and also systems with strong damping.

In all the cases considered our main focus is on understanding the role *nonlinearity of dissipation* plays in the theory. As mentioned before, nonlinearity of dissipation along with hyperbolicity of dynamic and the lack of compactness for nonlinear terms (sources) in the equations have been a source of considerable difficulties and obstacles in constructing comprehensive theory for this class of systems. In fact, majority of results and treatments deal with *linear dissipation* [**4, 19, 59, 61, 101**]. In the case of linear damping, difficulties related to the loss of regularity of linearized solutions and time-dependency of the principal part of the operator do not enter the picture. An excellent review of the theory pertaining to wave equation with *linear* damping is given in [**5**]. Thus, in treating nonlinear dissipation, our aim is to extract general properties of such systems that can be formulated at an abstract level, and at the same time are specific enough to capture the key characteristics of the models serving as motivating examples for the theory. In doing this, we were able to "abstract" systems with nonlinear damping and noncompact nonlinear terms that are also non-conservative (i.e. the energy is not necessarily decreasing) for which existence of global attractors and their properties can be asserted (see Chapter 3 and Chapter 4). In dealing with nonlinear dissipation we have made a systematic effort to work within minimal set of assumptions that are imposed on behaviour of nonlinear functions describing dissipation both at the origin and at infinity. Indeed, behaviour at infinity affects topological considerations (unboundedness of the resulting operators acting on appropriate functional spaces), while behaviour at the origin affects the "strength" of the damping action. Dissipation with derivative that vanishes at the origin has limited power in stabilizing dynamics.

Similarly, we have made an effort to display and to analyze the consequences of "unstructured" terms in the dynamics such as "non-conservative" forces that destroy monotonicity of the energy function. Indeed, for these models the energy is no longer decreasing -an assumption made in most - if not all - treatise on long behaviour of PDE's models. It turns out that non-conservative static forces (affecting the displacement) can be compensated by certain behaviour (hidden superlinearity) of potential energy, while non-conservative dynamic forces (affecting the velocity) require certain superlinearity of kinetic energy or of the damping. The first phenomenon is strongly linked to the uniqueness property of nonlinear part of potential energy which, in turn, can be asserted in some cases via nonlinear elliptic theory. The ultimate goal and task is to be able to answer several open questions that have been asked in the context of these hyperbolic equations. In fact, the results of Chapter 5 (resp. Chapter 6) provide an answer to several questions that have been open for nonlinear wave (resp. Karman plate) equations. These include: (i) finite-dimensionality of attractors with strongly nonlinear dissipation, (ii) uniform stabilization of solution to multiple equilibria, with a nonlinear and weak (at the origin) damping, (iii) existence of exponentially attracting finite-dimensional inertial sets. Indeed, finite dimensionality of attractors with a nonlinear dissipation has been known only, until recently, for wave equation in one dimension. Other properties such as convergence to multiple equilibria or exponentially attracting inertial sets have been considered only in the context of *linear* dissipation.

Finally we should mention that technical inspiration for our results and methods of proofs comes from recent developments in control theory. Certain inverse type of estimates such as observability and stabilizability estimates that became an object of recent interest in that field turned out to have critical bearing on derivation of asymptotic estimates in the theory of dynamical systems. Thus there seem to be present a strong synergy between the two fields that we hope to explore even further.

In closing, we wish to mention that although our aim was to provide a rather comprehensive treatment of the subject that is also self-contained (essentially all the proofs are provided), we were unable to touch certain topics that are of significant interest to the field. These include: (i) problems with boundary dissipation or related (though easier) localized (geometrically) dissipation, (ii) problems involving non-unique solutions. The reason why we do not consider boundary dissipation is that techniques involved are very much model dependent. Presentation of these results at the abstract level seemed to us rather artificial and perhaps unnecessarily cumbersome. However, readers interested in the topic may consult [20, 21, 22, 26, 28, 31] where essentially all the questions asked in this manuscript are solved for wave and von Karman plate equations with boundary dissipation. Regarding the second item above- non-unique solutions- the techniques developed so far are rather specific and model dependent (see, e.g., [3] and [86] and also the recent paper [105] devoted to wave equations with supercritical nonlinearities). A comprehensive survey of results for wave equations with linear damping is given in [5]. Special cases of nonlinear (but heavily structured) damping in wave equation are in [50].

1.2. The model and basic assumption

We consider problem (1.1) within the following framework: $\mathcal{A}$ is a closed, linear positive selfadjoint operator acting on a Hilbert space $\mathcal{H}$ with $\mathcal{D}(\mathcal{A}) \subset \mathcal{H}$. We shall denote by $|\cdot|$ the norm of $\mathcal{H}$; $(\cdot,\cdot)$ will denote scalar product in $\mathcal{H}$. We shall use the same symbol to denote the duality pairing between $\mathcal{D}(\mathcal{A}^{1/2})$ and $\mathcal{D}(\mathcal{A}^{1/2})'$. Let V be another Hilbert space such that $\mathcal{D}(\mathcal{A}^{1/2}) \subset V \subset \mathcal{H} \subset V' \subset \mathcal{D}(\mathcal{A}^{1/2})'$, all injections being continuous and dense, $M \in L(V, V')$, the bilinear form (Mu, v) is symmetric and $(Mu, u) \geq \alpha_0 |u|_V^2$, where $\alpha_0 > 0$ and $(\cdot,\cdot)$ is understood as a duality pairing between V and V'. Hence, $M^{-1} \in L(V', V)$. Setting $\bar{M} = M|_{\mathcal{H}}$ with $\mathcal{D}(\bar{M}) = \{u \in V; Mu \in \mathcal{H}\}$ we have $D(\bar{M}^{1/2}) = V$.

The following standing hypotheses are imposed on the nonlinear terms $D(u_t)$ and $F(u, u_t)$ in (1.1).

ASSUMPTION 1.1. **(D)** The operator $D : \mathcal{D}(\mathcal{A}^{1/2}) \to [\mathcal{D}(\mathcal{A}^{1/2})]'$ is assumed monotone and hemicontinuous with $D(0) = 0$, i.e.

$$(1.2) \qquad (Du - Dv, u - v) \geq 0 \quad \text{for all} \quad u, v \in \mathcal{D}(\mathcal{A}^{1/2})$$

and $\lambda \mapsto (D(u + \lambda v), v)$ is a continuous function from $\mathbb{R}$ into itself. Moreover, we assume that there exists a set $W \subset \mathcal{D}(\mathcal{A}^{1/2})$ such that $Dw \in V'$ for every $w \in W$ and W is dense in V.

(F) The nonlinear operator $F : \mathcal{D}(\mathcal{A}^{1/2}) \times V \to V'$ is locally Lipschitz, i.e.

$$(1.3) \qquad |F(u_1, v_1) - F(u_2, v_2)|_{V'} \leq L(r) \left[|\mathcal{A}^{1/2}(u_1 - u_2)| + |v_1 - v_2|_V \right]$$

for all $|\mathcal{A}^{1/2} u_i|, |v_i|_V \leq r$, where $r > 0$ is arbitrary. We assume that F has the form

$$(1.4) \qquad F(u, v) = -\Pi'(u) + F^*(u, v),$$

where $\Pi(u) = \Pi_0(u) + \Pi_1(u)$ is a C^1-functional on $\mathcal{D}(\mathcal{A}^{1/2})$ and $\Pi'(u)$ stands for Frechet derivative. We also assume that (i) $\Pi_0(u) \geq 0$ is a locally bounded on $\mathcal{D}(\mathcal{A}^{1/2})$, (ii) for every $\eta > 0$ there exists a constant $C_\eta \geq 0$ such that

(1.5) $$|\Pi_1(u)| \leq \eta \cdot \left(|\mathcal{A}^{1/2}u|^2 + \Pi_0(u)\right) + C_\eta, \quad u \in \mathcal{D}(\mathcal{A}^{1/2}),$$

and (iii) the nonlinear mapping $F^*(u,v)$ satisfies the following bound:

(1.6) $$(F^*(u,v),v) \leq c_1 + c_2 \left[|\mathcal{A}^{1/2}u|^2 + \Pi_0(u) + |v|_V^2\right] + k_0(Dv,v)$$

for any $u, v \in \mathcal{D}(\mathcal{A}^{1/2})$, where $c_1, c_2 > 0$ and $0 \leq k_0 < k$ are constants.

The energy associated with the model is the following:
$$\mathcal{E}(u,v) = \frac{1}{2}\left[|\mathcal{A}^{1/2}u|^2 + |M^{1/2}v|^2\right] + \Pi(u).$$

"Formal" energy relation (to be discussed later) takes the form:
(1.7)
$$\mathcal{E}(u(t),u_t(t)) + k\int_0^t (D(u_t)(s),u_t(s))ds = \mathcal{E}(u_0,u_1) + \int_0^t (F^*(u(s),u_t(s)),u_t(s))ds.$$

We note that the first condition in Assumption 1.1 is just monotonicity (and some mild regularity) requirement imposed on the dissipation. Condition (1.3) states local Lipschitz continuity of F with respect to the topology generated by the state space. Conditions (1.5) and (1.6) are meant to imply the boundedness of ground states. Relation (1.4) means that the (nonlinear) force F in our system is splitted into conservative, or potential, ($\tilde{F} = -\Pi'(u)$) and non-conservative (F^*) parts. The conservative force contributes to a non-negative part of the energy $\Pi_0(u)$ while the second part of the energy $\Pi_1(u)$ is of non-definite signature. We note that the presence of non-conservative, even linear, forces (i.e. forces which cannot be represented by potential operators) makes the study of dynamics more complicated. The main reason is that, as seen from (1.7), the energy of the system may fail to be monotone along trajectories (see also Theorem 1.5 and Remark 1.9 below).

Canonical examples which motivate and illustrate Assumption 1.1 are the following two models which are studied in Chapters 5 and 6 with details.

EXAMPLE 1.2 (**Semilinear wave equation**). Let $\Omega \subset \mathbb{R}^n$, $n = 2, 3$, be a bounded, connected set with a smooth boundary Γ. The exterior normal on Γ is denoted by ν. We consider the following equation

(1.8) $$w_{tt} - \Delta w + kg(w_t) + f(w) = f^*(w,w_t) \quad \text{in} \quad Q = [0,\infty) \times \Omega$$

subject to the boundary condition either of Dirichlet type $w = 0$ on $\Sigma \equiv [0,\infty) \times \Gamma$ or else of Robin type $\partial_\nu w + w = 0$ in Σ and the initial conditions $w(0) = w_0$ and $w_t(0) = w_1$. We assume that k is a positive parameter and impose the following assumptions on the nonlinear damping $g(s)$, non-conservative term $f^*(w,w_t)$ and the function $f(s)$:

- g is a continuous, nondecreasing function such that $g(0) = 0$ and there exists a positive constant m such that

(1.9) $$|g(s)| \leq m|s|^p \quad \text{for all} \quad |s| \geq 1,$$

  where $1 \leq p \leq 5$ when $n = 3$ and $1 \leq p < \infty$ for $n = 2$;
- $f^* : H^1(\Omega) \times L_2(\Omega) \to L_2(\Omega)$ is a globally Lipschitz mapping.

- $f \in C^1(\mathbb{R})$ is of the following polynomial growth:

(1.10) $$|f'(s)| \leq M|s|^q, \quad |s| \geq 1,$$

where $q \leq 2$ when $n = 3$ and $q < \infty$ when $n = 2$. Moreover, the following dissipativity condition holds:

(1.11) $$\liminf_{|s| \to \infty} \frac{f(s)}{s} \equiv \mu > -\lambda_1,$$

where $\lambda_1 > 0$ is the first eigenvalue of the operator $-\Delta$ equipped with appropriate boundary conditions.

Problem (1.8) is a special case of evolutionary system (1.1). To see this we set $\mathcal{H} \equiv V = L_2(\Omega)$, $M = I$ is the identity operator, $\mathcal{A} = -\Delta - \mu_0$ with the corresponding boundary conditions, $D(u) = g(u)$ and $F(u,v) = -f(u) - \mu_0 u + f^*(u,v)$. Here above the parameter μ_0 is chosen such that $-\mu < \mu_0 < \lambda_1$.

It is clear that $\mathcal{A}$ is a positive selfadjoint operator and we also have topological identifications $\mathcal{D}(\mathcal{A}^{1/2}) \sim H_0^1(\Omega)$ in Dirichlet case and $\mathcal{D}(\mathcal{A}^{1/2}) \sim H^1(\Omega)$ in Robin case.

We shall verify that the abstract assumptions imposed on the damping and nonlinear terms are indeed satisfied. For this it suffices to consider the case $n = 3$, as $n = 2$ leads to a simpler analysis. By (1.9), we have $|g(s)| \leq C(1 + |s|^5)$ for all $s \in \mathbb{R}$. Therefore, using the embedding $H^1(\Omega) \subset L^6(\Omega)$, we obtain

$$|g(u)|_{[\mathcal{D}(\mathcal{A}^{1/2})]'} \leq C\|g(u)\|_{L^{6/5}(\Omega)} \leq C\left(1 + \|u\|_{L^6(\Omega)}^5\right) \leq C\left(1 + \|u\|_{H^1(\Omega)}^5\right).$$

Thus, the operator D is a mapping from $\mathcal{D}(\mathcal{A}^{1/2})$ into $[\mathcal{D}(\mathcal{A}^{1/2})]'$. It is monotone because $g(s)$ is nondecreasing. The hemicontinuity of D easily follows from Lebesgue convergence theorem and from the embedding $H^1(\Omega) \subset L^6(\Omega)$. In this case we can take $W = \mathcal{D}(\mathcal{A}) \subset H^2(\Omega)$. Thus the operator D satisfies Assumption 1.1(D).

Property (1.3) (with $V = \mathcal{H}$) follows promptly from (1.10) and from the embedding $H^1(\Omega) \subset L^6(\Omega)$. Clearly, that the operator $F(u,v)$ has the representation (1.4) with $F^*(u,v) \equiv f^*(u,v)$ and $\Pi(u) \equiv \int_\Omega \hat{f}(u)dx$, where $\hat{f}$ denotes the antiderivative of $f(u) + \mu_0 u$. It follows from (1.11) that

$$\hat{f}(u) \geq -M_f + \frac{\mu + \mu_0}{4}u^2 \text{ for all } u \in \mathbb{R},$$

where $M_f > 0$ is a constant depending on f. Therefore, the decomposition of Π into the positive part and non-positive is accomplished by setting

$$\Pi_0(u) \equiv \int_\Omega \hat{f}(u)dx + M_f, \quad \Pi_1(u) = -M_f.$$

Property (1.5) holds with $\eta = 0$. Property (1.6) follows from global Lipschitz property assumed for f^*. Thus, all the hypotheses in Assumption 1.1 are satisfied for the model in hand.

We note that the simplest choice of a non-conservative force $f^*(u, u_t)$ obeying (1.6) is to take $f^*(u,v) = D_x u + v$, where D_x is a general first order differential operators in the variables x. Other choices, including superlinear non-conservative forces, are of course possible. However superlinear non-conservative forces typically require more structured form of the energy (see Chapter 6 and Chapter 7).

Detailed analysis of model (1.8) as a dynamical system along with all the consequences of abstract theory will be given in Chapter 5.

EXAMPLE 1.3 (**Von Karman evolution equations**). Let $\Omega \subset \mathbb{R}^2$ be bounded domain with a sufficiently smooth boundary Γ. Consider the following von Karman model with clamped boundary conditions:

$$u_{tt} - \alpha \Delta u_{tt} + k \cdot [g_0(u_t) - \alpha \cdot \mathrm{div} g(\nabla u_t)] + \Delta^2 u$$
$$= [v(u) + F_0, u] + p + L(u, u_t) \text{ in } \Omega \times (0, \infty),$$

(1.12) $$u = \frac{\partial}{\partial \nu} u = 0 \text{ on } \Gamma \times (0, \infty).$$

The Airy stress function $v(u)$ solves the problem

(1.13) $$\Delta^2 v(u) + [u, u] = 0 \text{ in } \Omega, \quad \frac{\partial}{\partial \nu} v(u) = v(u) = 0 \text{ on } \Gamma.$$

The von Karman bracket $[u, v]$ is given by

$$[u, v] \equiv u_{x_1 x_1} v_{x_2 x_2} + u_{x_2 x_2} v_{x_1 x_1} - 2 u_{x_1 x_2} v_{x_1 x_2}.$$

The parameter $\alpha \geq 0$ represents rotational forces. The function $p \in L_2(\Omega)$ represents external transverse forces applied to the plate and $F_0 \in H^2(\Omega)$ describes in-plane forces acting on the plate. The damping parameter k is positive. The operator $L(u, u_t)$ models the non-conservative forces in the plate. The presence of this term is responsible for the fact that energy is *not decreasing*. In typical applications arising in aeroelasticity problems (see, e.g., [8] or [38]) one has $Lu = -\rho u_{x_1}$, where ρ is determined by the velocity of the gas. More generally the operator L may be defined as a first order differential operator in the variables x_i and t. However, other choices (including superlinear) will be discussed later in Chapter 7 for the Berger plate model.

Equations of von Karman (1.12) and (1.13) are well known in nonlinear elasticity and constitute a basic model describing nonlinear oscillations of a plate accounting for large displacements, see, e.g., [83] and references therein. The model which accounts for moments of inertia, i.e. $\alpha > 0$ is often referred as *modified von Karman equations*. The rotational case when $\alpha > 0$ represents a purely hyperbolic dynamics with finite speed of propagation. Instead, when $\alpha = 0$ we have an infinite speed of propagation. Thus, the mathematical properties of these two models are very different.

We impose the following assumptions on the damping functions g_0 and g:
- g_0 is a monotone nondecreasing and continuous function, $g_0(0) = 0$;
- The function $g = (g_1, g_2) : \mathbb{R}^2 \mapsto \mathbb{R}^2$ is a continuous monotone nondecreasing mapping in $\mathbb{R}^2$, i.e.

$$\sum_{i=1,2} (g_i(s_1, s_2) - g_i(s_1', s_2')) (s_i - s_i') \geq 0, \quad (s_1, s_2), (s_1', s_2') \in \mathbb{R}^2.$$

Moreover, we assume that $g(0) = 0$ and g is of polynomial growth at infinity, i.e. we have estimate

$$|g(s)| \leq C(1 + |s|^p), \quad s = (s_1, s_2) \in \mathbb{R}^2, \ |s| = \sqrt{s_1^2 + s_2^2},$$

with some constants $C > 0$ and $p \geq 1$.

In order to rewrite von Karman equations (1.12) and (1.13) as a second order abstract equation of the form (1.1) we introduce the following spaces and operators:
- $\mathcal{H} \equiv L_2(\Omega)$, $V \equiv V_\alpha$, where $V_\alpha = H_0^1(\Omega)$ for $\alpha > 0$ and $V_\alpha = L_2(\Omega)$ for $\alpha = 0$.

- $\mathcal{A}u \equiv \Delta^2 u$, $u \in \mathcal{D}(\mathcal{A})$: $\mathcal{D}(\mathcal{A}) \equiv H_0^2(\Omega) \cap H^4(\Omega)$.
- $Mu \equiv u - \alpha\Delta u$, $u \in \mathcal{D}(M)$, where $\mathcal{D}(M) \equiv H_0^1(\Omega) \cap H^2(\Omega)$ in the case $\alpha > 0$ and $\mathcal{D}(M) \equiv L_2(\Omega)$ when $\alpha = 0$.
- $F(u,w) \equiv [u, v(u) + F_0] + L(u,w) + p$, where $v(u) \in H_0^2(\Omega)$ solves (1.13).
- $D(u) \equiv g_0(u) - \alpha\operatorname{div}(g(\nabla u))$.

It is clear that $\mathcal{A}$ and M satisfy the conditions introduced above.

We also have $F(u,w) = -\Pi'(u) + F^*(u,w)$ with $F^*(u,w) = L(u,w)$ and

$$\Pi(u) = \Pi_0(u) + \Pi_1(u),$$

where

$$\Pi_0(u) = \frac{1}{4}\int_\Omega |\Delta v(u)|^2 dx,\ \Pi_1(u) = -\frac{1}{2}\int_\Omega F_0(x)[u,u](x)dx - \int_\Omega p(x)u(x)dx.$$

It will be verified later that Assumption 1.1(F) in both cases $\alpha > 0$ and $\alpha = 0$ are satisfied. This is a non-trivial fact that has to do with regularity of Airy's stress function and uniqueness property of von Karman bracket (see [**13, 15, 47, 77**]).

As to the damping operator we obviously have that

$$|D(u)|_{V'} \leq \|g_0(u)\|_{L_2(\Omega)} + \alpha\|g(\nabla u)\|_{L_2(\Omega)}.$$

Using the embeddings $H^2(\Omega) \subset L_\infty(\Omega)$ and $H^1(\Omega) \subset L_q(\Omega)$ for every $1 \leq q < \infty$, we find that

$$\|g_0(u)\|_{L_2(\Omega)} \leq C_1 \cdot \sup\left\{|g_0(s)| : |s| \leq C_2\|u\|_{H^2(\Omega)}\right\}$$

and

$$\|g(\nabla u)\|_{L_2(\Omega)} \leq C\left(1 + \|\nabla u\|_{L_{2p}(\Omega)}^p\right) \leq C\left(1 + \|u\|_{H^2(\Omega)}^p\right).$$

Therefore the operator D maps $\mathcal{D}(\mathcal{A}^{1/2})$ into V'. It is monotone, because g_0 and g are monotone and we have the relation

$$(Du, v) = \int_\Omega g_0(u)v dx + \alpha\int_\Omega g(\nabla u)\nabla v dx, \quad u, v \in H_0^2(\Omega) \equiv \mathcal{D}(\mathcal{A}^{1/2}).$$

As in Example 1.2 the hemicontinuity of D follows from the Lebesgue convergence theorem. Thus Assumption 1.1(D) is satisfied with $W = \mathcal{D}(\mathcal{A}^{1/2})$.

Detailed analysis of this model within the context of dynamical system is given in Chapter 6.

The two models introduced above should serve as a guiding prototypes for the type of dynamics and type of abstract assumptions to be considered. In particular, features such as strongly nonlinear dissipation without assumptions on the growth at the origin, non-compact nonlinear forcing terms along with a presence of non-conservative forces are fully displayed in these two cases of wave and plate equations considered.

1.3. Well-posedness

In this section we discuss existence and uniqueness of solutions to problem (1.1) under Assumption 1.1. We rely on the theory of monotone operators (see, e.g., [**6**] and [**95**]).

Below we use the following notations.

Let H be a separable Banach space. We denote by $L_p(a,b;H)$ with $1 \leq p \leq \infty$ the space of (equivalence classes of) Bochner measurable functions $f : [a,b] \mapsto H$ such that $\|f(\cdot)\|_H \in L_p(a,b)$. Each $L_p(a,b;H)$ is a Banach space with the norms

$$\|f\|_{L_p(a,b;H)} = \left(\int_a^b \|f(t)\|_H^p dt\right)^{1/p}, \quad 1 \leq p < \infty,$$

$$\|f\|_{L_\infty(a,b;H)} = \text{esssup}\left\{\|f(t)\|_H \,:\, t \in [a,b]\right\}.$$

We also denote by $C(a,b;H)$ the space of strongly continuous functions with values in H and use the space

(1.14) $$W_p^1(a,b;H) = \{f \in C(a,b;H) \,:\, f' \in L_p(a,b;H)\},$$

where $f'(t)$ is a distributional derivative of $f(t)$ with respect to t. We note that the space $W_1^1(a,b;H)$ coincides with the set absolutely continuous functions from $[a,b]$ into H (see, e.g., [**95**, Theorem 1.7, p.105]).

We will use the following definitions of solutions.

DEFINITION 1.4. A function $u(t) \in C([0,T]; \mathcal{D}(\mathcal{A}^{1/2})) \cap C^1([0,T]; V)$ possessing the properties $u(0) = u_0$ and $u_t(0) = u_1$ is said to be

(S) *strong solution* to problem (1.1) on the interval $[0,T]$, iff
- $u \in W_1^1(a,b; \mathcal{D}(\mathcal{A}^{1/2}))$ and $u_t \in W_1^1(a,b; V)$ for any $0 < a < b < T$;
- $\mathcal{A}u(t) + kDu_t(t) \in V'$ for almost all $t \in [0,T]$;
- equation (1.1) is satisfied in V' for almost all $t \in [0,T]$;

(G) *generalized solution* to problem (1.1) on the interval $[0,T]$, iff there exists a sequence of strong solutions $\{u_n(t)\}$ to problem (1.1) with initial data (u_{0n}, u_{1n}) instead of (u_0, u_1) such that

(1.15) $$\lim_{n\to\infty} \max_{t\in[0,T]} \left\{|M^{1/2}(\partial_t u(t) - \partial_t u_n(t))| + |\mathcal{A}^{1/2}(u(t) - u_n(t))|\right\} = 0.$$

The main result related to well-posedness is the following theorem.

THEOREM 1.5. *Let $T > 0$ be arbitrary. Under Assumption 1.1 the following statements hold.*

- **strong solutions** *For every $(u_0, u_1) \in \mathcal{D}(\mathcal{A}^{1/2}) \times \mathcal{D}(\mathcal{A}^{1/2})$, such that $\mathcal{A}u_0 + kDu_1 \in V'$ there exists unique strong solution to problem (1.1) on the interval $[0,T]$ such that*

(1.16) $$(u_t, u_{tt}) \in L_\infty(0, T; \mathcal{D}(\mathcal{A}^{1/2}) \times V), \quad u_t \in C_r([0,T); \mathcal{D}(\mathcal{A}^{1/2})),$$

$$u_{tt} \in C_r([0,T); V) \quad \text{and} \quad \mathcal{A}u(t) + kDu_t(t) \in C_r([0,T); V'),$$

where we denote by C_r the space of right continuous functions. This solution satisfies the energy relation

(1.17)
$$\mathcal{E}(u(t), u_t(t)) + k\int_0^t (Du_t(\tau), u_t(\tau))d\tau = \mathcal{E}(u_0, u_1) + \int_0^t (F^*(u(\tau), u_t(\tau)), u_t(\tau))d\tau,$$

where

(1.18) $$\mathcal{E}(u_0, u_1) = E(u_0, u_1) + \Pi_1(u_0)$$

and the positive part of the energy $E(u_0, u_1)$ is defined by

$$(1.19) \quad E(u_0, u_1) = \frac{1}{2}\left((Mu_1, u_1) + (\mathcal{A}u_0, u_0)\right) + \Pi_0(u_0) \equiv E_0(u_0, u_1) + \Pi_0(u_0).$$

- **generalized solutions** For every $(u_0; u_1) \in \mathcal{D}(\mathcal{A}^{1/2}) \times V$ there exists unique generalized solution such that
$$(u, u_t) \in C([0, T]; \mathcal{D}(\mathcal{A}^{1/2}) \times V).$$

Both strong and generalized solutions satisfy the estimate

$$(1.20) \quad E(u(t), u_t(t)) \leq c_0\left(1 + E(u_0, u_1)\right)e^{c_1 t}, \quad t \in [0, T].$$

with some constants c_0 and c_1. Under the Lipschitz condition

$$(1.21) \quad |F^*(u_1, v_1) - F^*(u_2, v_2)|_{V'} \leq L(r)\left[|\mathcal{A}^{1/2}(u_1 - u_2)| + |v_1 - v_2|_V\right]$$

for all $|\mathcal{A}^{1/2}u_i|, |v_i|_V \leq r$, where $r > 0$ is arbitrary, these solutions satisfy the energy inequality

$$(1.22) \quad \mathcal{E}(u(t), u_t(t)) \leq \mathcal{E}(u_0, u_1) + \int_0^t (F^*(u(\tau), u_t(\tau)), u_t(\tau))d\tau.$$

This theorem follows from rather standard application of monotone operator theory with locally Lipschitz perturbations. (See also [76] and [78, Chap.2], where general evolutions with boundary dissipations are considered). For the sake of completeness we give a sketch of the main steps of the proof relying on the theory of monotone operators (see [6] or [95]).

PROOF. Let us rewrite problem (1.1) as a first order equation. In order to accomplish this we introduce the following operator $A : \mathcal{D}(A) \subset H \mapsto H$, where $H \equiv \mathcal{D}(\mathcal{A}^{1/2}) \times V$, defined by

$$(1.23) \quad A = \begin{pmatrix} 0 & -I \\ M^{-1}\mathcal{A} & kM^{-1}D \end{pmatrix}$$

with

$$\mathcal{D}(A) = \left\{u = (x, y) \in \mathcal{D}(\mathcal{A}^{1/2}) \times \mathcal{D}(\mathcal{A}^{1/2}) : \mathcal{A}x + kDy \in V'\right\}.$$

With the above notation the original evolution problem (1.1) is equivalent to the equation

$$(1.24) \quad \begin{cases} \frac{d}{dt}U(t) + AU(t) = \begin{pmatrix} 0 \\ M^{-1}F(u(t), u_t(t)) \end{pmatrix} \\ U(0) = U_0 \equiv (u_0, u_1), \end{cases}$$

where $U(t) = (u(t), u_t(t))$.

LEMMA 1.6. *Under Assumption 1.1, the operator A defined in (1.23) is m-accretive.*

We recall (see, e.g., [95, p.18]) that an operator $A : \mathcal{D}(A) \subseteq H \mapsto H$ is called *accretive* iff we have

$$(Ax_1 - Ax_2, x_1 - x_2)_H \geq 0 \quad \text{for any} \quad x_1, x_2 \in \mathcal{D}(A).$$

It is *maximal accretive (m-accretive)* if, in addition, $R(I + A) = H$, where $R(I + A)$ is the range of $I + A$.

1.3. WELL-POSEDNESS

PROOF. STEP 1. We first prove that the operator A is accretive.

We take arbitrary elements $u = (u_1, u_2)$, $w = (w_1, w_2) \in \mathcal{D}(A)$. Let $\xi = (\xi_1, \xi_2) = A(u)$ and $\eta = (\eta_1, \eta_2) = A(w)$. Thus, in particular, $\xi_1 = -u_2, \eta_1 = -w_2$ and
$$\xi_2 = M^{-1}[\mathcal{A}u_1 + kDu_2] \text{ and } \eta_2 = M^{-1}[\mathcal{A}w_1 + kDw_2].$$
Since
$$(A(u) - A(w), u - w)_H = (\xi - \eta, u - w)_H$$
$$= (\mathcal{A}^{1/2}(\xi_1 - \eta_1), \mathcal{A}^{1/2}(u_1 - w_1))_{\mathcal{H}} + (M(\xi_2 - \eta_2), (u_2 - w_2))_{\mathcal{H}},$$
we obtain that
$$(A(u) - A(w), u - w)_H$$
$$= -(\mathcal{A}(u_2 - w_2), u_1 - w_1)_{\mathcal{H}} + (\mathcal{A}(u_1 - w_1) + k(Du_2 - Dw_2), u_2 - w_2)_{\mathcal{H}}$$
$$= k(Du_2 - Dw_2, u_2 - w_2)_{\mathcal{H}} \geq 0,$$
where we have used monotonicity assumptions of D. Thus A is accretive.

STEP 2. To prove maximality of A we need to show that $R(I + A) = H$. The above entails to solving the following system of equations: given $f_0 \in \mathcal{D}(\mathcal{A}^{1/2})$ and $f_1 \in V$ find $(x, y) \in \mathcal{D}(A)$ such that

(1.25) $$-y + x = f_0,$$
$$\mathcal{A}x + kDy + My = Mf_1.$$

Eliminating x yields

(1.26) $$\mathcal{A}y + kDy + My = Mf_1 - \mathcal{A}f_0 \in [\mathcal{D}(\mathcal{A}^{1/2})]'.$$

If we denote $v = \mathcal{A}^{1/2}y$, then we obtain the relation

(1.27) $$v + Sv = \mathcal{A}^{-1/2}(Mf_1 - \mathcal{A}f_0) \in \mathcal{H},$$

where
$$Sv = k\mathcal{A}^{-1/2}D\left(\mathcal{A}^{-1/2}v\right) + \mathcal{A}^{-1/2}M\mathcal{A}^{-1/2}v.$$

It follows from Assumption 1.1(D) that $k\mathcal{A}^{-1/2}D\left(\mathcal{A}^{-1/2}v\right)$ is m-accretive in $\mathcal{H}$. Since $\mathcal{D}(\mathcal{A}^{1/2}) \subset V$, it is also clear that $\mathcal{A}^{-1/2}M\mathcal{A}^{-1/2}$ is a bounded linear positive operator in $\mathcal{H}$. Consequently by [95, Lemma 2.1, p.165] the operator S is m-accretive in $\mathcal{H}$, i.e. $R(I + S) = \mathcal{H}$. Therefore, there exists $v \in \mathcal{H}$ which solves (1.27). Consequently $y = \mathcal{A}^{-1/2}v \in \mathcal{D}(\mathcal{A}^{1/2})$ is solution to (1.26) for the variable $y \in \mathcal{D}(\mathcal{A}^{1/2})$. Going back to (1.25) we obtain $x \in \mathcal{D}(\mathcal{A}^{1/2})$. It is clear that $(x, y) \in \mathcal{D}(A)$. The proof of maximality property has been completed. □

It follows from Assumption 1.1(F) that the operator
$$B(U) = \begin{pmatrix} 0 \\ M^{-1}F(U) \end{pmatrix}, \quad U = (x; y) \in H,$$
is locally Lipschitz on H. Indeed, the above follows from local Lipschitz property of $M^{-1/2}F$ acting between $\mathcal{D}(\mathcal{A}^{1/2}) \times V$ and $\mathcal{H}$ and the definition of the norm in V.

Therefore, using [95, Theorems 4.1 and 4.1A, Chapt.4] and applying the same argument as in the proof of Theorem 7.2[20] we conclude for any $U_0 \in \mathcal{D}(A)$ there exists $t_{max} \leq \infty$ such that (1.24) has a unique strong solution U on the interval

$[0, t_{max})$. Furthermore, if we assume only $U_0 \in \overline{\mathcal{D}(\mathcal{A})}$ we obtain a unique generalized solution $U \in C([0, t_{max}); H)$ to (1.24). In both cases we have

$$\lim_{t \nearrow t_{max}} \|U(t)\|_H = \infty \quad \text{provided} \quad t_{max} < \infty. \tag{1.28}$$

We note that, by Assumption 1.1(D), $\overline{\mathcal{D}(\mathcal{A})} = H$ because the set $\mathcal{D}(\mathcal{A}) \times W$ lies in in $\mathcal{D}(A)$ and W is dense in V.

Now we prove the global existence of strong and generalized solutions.

Let $u(t)$ be a strong solution on some interval $[0, t_{max})$. Multiplying (1.1) by u_t in $\mathcal{H}$ we obtain

$$E_0(u(t), u_t(t)) + k \int_0^t (Du_t(\tau), u_t(\tau)) d\tau$$
$$= E_0(u_0, u_1) + \int_0^t (F(u(\tau), u_t(\tau)), u_t(\tau)) d\tau \tag{1.29}$$

for $t \in [0, t_{max})$. A simple calculation shows that

$$(F(u(t)), u_t(t)) = -\frac{d}{dt}\Pi(u(t)) + (F^*(u(t), u_t(t)), u_t(t)), \quad t \in [0, t_{max}).$$

Thus from (1.29) we obtain the energy relation (1.17) for the strong solution $u(t)$ and for $t \in [0, t_{max})$.

We note that relation (1.5) implies that there exists a constant $c > 0$ such that

$$\frac{1}{2} \cdot E(u_0, u_1) - c \leq \mathcal{E}(u_0, u_1) \leq 2 \cdot E(u_0, u_1) + c, \quad u_0 \in \mathcal{D}(\mathcal{A}^{1/2}), \; u_1 \in V. \tag{1.30}$$

Therefore using (1.17) and (1.30) we obtain that

$$E(u(t), u_t(t)) + 2k \int_0^t (Du_t(\tau), u_t(\tau)) d\tau$$
$$\leq C(1 + E(u_0, u_1)) + 2 \int_0^t (F^*(u(\tau), u_t(\tau)), u_t(\tau)) d\tau.$$

Thus by (1.6) we have that

$$E(u(t), u_t(t)) \leq c_1 (1 + E(u_0, u_1)) + c_2 \int_0^t E(u(\tau), u_t(\tau)) d\tau$$

with some constants $c_i > 0$. This implies, after rescaling of constants, that

$$E(u(t), u_t(t)) \leq c_0 (1 + E(u_0, u_1)) e^{c_1 t}, \quad t \in [0, t_{max}).$$

with some constants c_0 and c_1. Using (1.15) it is easy to find that the last relation is also true for generalized solutions. Thus, by property (1.28), we have the global existence (and uniqueness) both strong and generalized solutions. It is clear that generalized solutions satisfy (1.20) and also (1.22) under condition (1.21). □

REMARK 1.7. Assume that the mapping D have additional regularity properties. Namely we assume that $D : \mathcal{D}(\mathcal{A}^{1/2}) \mapsto V'$ is a continuous mapping. In this case the compatibility condition can be written in the form $\mathcal{A}u_0 \in V'$ and the strong solution $u(t)$ possesses the property $\mathcal{A}u \in C_r([0, t_{max}); V')$.

REMARK 1.8. Condition (F) in Assumption 1.1 requires that the nonlinear term $F(u, v)$ be locally Lipschitz. This assumption is used in order to prove local existence of solutions. When locally Lipschitz condition is violated and some other structural hypotheses are imposed one can still prove existence and uniqueness

of solutions. For example, we can assume that Assumption 1.1 holds with (1.3) replaced by the following condition
(1.31)
$$|F(u,v) - F(\hat{u},\hat{v})|^2_{V'} \leq L(r) \left[|\mathcal{A}^{1/2}(u-\hat{u})|^2 + |v-\hat{v}|^2_V + (D(v)-D(\hat{v}), v-\hat{v}) \right]$$

for $u, \hat{u}, v, \hat{v} \in \mathcal{D}(\mathcal{A}^{1/2})$ such that

(1.32) $$|\mathcal{A}^{1/2}u|^2 + |v|^2_V \leq r \quad \text{and} \quad |\mathcal{A}^{1/2}\hat{u}|^2 + |\hat{v}|^2_V \leq r.$$

In this case the well-posedness statement of Theorem 1.5 remains true. The argument is the same as in the proof of Theorem 1.5 after noticing that the potentially non-Lipschitz part of $F(u,v)$ can be factored into the monotone part of the operator. More specifically, it follows from from (1.31) that the operator

$$D_*(u,v) = k \cdot D(v) - F(u,v)$$

possesses the property

$$(D_*(u,v) - D_*(\hat{u},\hat{v}), v-\hat{v}) \geq \frac{k}{2} \cdot (D(v) - D(\hat{v}), v-\hat{v}) \\ -C(r)\left[|\mathcal{A}^{1/2}(u-\hat{u})|^2 + |v-\hat{v}|^2_V\right]$$

for some $C(r) > 0$ provided that (1.32) is satisfied. This relation makes it possible to prove that the operator $A+C(r)I$, where A is given by (1.23) with D_* instead of D, is accretive on all elements possessing property (1.32) Thus we are able to fall into the same framework as in the locally Lipschitz case of Theorem 1.5. Using the same regularizing procedure as in [20, Theorem 7.2] and relying on [95, Theorems 4.1 and 4.1A, Chapt.4] we can obtain the well-posedness result in the case considered.

REMARK 1.9. In the case $F^* \equiv 0$ the full energy $\mathcal{E}(u(t), u_t(t))$ is non-increasing along trajectories. However, in general, the dynamical system may not possess this property, as it is most often the case when the term F^* is present in the model.

The following assertion shows that under some conditions the energy inequality (1.22) valid for generalized solutions can be stated in a stronger form.

PROPOSITION 1.10. *In addition to Assumption 1.1 assume that (1.21) holds and there exists a convex function* $\varphi : V \mapsto \mathbb{R} \cup \{+\infty\}$ *such that*

- φ *is lower semicontinuous on V, i.e.*

$$\{v_n \to v \ \text{in} \ V\} \implies \liminf_{n\to\infty} \varphi(v_n) \geq \varphi(v);$$

- $\varphi(v) \leq (Dv, v)$ *for any $v \in \mathcal{D}(\mathcal{A}^{1/2})$.*

Then any generalized solution $u(t)$ to problem (1.1) satisfies the energy inequality of the form
(1.33)
$$\mathcal{E}(u(t), u_t(t)) + k\int_s^t \varphi(u_t(\tau))d\tau \leq \mathcal{E}(u(s), u_t(s)) + \int_s^t (F^*(u(\tau), u_t(\tau)), u_t(\tau))d\tau$$

for all $0 \leq s \leq t \leq T$.

PROOF. Let $u(t)$ be a generalized solution and $\{u^n(t)\}$ be a sequence of strong solutions such that (1.15) holds. We use the energy relation

$$\mathcal{E}(u^n(t), u_t^n(t)) + k \int_s^t (Du_t^n(\tau), u_t^n(\tau)) d\tau$$
(1.34)
$$= \mathcal{E}(u^n(s), u_t^n(s)) + \int_s^t (F^*(u^n(\tau), u_t^n(\tau)), u_t^n(\tau)) d\tau .$$

Since the energy $\mathcal{E}(u, u_t)$ is continuous on $\mathcal{D}(\mathcal{A}^{1/2}) \times V$, relation (1.15) implies that

$$\lim_{n \to \infty} \mathcal{E}(u^n(t), u_t^n(t)) = \mathcal{E}(u(t), u_t(t)) \quad \text{for every} \quad t \in [0, T].$$

Using property (1.21) of F^* and relation (1.15) again it is also easy to see that

$$\lim_{n \to \infty} \int_s^t (F^*(u^n(\tau), u_t^n(\tau)), u_t^n(\tau)) d\tau = \int_s^t (F^*(u(\tau), u_t(\tau)), u_t(\tau)) d\tau .$$

Since φ is lower semicontinuous, by Fatou's lemma we have that

(1.35) $\displaystyle\liminf_{n \to \infty} \int_s^t (Du_t^n(\tau), u_t^n(\tau)) d\tau \geq \int_s^t \liminf_{n \to \infty} \varphi(u_t^n(\tau)) d\tau \geq \int_s^t \varphi(u_t(\tau)) d\tau .$

Thus, relation (1.33) follows from (1.34). □

REMARK 1.11. We note that in the case of wave equation (1.8) the hypothesis of Proposition 1.10 holds provided $s \mapsto sg(s)$ is a continuous convex function on $\mathbb{R}$. In this case $\varphi : L_2(\Omega) \mapsto \mathbb{R} \cup \{+\infty\}$ has the form

(1.36) $\quad \varphi(v) = \displaystyle\int_\Omega v(x) g(v(x)) dx \text{ if } vg(v) \in L_1(\Omega) \quad \text{and} \quad \varphi(v) = +\infty \text{ otherwise,}$

(see [**95**, Proposition 8.1, p.85]). To apply Proposition 1.10 in the case of a general damping function one can choose

$$\varphi(v) = \int_\Omega j_g(v(x)) dx \quad \text{with} \quad j_g(s) = \int_0^s g(\xi) d\xi.$$

However, it should be noted, that using appropriate approximations and properties of convergence in $L_2(\Omega)$ we can prove (1.33) with φ of the form (1.36) in the general case (see the argument given in the proof of Theorem 5.10 in Chapter 5).

The assertion stated below gives a condition under which generalized solutions satisfy weak (variational) form of equation (1.1) and also energy equality (1.17). The conditions required are rather strong and for this reason this particular result will not be of much use used in the sequel. However, for sake of completeness we provide the formulation and a brief proof.

PROPOSITION 1.12. *Let the hypotheses of Theorem 1.5 hold. Assume additionally that the operator D maps V into V' and is a monotone hemicontinuous operator which is bounded on bounded sets, i.e.*

(1.37) $\quad \sup\{|D(v)|_{V'} : v \in V, |v|_V \leq \rho\} < \infty \quad \text{for any} \quad \rho > 0.$

Then every generalized solution is also weak, i.e., the relation

(1.38)
$$(Mu_t(t), \psi) = (Mu_1, \psi) - \int_0^t ((\mathcal{A}u(\tau), \psi) +$$
$$-k(Du_t(\tau), \psi) + (F(u(\tau), u_t(\tau)), \psi)) d\tau$$

holds for every $\psi \in \mathcal{D}(\mathcal{A}^{1/2})$ and for almost all $t \in [0,T]$. Moreover, generalized solutions satisfy the energy relation (1.17) under conditions (1.21) and (1.37).

PROOF. Since any strong solution satisfies (1.38), we can use relation (1.15) to make limit transition in (1.38). By (1.37) we can apply the Lebesgue convergence theorem to the integral with the damping term. Hence, every generalized solution is weak.

Let us prove the energy inequality (1.17) for generalized (weak) solutions.

As the argument given in the proof of Proposition 1.10 shows, to obtain (1.17) we need only check that

$$(1.39) \qquad \lim_{n \to \infty} \int_0^t (Du_t^n(\tau), u_t^n(\tau)) d\tau = \int_0^t (Du_t(\tau), u_t(\tau)) d\tau ,$$

where $u(t)$ is a generalized solution and $\{u^n(t)\}$ is a sequence of strong solutions such that (1.15) holds.

By [**95**, Lemmata 2.1 and 2.2, p.38] $D : V \mapsto V'$ is demicontinuous, i.e. $Du^n(t) \to Du(t)$ weakly in V' for every t and hence

$$\lim_{n \to \infty} (Du_t^n(\tau), u_t^n(\tau)) = (Du_t(\tau), u_t(\tau)), \quad \tau \in [0,T].$$

Thus property (1.37) and the Lebesgue convergence theorem imply (1.39). $\square$

REMARK 1.13. Assume that the operator D has the form $D = D_1 + D_2$, where D_1 possesses the properties described in the statement of Proposition 1.10 with some convex lower semicontinuous function φ and D_2 enjoys requirements of Proposition 1.12. Then the same arguments as in Propositions 1.10 and 1.12 show that under condition (1.21) any generalized solution solution satisfies the energy inequality of the form

$$\mathcal{E}(u(t), u_t(t)) \;+\; k \int_s^t \varphi(u_t(\tau)) d\tau + k \int_0^t (D_2 u_t(\tau), u_t(\tau)) d\tau$$
$$\leq \;\; \mathcal{E}(u(s), u_t(s)) + \int_s^t (F^*(u(\tau), u_t(\tau)), u_t(\tau)) d\tau.$$

REMARK 1.14. Assumption (1.37) is too restrictive. For instance, in the case of the wave equation considered in Example 1.2 relation (1.37) is equivalent to the requirement that $g(s)$ is linearly bounded, i.e.

$$(1.40) \qquad\qquad |g(s)| \leq C(1 + |s|) \quad \text{for all} \quad s \in \mathbb{R}.$$

Indeed, if $\sup_{s \in \mathbb{R}} \left[|g(s)| \cdot (1 + |s|)^{-1} \right] = +\infty$ then there exists a sequence $\{a_n\}$ such that

$$|a_n| \to \infty \quad \text{and} \quad |g(a_n)| \geq n(1 + |a_n|), \quad n = 1, 2, \ldots$$

Let Ω_n be a measurable subset of Ω such that $\text{Leb}(\Omega_n) = (1 + |a_n|)^{-2}$. We define a function $u_n \in L_2(\Omega)$ by the relations: $u_n(x) = a_n$ if $x \in \Omega_n$ and $u_n(x) = 0$ otherwise. Then we have that

$$\|u_n\|_{L_2(\Omega)} \leq 1 \quad \text{and} \quad \int_\Omega |g(u_n(x))|^2 dx \geq \int_{\Omega_n} |g(u_n(x))|^2 dx \geq n^2$$

Thus, (1.37) implies (1.40). It is also obvious that (1.40) implies (1.37).

Later we shall develop other versions of additional assumptions concerning D which guarantee the validity of the energy relation (1.17) for generalized solutions. However these conditions will be more problem specific and for sake of clarity of exposition we relegate these to Chapters 5 and 6 where concrete models are considered.

Below we also need the following simple assertion.

PROPOSITION 1.15. *Let Assumption 1.1 hold and $0 < T \leq \infty$. Then for any $R > 0$ we have the following estimate*

$$
\begin{aligned}
&|M^{1/2}(u_{1t}(t) - u_{2t}(t))|^2 + |\mathcal{A}^{1/2}(u_1(t) - u_2(t))|^2 \\
&\qquad \leq e^{L(R)t}\left(|M^{1/2}(u_{11} - u_{12})|^2 + |\mathcal{A}^{1/2}(u_{01} - u_{02})|^2\right), \quad t \in [0, T)
\end{aligned}
\tag{1.41}
$$

where $u_1(t)$ and $u_2(t)$ are two solutions to problem (1.1) with initial data $(u_{01}; u_{11})$ and $(u_{02}; u_{12})$ respectively such that $|\mathcal{A}^{1/2}(u_i(t)| + |M^{1/2}(u_{it}(t)|^2 \leq R^2$ for all $t \in [0, T)$, $i = 1, 2$. Here $L(R)$ is the Lipschitz constant from (1.3).

PROOF. The proof is standard and based on a classical energy type of argument. Let $u_1(t)$ and $u_2(t)$ be strong solutions to problem (1.1). From (1.1) for the difference $u(t) = u_1(t) - u_2(t)$ we have the relation

$$Mu_{tt} + k\left[Du_{1t} - Du_{2t}\right] + \mathcal{A}u = F(u_1, u_{1t}) - F(u_2, u_{2t}).$$

This implies the following energy relation

$$
\begin{aligned}
&E_0(u(t), u_t(t)) + k\int_0^t (Du_{1t}(\tau) - Du_{2t}(\tau), u_t(\tau))d\tau \\
&= E_0(u(0), u_t(0)) + \int_0^t (F(u_1(\tau), u_{1t}(\tau)) - F(u_2(\tau), u_{2t}(\tau)), u_t(\tau))d\tau,
\end{aligned}
$$

where $E_0(u, v) = \frac{1}{2}\left((Mv, v) + (\mathcal{A}u, u)\right)$. Therefore using (1.2), (1.3) and Gronwall's lemma we obtain (1.41) valid first for strong solutions. The limit process along with standard density argument allows to extend the validity of (1.41) to all generalized solutions. □

In closing this section we conclude that under Assumption 1.1 equation (1.1) generates a dynamical system (H, S_t) in the Hilbert space $H = \mathcal{D}(\mathcal{A}^{1/2}) \times V$ by the formula $S_t(u_0, u_1) = (u(t); u_t(t))$, where $u(t)$ is the solution to (1.1). Our main goal is the study of long time behaviour of this system. In the next Chapter 2 we formulate and prove several abstract results which pertain to general criteria for asserting the existence, compactness and finite-dimensionality of attractors. These abstract results will be then applied to evolutions in question. Concrete applications are given in Chapters 5, 6 and 7.

CHAPTER 2

Abstract results on global attractors

In this chapter we deal with general results on asymptotic behaviour of abstract infinite-dimensional dissipative dynamical systems. The main issue we address are: asymptotic smoothness (in the sense of [**59**]) of the systems, finite dimension of invariant sets and the existence of fractal exponential attractors (inertial sets). The main assumption in all these results is some kind of squeezing property which, as we shall see in Chapters 3 and 4, is closely related to stabilizability estimates from control theory. We also discuss properties of gradient systems, i.e. the systems possessing a functional which does not increase along trajectories.

By definition a dynamical system is a pair of objects (X, S_t) consisting of a complete metric space X and a family of continuous mappings $\{S_t : t \in \mathbb{R}_+\}$ of X into itself with the semigroup properties:

$$S_0 = I, \quad S_{t+\tau} = S_t \circ S_\tau.$$

We also assume that the semigroup S_t is strongly continuous, this is to say, for any fixed $y_0 \in X$ $y(t) = S_t y_0$ is continuous for any $t \geq 0$. Therewith X is called a *phase space* (or state space) and S_t is called *evolution semigroup* (or evolution operator).

DEFINITION 2.1. Let (X, S_t) be a dynamical system.
- A closed set $B \subset X$ is said to be *absorbing* for (X, S_t) iff for any bounded set $D \subset X$ there exists $t_0(D)$ such that $S_t D \subset B$ for all $t \geq t_0(D)$.
- (X, S_t) is said to be *(ultimately) dissipative* iff it possesses a bounded absorbing set B. If X is a Banach space, then a value $R > 0$ is said to be a *radius of dissipativity* of (X, S_t) iff $B \subset \{x \in X : \|x\|_X \leq R\}$
- (X, S_t) is said to be *asymptotically smooth* iff for any bounded set D such that $S_t D \subset D$ for $t > 0$ there exists a compact set K in the closure $\overline{D}$ of D, such that

(2.1) $$\lim_{t \to +\infty} d_X\{S_t D \,|\, K\} = 0.$$

Here and below $d_X\{A|B\} = \sup_{x \in A} \mathrm{dist}_X(x, B)$ is the Hausdorff semidistance.

DEFINITION 2.2. A bounded closed set $A \subset X$ is said to be a *global attractor* of the dynamical system (X, S_t) iff (i) A is an invariant set, i.e. $S_t A = A$ for $t \geq 0$, and (ii) A is uniformly attracting, i.e. for all bounded set $D \subset X$

(2.2) $$\lim_{t \to +\infty} d_X\{S_t D \,|\, A\} = 0.$$

It is well known that the properties of *ultimate dissipativity* and *asymptotic smoothness* are critical for proving existence of global attractors. In fact, the following result is well known [**4, 19, 59, 73, 101**].

THEOREM 2.3. *Let (X, S_t) be a dissipative dynamical system in a complete metric space X. Then (X, S_t) possesses a compact global attractor A if and only if (X, S_t) is asymptotically smooth.*

2.1. Criteria for asymptotic smoothness of dynamical systems

In this section we shall provide several assertions which give a convenient criteria for asymptotic smoothness of a dynamical system. Two of them generalize the results presented in [**59**] and [**12**]. We shall see later that, these particular generalizations are critical for establishing asymptotic smoothness in hyperbolic like flows with nonlinear damping.

THEOREM 2.4. *Let (X, S_t) be a dynamical system on a complete metric space X endowed with a metric d. Assume that for any bounded positively invariant set B in X there exist $T > 0$, a continuous non-decreasing function $q : \mathbb{R}_+ \mapsto \mathbb{R}_+$ and a pseudometric ϱ_B^T on $C(0, T; X)$ such that*

(i) *$q(0) = 0$ and $q(s) < s$ for $s > 0$;*
(ii) *the pseudometric ϱ_B^T is precompact (with respect to X) in the following sense: any sequence $\{x_n\} \subset B$ has a subsequence $\{x_{n_k}\}$ such that the sequence $\{y_k\} \subset C(0, T; X)$ of elements $y_k(\tau) = S_\tau x_{n_k}$ is Cauchy with respect to ϱ_B^T;*
(iii) *the following inequality holds*

$$(2.3) \quad d(S_T y_1, S_T y_2) \leq q\left(d(y_1, y_2) + \varrho_B^T(\{S_\tau y_1\}, \{S_\tau y_2\})\right),$$

for every $y_1, y_2 \in B$, where we denote by $\{S_\tau y_i\}$ the element in the space $C(0, T; X)$ given by function $y_i(\tau) = S_\tau y_i$.

Then (X, S_t) is an asymptotically smooth dynamical system.

REMARK 2.5. Instead of (2.3) we can also assume that

$$(2.4) \quad d(S_T y_1, S_T y_2) \leq q(d(y_1, y_2)) + \varrho_B^T(\{S_\tau y_1\}, \{S_\tau y_2\}),$$

(pseudometric ϱ_B^T outside q).

PROOF. We follow the same line of arguments as in [**27**], where this theorem is proved for X being a Banach space. As in [**27**] (cf. also [**59**, Lemma 2.3.6]) we involve the Kuratowski α-measure of noncompactness which is defined by the formula

$$(2.5) \quad \alpha(B) = \inf\{\delta : B \text{ has a finite cover of diameter } < \delta\}$$

for every bounded set B of X. The α-measure has the following properties (see, e.g., [**59**] or [**92**, Lemma 22.2]):

(i) $\alpha(B) = 0$ if and only if B is precompact;
(ii) $\alpha(A \cup B) = \max\{\alpha(A), \alpha(B)\}$;
(iii) $\alpha(B) = \alpha(\overline{B})$, where $\overline{B}$ is the closure of B;
(iv) if $B_1 \supset B_2 \supset B_3 \ldots$ are nonempty closed sets in X such that $\alpha(B_n) \to 0$ as $n \to \infty$, then $\cap_{n \geq 1} B_n$ is nonempty and compact.

We first prove that

$$(2.6) \quad \alpha(S_T B) \leq q(\alpha(B)) \quad \text{for every bounded positively invariant set } B.$$

For any $\varepsilon > 0$ there exist sets $F_1, \ldots, F_n$ such that

$$B = F_1 \cup \ldots \cup F_n, \quad \text{diam } F_i < \alpha(B) + \varepsilon.$$

It follows from assumption (ii) that there exists a finite set $\mathcal{N} = \{x_i : i = 1, 2 \ldots m\} \subset B$ such that for every $y \in B$ there is $x_i \in \mathcal{N}$ with the property $\varrho_B^T(\{S_\tau y\}, \{S_\tau x_i\}) \leq \varepsilon$. It means that
$$B = \cup_{i=1}^m C_i, \quad C_i = \{y \in B : \varrho_B^T(\{S_\tau y\}, \{S_\tau x_i\}) \leq \varepsilon\}.$$
Now we can write the representations
$$(2.7) \qquad B = \cup_{i,j}(C_i \cap F_j), \quad \text{and} \quad S_T B = \cup_{i,j}(S_T(C_i \cap F_j)).$$
Since $\operatorname{diam} F_j < \alpha(B) + \varepsilon$ and
$$\varrho_B^T(\{S_\tau y_1\}, \{S_\tau y_2\}) \leq \varrho_B^T(\{S_\tau y_1\}, \{S_\tau x_i\}) + \varrho_B^T(\{S_\tau x_i\}, \{S_\tau y_2\}),$$
it follows from (2.3) that
$$d(S_T y_1, S_T y_2) \leq q\left([\alpha(B) + \varepsilon] + 2\varepsilon\right),$$
for any $y_1, y_2 \in C_i \cap F_j$. Thus
$$\operatorname{diam}(S_T(C_j \cap F_i)) \leq q\left(\alpha(B) + 3\varepsilon\right).$$
Therefore using (2.7) and the definition of α-measure we obtain (2.6).

Since $S_t B \subset B$, we obviously have the representation
$$\omega(B) = \bigcap_{k=1}^\infty B_k, \quad B_k \equiv \overline{S_{kT} B},$$
for the ω-limit set $\omega(B)$. It is also clear that $B_k \supset B_{k+1}$ for every k and
$$(2.8) \qquad \alpha(B_k) = \alpha(S_T B_{k-1}) \leq q(\alpha(B_{k-1})) \quad k = 1, 2, \ldots.$$
Since $q(s) < s$, the sequence $\{\alpha(B_k)\}$ is decreasing. Therefore there exists $\alpha_0 = \lim_{k \to \infty} \alpha(B_k)$. From (2.8) we have that $\alpha_0 \leq q(\alpha_0)$ which is possible only if $\alpha_0 = 0$. Thus by property (iv) of the α-measure, $\omega(B)$ is a nonempty compact set. The standard argument (see, e.g., [**59**, p.13,14]) or [**19**, p.30]) allows us to conclude that $\omega(B)$ attracts B uniformly. Thus (X, S_t) is asymptotically smooth. $\square$

REMARK 2.6. Property (2.6) implies that for every bounded forward invariant set B there exists $T > 0$ such that $\alpha(S_T B) < \alpha(B)$ provided $\alpha(B) > 0$. If for some fixed $T > 0$ this property holds for every bounded set B such that $S_T B$ is also bounded, then, by the definition (see [**59**]), S_T is a conditional α-condensing mapping. It is known [**59**] that conditional α-condensing mappings are asymptotically smooth. Therefore Theorem 2.4 can be considered as a generalization of the results presented in [**59**].

Theorem 2.4 is tailored to problems where the convergence to a compact set is not necessarily exponential. This corresponds to situations when the damping at the origin is nonlinear. If the damping at the origin is linear, then a more restrictive version of Theorem 2.4 suffices and is more convenient in applications. In fact, the following result, which is a "linear" version of Theorem 2.4 emerges as the Corollary of Theorem 2.4.

COROLLARY 2.7. *Let (X, S_t) be a dynamical system on a Banach space X. Assume that for any bounded positively invariant set B in X and for any $t \geq t_0 = t_0(B) \geq 0$ there exist a function $K_B(t)$ on $[t_0, +\infty)$ and a pseudometric ϱ_B^t on $C(0, t; X)$ such that*

(i) *$K_B(t) \geq 0$ and $\lim_{t \to \infty} K_B(t) = 0$;*

(ii) *the pseudometric ϱ_B^t is precompact (with respect to the norm of X).*
(iii) *the estimate*

(2.9) $$\|S_t y_1 - S_t y_2\| \leq K_B(t) \cdot \|y_1 - y_2\| + \varrho_B^t(\{S_\tau y_1\}, \{S_\tau y_2\}), \quad t \geq t_0,$$

holds for every $y_1, y_2 \in B$, where we denote by $\{S_\tau y_i\}$ the element in the space $C(0, t; X)$ given by function $y_i(\tau) = S_\tau y_i$.

Then (X, S_t) is an asymptotically smooth dynamical system.

PROOF. It suffices to apply Theorem 2.4 with $d(x, y) = \|x - y\|$ and $q(s) = K_B(T) \cdot s$, where T is chosen sufficiently large so that $K_B(T) \leq \gamma < 1$. □

Corollary 2.7 implies the following result which was proved earlier in [**12**].

PROPOSITION 2.8. *Assume that a dynamical system (X, S_t) on a Banach space X possesses the following property: for any bounded positively invariant set B in X there exist functions $C_B(t) \geq 0$ and $K_B(t) \geq 0$ such that $\lim_{t \to \infty} K_B(t) = 0$, a time $t_0 = t_0(B)$ and a precompact pseudometric ϱ on X such that*

(2.10) $$\|S_t y_1 - S_t y_2\| \leq K_B(t) \cdot \|y_1 - y_2\| + C_B(t) \cdot \varrho(y_1, y_2), \quad t \geq t_0,$$

for every $y_1, y_2 \in B$. Then (X, S_t) is an asymptotically smooth dynamical system.

We recall that a pseudometric ϱ on a Banach space X is said to be *precompact* (with respect to the norm of X) if any bounded sequence (in norm) has a subsequence which is Cauchy with respect to ϱ.

PROOF. It is clear that a pseudometric ϱ_B^t defined on $C(0, t; X)$ by the formula

$$\varrho_B^t(\{S_\tau y_1\}, \{S_\tau y_2\}) = C_B(t) \cdot \varrho(y_1, y_2)$$

satisfies assumptions (ii) and (iii) in Corollary 2.7. □

REMARK 2.9. The main difference between (2.9) and (2.10) is the fact that the term in (2.9) involving compact pseudometric is evaluated on trajectories, rather than on a specific point in the phase space, as in (2.10). Thus, (2.9) is definitely a weaker requirement than that of [**12**]. While in the case of linear damping it is relatively simple to show that inequality in (2.9) implies that of (2.10), this is not the case in hyperbolic flows with a nonlinear damping. Thus, the above generalization is critical for models considered in this paper.

The following version of the compactness criterion provides more flexibility, with respect to method based on Theorem 2.4 by allowing taking sequential limits rather then the simultaneous limits. This was an observation made for the first time in [**66**].

PROPOSITION 2.10. *Let (X, S_t) be a dynamical system on a complete metric space X endowed with a metric d. Assume that for any bounded positively invariant set B in X and for any $\epsilon > 0$ there exists $T \equiv T(\epsilon, B)$ such that*

(2.11) $$d(S_T y_1, S_T y_2) \leq \epsilon + \Psi_{\epsilon, B, T}(y_1, y_2), y_i \in B,$$

where $\Psi_{\epsilon, B, T}(y_1, y_2)$ is a function defined on $B \times B$ such that

(2.12) $$\liminf_{m \to \infty} \liminf_{n \to \infty} \Psi_{\epsilon, B, T}(y_n, y_m) = 0$$

for every sequence $\{y_n\}$ from B. Then (X, S_t) is an asymptotically smooth dynamical system.

The result stated in Proposition 2.10 is an abstract version of Theorem 2 in [**66**] that can be derived from the arguments given in [**66**]. For the reader's convenience an independent and shorter proof of the same result is given.

PROOF. As in the proof of Theorem 2.4 it is sufficient to prove that
$$\lim_{t \to \infty} \alpha(S_t B) = 0,$$
where $\alpha(B)$ is the Kuratowski α-measure defined (2.5).

Since $S_{t_1} B \subset S_{t_2} B$ for $t_1 > t_2$, the function $\alpha(t) \equiv \alpha(S_t B)$ is non-increasing, Therefore it is sufficient to prove that for any $\varepsilon > 0$ there exists $T > 0$ such that $\alpha(S_T B) \leq \varepsilon$. If this is not true, then there is $\varepsilon_0 > 0$ such that
$$\alpha(S_T B) \geq 6\varepsilon_0 \quad \text{for all} \quad T > 0.$$
For this ε_0 we choose T_0 such that (2.11) and (2.12) hold. The relation $\alpha(S_{T_0} B) \geq 6\varepsilon_0$ implies that there exists an infinite sequence $\{y_n\}_{n=1}^\infty$ such that
$$d(S_{T_0} y_n, S_{T_0} y_m) \geq 2\varepsilon_0 \quad \text{for all} \quad n \neq m, \ n, m = 1, 2, \ldots$$
Therefore from (2.11) we have that
$$\Psi_{\varepsilon_0, B, T_0}(y_n, y_m) \geq \varepsilon_0 \quad \text{for all} \quad n \neq m, \ n, m = 1, 2, \ldots$$
This contradicts (2.12). The proof is complete. □

We conclude this section with the following well-known assertion (see, e.g., [**59**, Lemma 3.2.3] or [**92**, Lemma 22.4]). We put it here for the sake of further references only.

PROPOSITION 2.11. *Let (X, S_t) be a dynamical system on a Banach space X. Assume that for any bounded positively invariant set B in X the evolution operator S_t admits the splitting*
$$S_t y = U_t y + K_t y, \quad y \in B, \ t \geq 0,$$
such that
- $\sup \{\|U_t y\| : y \in B\} \to 0$ *as $t \to \infty$, and*
- *there exists $t_0 = t_0(B) \geq 0$ such that $K_t B$ is a precompact set in X for every $t \geq t_0$.*

Then (X, S_t) is an asymptotically smooth dynamical system.

PROOF. Since α-measure possesses the property $\alpha(A + B) \leq \alpha(A) + \alpha(B)$ and $\alpha(K) = 0$ for every precompact set (see, e.g., [**92**, Lemma 22.2]), for $t \geq t_0$ we have that
$$\alpha(S_t B) \leq \alpha(U_t B) + \alpha(K_t B) \leq \alpha(U_t B) \leq 2\mathrm{diam}\{U_t B\} \leq 4 \sup_{y \in B} \|U_t y\|.$$
Thus $\alpha(S_t B) \to 0$ as $t \to \infty$. Since $S_t B \subset B$, we can apply the same argument as in the proof of Theorem 2.4. □

2.2. Criteria for finite dimensionality of attractors

Finite dimensionality is an important property of global attractors which can be established for several dynamical systems, including these arising in significant applications. There are several approaches which provide effective estimates for the dimension of attractors of dynamical systems generated by PDE's (see, e.g., [**4, 73, 101**]). Here we present an approach which do not require C^1-smoothness of the evolutionary operator. The reason for this focus is that dynamical systems of hyperbolic-like nature do not display smoothing effects - unlike parabolic equations. Therefore, the C^1-smoothness of the flows is most often out of question, particularly in problems with a nonlinear dissipation. On one hand our Theorem 2.15 generalizes Ladyzhenskaya's theorem (see, e.g., [**73**]) on finite dimensionality of invariant sets. On the other hand it is related to squeezing property in the form considered in [**89**]. However, as in the case of Ladyzhenskaya's theorem we wish to point out that the estimate of the dimension based on the theorems below usually turn out to be conservative.

Fractal dimension is the most commonly used measures in the theory of infinite dimensional dynamical systems.

DEFINITION 2.12. Let M be a compact set in a metric space X. The *fractal (box-counting) dimension* $\dim_f M$ of M is defined by

$$\dim_f M = \limsup_{\varepsilon \to 0} \frac{\ln n(M, \varepsilon)}{\ln(1/\varepsilon)},$$

where $n(M, \varepsilon)$ is the minimal number of closed balls of the radius ε which cover the set M.

The importance of the notion of finite fractal dimension is illustrated by the following property (see [**53**]): if M compact set in X such that $\dim_f M < n/2$ for some $n \in \mathbb{N}$, then there exists an injective Lipschitz mapping $L : M \mapsto \mathbb{R}^n$ such that its inverse is Hölder continuous. This means that M can be placed in the graph of Hölder continuous mapping which maps a compact subset of $\mathbb{R}^n$ onto M.

We refer to [**46**] for other properties of fractal dimension. Here we only note that the fractal dimension $\dim_f M$ of the set M can be represented by the formula

$$(2.13) \qquad \dim_f M = \limsup_{\varepsilon \to 0} \frac{\ln N(M, \varepsilon)}{\ln(1/\varepsilon)},$$

where $N(M, \varepsilon)$ is the minimal number of *closed sets* of the diameter 2ε which cover M.

Below we need the following definition.

DEFINITION 2.13. Let X be a complete metric space endowed with the metric d and M be a bounded closed set in X. Assume that ϱ is a pseudometric defined on M. Let $B \subset M$ and $\varepsilon > 0$.

- A subset $\mathcal{U}$ in B is said to be (ε, ϱ)-*distinguishable* if $\varrho(x, x') > \varepsilon$ for any $x, x' \in \mathcal{U}$, $x \neq x'$. We denote by $m_\varrho(B, \varepsilon)$ the maximal cardinality of an (ε, ϱ)-distinguishable subset of B.
- The pseudometric ϱ is said to be *compact* on M iff $m_\varrho(M, \varepsilon)$ is finite for every $\varepsilon > 0$.

- Similarly to [**68**], we shall call the value $\ln m_\varrho(B,\varepsilon)$ by ε-*capacity* of the set B (with respect to the pseudometric ϱ), or shortly, (ε,ϱ)-*capacity*. For any $r > 0$ we define a *local* (r,ε,ϱ)-*capacity* of the set M by the formula

(2.14) $$\mathcal{C}_\varrho(M;r,\varepsilon) = \sup\{\ln m_\varrho(B,\varepsilon) \,:\, B \subseteq M,\ \mathrm{diam}\,B \leq 2r\}.$$

Now we are in position to state our basic assertion of this section. We shall use it below to obtain appropriate criteria which are convenient for applications to second order in time evolution equations.

THEOREM 2.14. *Let (X,d) be a complete metric space and M be a bounded closed set in X. Assume that there exists a mapping $V : M \mapsto X$ such that*
  (i) $M \subseteq VM$;
  (ii) *there exist a compact pseudometric ϱ on M and a number $0 < \eta < 1$ such that*

(2.15) $$d(Vv_1, Vv_2) \leq \eta \cdot d(v_1, v_2) + \varrho(v_1, v_2),\ v_1, v_2 \in M.$$

Then M is a compact set in X with the fractal dimension $\dim_f M$ estimated as follows

(2.16) $$\dim_f M \leq \left[\ln\frac{1}{\delta + \eta}\right]^{-1} \cdot \limsup_{\varepsilon \to 0} \mathcal{C}_\varrho(M;\varepsilon, \delta\varepsilon)$$

for every $\delta \in (0, 1 - \eta)$, where $\mathcal{C}_\varrho(M;r,\varepsilon)$ is the local (r,ε,ϱ)-capacity of the set M defined by (2.14).

The right hand side in estimate (2.16) is allowed to be infinity. Therefore it is important to find conditions under which the parameter

$$c_\varrho(M,\delta) \equiv \limsup_{\varepsilon \to 0} \mathcal{C}_\varrho(M;\varepsilon, \delta\varepsilon)$$

is finite. We do it in the following assertion which is proved to be useful in problems with nonlinear damping.

THEOREM 2.15. *Let X be a Banach space and M be a bounded closed set in X. Assume that there exists a mapping $V : M \mapsto X$ such that*
  (i) $M \subseteq VM$;
  (ii) V *is Lipschitz on M, i.e, there exists $L > 0$ such that*

(2.17) $$\|Vv_1 - Vv_2\| \leq L\|v_1 - v_2\|,\ v_1, v_2 \in M;$$

  (iii) *there exist compact seminorms $n_1(x)$ and $n_2(x)$ on X such that*

(2.18) $$\|Vv_1 - Vv_2\| \leq \eta\|v_1 - v_2\| + K \cdot [n_1(v_1 - v_2) + n_2(Vv_1 - Vv_2)]$$

  for any $v_1, v_2 \in M$, where $0 < \eta < 1$ and $K > 0$ are constants (a seminorm $n(x)$ on X is said to be compact iff for any bounded set $B \subset X$ there exists a sequence $\{x_n\} \subset B$ such that $n(x_m - x_n) \to 0$ as $m, n \to \infty$).

Then M is a compact set in X of a finite fractal dimension. Moreover, we have the estimate

(2.19) $$\dim_f M \leq \left[\ln\frac{2}{1+\eta}\right]^{-1} \cdot \ln m_0\left(\frac{4K(1+L^2)^{1/2}}{1-\eta}\right),$$

where $m_0(R)$ is the maximal number of pairs (x_i, y_i) in $X \times X$ possessing the properties

$$\|x_i\|^2 + \|y_i\|^2 \leq R^2,\ n_1(x_i - x_j) + n_2(y_i - y_j) > 1,\ i \neq j.$$

REMARK 2.16. It is easy to see that under the hypothesis concerning the seminorms n_1 and n_2 the characteristic $m_0(R)$ is finite for any fixed R (see, e.g., the argument given in the proof of Lemma 2.19 below). We also note that it was proved in [27] that if X is a separable Hilbert space and the seminorms n_1 and n_2 have the form $n_i(v) = \|P_i v\|$, $i = 1, 2$, where P_1 and P_2 are finite-dimensional orthoprojectors, then

$$(2.20) \quad \dim_f M \leq (\dim P_1 + \dim P_2) \cdot \ln\left(1 + \frac{8(1+L)\sqrt{2}K}{1-\eta}\right) \cdot \left[\ln \frac{2}{1+\eta}\right]^{-1}.$$

We also refer to [28] for an analysis of the dimension problem in the case of more general (nonlinear) relations of the type (2.18).

REMARK 2.17. The result given in Theorem 2.15 generalizes squeezing properties given in [71, 39] and most recently [89]. Indeed, Ladyzhenskaya's theorem [71] on finite dimension of invariant sets follows from Theorem 2.15. To see this we assume that X is a separable Hilbert space and take $n_1 \equiv 0$ and $n_2(v) = \|Pv\|$ in relation (2.18), where P is a finite dimensional projector. In this case relation (2.18) easily follows from Ladyzhenskaya's squeezing assumption:

$$\|(I-P)(Vv_1 - Vv_2)\| \leq \eta \|v_1 - v_2\|, \ v_1, v_2 \in M.$$

Theorem 2.15 also generalizes the result by Prazak [89], which relies on the so-called "generalized squeezing property". To obtain the conclusion of Lemma 4.1 [89] on dimension we need only apply Theorem 2.15 in a separable Hilbert space X with $n_1(a) = n_2(a) = \|Pa\|$, where P is a finite dimensional projector. One of the main advantages of our approach in comparison with results in [71] and [89] is that Theorem 2.15 does not contain finite-dimensional projectors in explicit form. This fact is very handy in applications to hyperbolic problems with boundary nonlinear damping.

For the proof of Theorem 2.14 we need the following lemma which is also useful in the construction of inertial sets in Sect. 2.3.

LEMMA 2.18. *Let M be a bounded closed set in a complete metric space (X, d). Assume that $V : M \mapsto X$ is a mapping such that (2.15) hold with some $\eta > 0$. Then*

$$(2.21) \quad \alpha(VB) \leq \eta \cdot \alpha(B) \text{ for any } B \subseteq M,$$

where $\alpha(B)$ is the Kuratowski α-measure of noncompactness of the set B defined by (2.5). Thus V is α-contraction [59] on M provided $\eta < 1$.

PROOF. We use the same idea as in the proof of Theorem 2.4. By the definition of $\alpha(B)$ (see (2.5)), for any $\varepsilon > 0$ there exist sets $F_1, \ldots, F_n$ such that

$$B = F_1 \cup \ldots \cup F_n, \quad \operatorname{diam} F_i < \alpha(B) + \varepsilon.$$

Since ϱ is a compact pseudometric on M, the maximal (ε, ϱ)-distinguishable subset of B is finite. Thus there exists a finite set $\mathcal{N} = \{x_i : i = 1, 2 \ldots m\} \subset B$ such that for every $y \in B$ there is $i \in \{1, 2 \ldots m\}$ with the property $\varrho(y, x_i) \leq \varepsilon$. It means that

$$B = \cup_{i=1}^m C_i, \quad C_i = \{y \in B : \varrho(y, x_i) \leq \varepsilon\}.$$

No we can write the representations

$$B = \cup_{i,j}(C_i \cap F_j), \quad \text{and} \quad VB = \cup_{i,j}(V(C_i \cap F_j)).$$

Using (2.15) it is easy to see that
$$\operatorname{diam}(V(C_j \cap F_i)) \leq \eta \cdot \alpha(B) + \varepsilon \cdot [2 + \eta].$$
This implies (2.21). □

Proof of Theorem 2.14. Lemma 2.18 implies that
$$\alpha(M) \leq \alpha(VM) \leq \eta \cdot \alpha(M).$$
Since $0 < \eta < 1$, this is possible only if $\alpha(M) = 0$. Thus M is compact.

Assume that $\{F_i : i = 1, \ldots, N(M, \varepsilon)\}$ is the minimal covering of M by its closed subsets with a diameter equal or less than 2ε. Let $0 < \delta < 1$ and $\{x_j^i : j = 1, \ldots, n_i\} \subset F_i$ be a maximal $(\delta\varepsilon, \varrho)$-distinguishable subset of F_i. Since ϱ is compact, this finite set exists, and by Definition 2.13 we have that
$$n_i = m_\varrho(F_i, \delta\varepsilon) \leq \exp\{\mathcal{C}_\varrho(M; \varepsilon, \delta\varepsilon)\},$$
where $\mathcal{C}_\varrho(M; r, \varepsilon)$ is the local $(r, \varepsilon, \varrho)$-capacity of the set M defined by (2.14). We also have that
$$F_i \subset \bigcup_{j=1}^{n_i} B_j^i, \ B_j^i \equiv \{v \in F_i : \rho(v, x_j^i) \leq \delta \cdot \varepsilon\}.$$
Therefore
$$VM \subset \bigcup_{i=1}^{N(M,\varepsilon)} \bigcup_{j=1}^{n_i} VB_j^i.$$
If $y_1, y_2 \in B_j^i$, then from (2.15) we have
$$d(Vy_1, Vy_2) \leq \eta d(y_1, y_2) + \rho(y_1, x_j^i) + \rho(y_2, x_j^i) \leq 2(\eta + \delta)\varepsilon.$$
Thus $\operatorname{diam} VB_j^i \leq 2(\eta + \delta)\varepsilon$ for any $\varepsilon > 0$ and $0 < \delta < 1$. Therefore
$$(2.22) \qquad N(VM, (\eta + \delta)\varepsilon) \leq \exp\{\mathcal{C}_\varrho(M; \varepsilon, \delta\varepsilon)\} \cdot N(M, \varepsilon).$$
For further use we note that relation (2.22) remains true *without* the hypotheses $M \subseteq VM$.

If we choose $0 < \delta < 1 - \eta$, then from (2.22) under the assumption $M \subseteq VM$ we obtain that
$$\ln N(M, q\varepsilon) \leq \mathcal{C}_\varrho(M; \varepsilon, \delta\varepsilon) + \ln N(M, \varepsilon)$$
for any $\varepsilon > 0$, where $q = \eta + \delta < 1$. Let $\varepsilon_n = q^n \varepsilon_0$ for some $\varepsilon_0 > 0$. It is clear that
$$(2.23) \qquad \ln N(M, \varepsilon_n) \leq \sum_{k=0}^{n-1} \mathcal{C}_\varrho(M; \varepsilon_k, \delta\varepsilon_k) + \ln N(M, \varepsilon_0), \quad n = 1, 2, \ldots.$$
Let $\kappa > 0$ be arbitrary. We can choose $\varepsilon_0 = \varepsilon_0(\kappa) > 0$ such that
$$\mathcal{C}_\varrho(M; \varepsilon, \delta\varepsilon) \leq c_\varrho(M, \delta) + \kappa \quad \text{for all} \quad 0 < \varepsilon \leq \varepsilon_0,$$
where
$$c_\varrho(M, \delta) = \limsup_{\varepsilon \to 0} \mathcal{C}_\varrho(M; \varepsilon, \delta\varepsilon).$$
Therefore from (2.23) we obtain that
$$\ln N(M, \varepsilon_n) \leq n \cdot (c_\varrho(M, \delta) + \kappa) + \ln N(M, \varepsilon_0), \ n = 1, 2, \ldots.$$
Now for any $\varepsilon < \varepsilon_0$ we can find $n = n_\varepsilon$ and $\widetilde{\varepsilon} \in [\varepsilon_1, \varepsilon_0)$ such that
$$(2.24) \qquad \varepsilon_{n+1} \leq \varepsilon < \varepsilon_n, \ \varepsilon = q^n \widetilde{\varepsilon}.$$

Hence we have that
$$\ln N(M,\varepsilon) \leq n_\varepsilon \cdot (c_\varrho(M,\delta) + \kappa) + \ln N(M,\widetilde{\varepsilon})$$
$$\leq n_\varepsilon \cdot (c_\varrho(M,\delta) + \kappa) + \ln N(M,\varepsilon_1).$$
Thus, by (2.13) we obtain that
$$\dim_f M \leq (c_\varrho(M,\delta) + \kappa) \cdot \limsup_{\varepsilon \to 0} \frac{n_\varepsilon}{\ln(1/\varepsilon)}.$$
It follows from (2.24) that
$$n_\varepsilon \leq \frac{\ln(\widetilde{\varepsilon}/\varepsilon)}{\ln(1/q)} \leq \frac{\ln(\varepsilon_0/\varepsilon)}{\ln(1/q)}.$$
Therefore
$$\dim_f M \leq (c_\varrho(M,\delta) + \kappa) \cdot \frac{1}{\ln(1/q)} \quad \text{for every} \quad \kappa > 0.$$
This implies (2.16). The proof of Theorem 2.14 is complete. □

Proof of Theorem 2.15. It is sufficient to prove the following lemma.

LEMMA 2.19. *Under conditions of Theorem 2.15 the pseudometric*
$$(2.25) \qquad \varrho(x,y) = K \cdot (n_1(x-y) + n_2(Vx - Vy))$$
is compact on M in the sense of Definition 2.13. The local (r,ε,ϱ)-capacity of the set M with respect to this pseudometric admits the estimate
$$(2.26) \qquad C_\varrho(M;r,\varepsilon) \leq \ln m_0\left(\frac{2K(1+L^2)^{1/2}r}{\varepsilon}\right),$$
where $m_0(R)$ is the maximal number of pairs (x_i, y_i) in $X \times X$ possessing the properties
$$\|x_i\|^2 + \|y_i\|^2 \leq R^2, \quad n_1(x_i - x_j) + n_2(y_i - y_j) > 1, \ i \neq j.$$

PROOF. If ϱ is not compact, then for some $\varepsilon > 0$, there exists a sequence $\{z_n\} \subset M$ such that
$$(2.27) \qquad n_1(z_n - z_m) + n_2(Vz_n - Vz_m) \geq \varepsilon \quad \text{for all} \quad n \neq m.$$
Since M is bounded, the sequences $\{z_n\}$ and $\{Vz_n\}$ are bounded. Since the seminorms n_1 and n_2 are compact, this implies that there exists a subsequence $\{z_{n_k}\}$ such that
$$n_1(z_{n_k} - z_{n_l}) + n_2(Vz_{n_k} - Vz_{n_l}) \to 0 \quad \text{when} \quad k,l \to \infty,$$
which is impossible due to (2.27). Thus the pseudometric ϱ is compact on M.

Now we prove (2.26). Let $B \subset M$ and $m_\varrho(B,\varepsilon)$ be the maximal cardinality of an (ε,ϱ)-distinguishable subset of B. In the space $X \times X$ endowed with the norm
$$|Z| = \left(\|x\|^2 + \|y\|^2\right)^{1/2}, \quad Z = (x,y) \in X \times X,$$
we consider the set $\mathcal{B} = \{Z = (x, Vx) : x \in B\}$. It is clear that
$$m_\varrho(B,\varepsilon) = \aleph\left\{Z_i \in \mathcal{B} : N(Z_i - Z_j) > \varepsilon \cdot K^{-1}, \ i \neq j\right\},$$
where $N(Z) = n_1(x) + n_2(y)$, $Z = (x,y)$, and $\aleph\{...\}$ denotes the maximal number of elements with the given properties. Since by (2.17)
$$\operatorname{diam}\mathcal{B} = \sup_{x,y \in B}\left(\|x-y\|^2 + \|Vx - Vy\|^2\right)^{1/2} \leq R \equiv \left(1 + L^2\right)^{1/2}\operatorname{diam}B,$$

there exist $Y_0 \in \mathcal{B}$ such that
$$\mathcal{B} \subset B_R(Y_0) \equiv \{Z \in X \times X : |Z - Y_0| \leq R\}$$
Therefore, using the property $N(\lambda Z) = \lambda N(Z)$ for any $\lambda > 0$ and $Z = (x, y)$, we obtain that
$$\begin{aligned} m_\varrho(B, \varepsilon) &\leq \aleph\{Z_i \in B_R(Y_0) : N(Z_i - Z_j) > \varepsilon \cdot K^{-1}, \; i \neq j\} \\ &= \aleph\{Z_i \in B_R(0) : N(Z_i - Z_j) > \varepsilon \cdot K^{-1}, \; i \neq j\} \\ &= \aleph\{Z_i \in B_{RK/\varepsilon}(0) : N(Z_i - Z_j) > 1, \; i \neq j\} = m_0(RK/\varepsilon). \end{aligned}$$
By Definition (2.14) this implies (2.26). $\square$

To conclude the proof of Theorem 2.15 we need only apply Theorem 2.14 with ϱ given by (2.25) and $\delta = (1-\eta)/2$ and use Lemma 2.19.

Theorem 2.14 also implies the following assertion.

COROLLARY 2.20. *Let X be a Banach space and M be a bounded closed set in X. Assume that there exists a mapping $V : M \mapsto X$ such that*
  (i) *$M \subseteq VM$;*
  (ii) *the mapping V admits the splitting*
(2.28) $$V = S + K,$$
  where S is Lipschitz and stable on M, i.e, there exists $0 < \eta < 1$ such that
(2.29) $$\|Sv_1 - Sv_2\| \leq \eta\|v_1 - v_2\|, \quad v_1, v_2 \in M,$$
  and K is a Lipschitz mapping from M into Y, where Y is a Banach space which is compactly embedded in X, i.e.,
(2.30) $$\|Kv_1 - Kv_2\|_Y \leq L_K\|v_1 - v_2\|, \quad v_1, v_2 \in M.$$
Then M is a compact set in X of a finite fractal dimension. Moreover, we have the estimate
(2.31) $$\dim_f M \leq \left[\ln \frac{2}{1+\eta}\right]^{-1} \cdot \ln m_{Y,X}\left(\frac{4L_K}{1-\eta}\right),$$
where $m_{Y,X}(R)$ is the maximal number of points x_i in the ball of the radius R in Y possessing the properties $\|x_i - x_j\| > 1$, $i \neq j$.

PROOF. It follows from (2.29) and (2.30) that V is Lipschitz on M and that relation (2.15) holds with
$$d(v_1, v_2) = \|v_1 - v_2\| \quad \text{and} \quad \varrho(v_1, v_2) = \|Kv_1 - Kv_2\|.$$
A simple argument shows that this pseudometric ϱ is compact.

Thus we need only check relation (2.31). We shall use the same method as in Lemma 2.19. Let $B \subset M$ and $m_\varrho(B, \varepsilon)$ be the maximal cardinality of an (ε, ϱ)-distinguishable subset of B, i.e,
$$\begin{aligned} m_\varrho(B, \varepsilon) &= \aleph\{x_i \in B : \|Kx_i - Kx_j\| > \varepsilon, \; i \neq j\} \\ &= \aleph\{z_i \in KB : \|z_i - z_j\| > \varepsilon, \; i \neq j\}, \end{aligned}$$
where, as in Lemma 2.19, $\aleph\{...\}$ denotes the maximal number of elements with the given properties. It is easy to see from (2.30) that
$$KB \subset B_R^Y(y_0) = \{y \in Y : \|y - y_0\|_Y \leq R \equiv L_K \mathrm{diam} B\}$$

for some $y_0 \in KB$. Therefore, as in Lemma 2.19, we can conclude that
$$m_\varrho(B,\varepsilon) \le \aleph\{z_i \in B_R^Y(y_0) : \|z_i - z_j\| > \varepsilon,\ i \ne j\}$$
$$= \aleph\{z_i \in B_R^Y(0) : \|z_i - z_j\| > \varepsilon,\ i \ne j\}$$
$$= \aleph\{z_i \in B_{R/\varepsilon}^Y(0) : \|z_i - z_j\| > 1,\ i \ne j\} = m_{Y,X}(R/\varepsilon).$$

Therefore by Definition (2.14) we have
$$(2.32) \qquad \mathcal{C}_\varrho(M;r,\varepsilon) \le \ln m_{Y,X}\left(\frac{2L_K r}{\varepsilon}\right).$$

Hence estimate (2.31) follows from (2.16) with $\delta = (1-\eta)/2$. □

If in the splitting (2.28) we have $S \equiv 0$, then we easily arrive to the following assertion which was proved in [**85**, Lemma 1.3].

COROLLARY 2.21. *Let X and Y be Banach spaces such that Y is compactly embedded in X. Let M be a bounded closed set in X. Assume that $V : M \mapsto Y$ is a Lipschitz mapping from M into Y, i.e.,*
$$\|Vv_1 - Vv_2\|_Y \le L\|v_1 - v_2\|_X,\ v_1, v_2 \in M.$$
If $M \subset VM$, then M is a compact set in X and its fractal dimension (in X) admits the estimate
$$\dim_f M \le \frac{\ln m_{Y,X}(4L)}{\ln 2},$$
where $m_{Y,X}(R)$ is the same as in the Corollary 2.20.

PROOF. We apply Corollary 2.20 with $S \equiv 0$ and $\eta = 0$. □

2.3. Exponentially attracting positively invariant sets

In Theorem 2.14 we dealt with a negatively invariant set M. As for positively invariant set, their finite dimensionality is not guaranteed by condition (2.15). However, as the following theorem states, they are attracted by finite-dimensional compacts with exponential speed.

THEOREM 2.22. *Let (X,d) be a complete metric space and M be a bounded closed set in X. Assume that the mapping $V : M \mapsto M$ possesses the properties:*

(i) *there exist a compact pseudometric ϱ on M and a number $0 < \eta < 1$ such that*
$$(2.33) \qquad d(Vv_1, Vv_2) \le \eta \cdot d(v_1, v_2) + \varrho(v_1, v_2),\ v_1, v_2 \in M;$$

(ii) *for some $\delta \in (0, 1-\eta)$ we have that*
$$(2.34) \qquad c_\varrho(M, \delta) \equiv \limsup_{\varepsilon \to 0} \mathcal{C}_\varrho(M; \varepsilon, \delta\varepsilon) < \infty,$$

where $\mathcal{C}_\varrho(M; r, \varepsilon)$ is the local $(r, \varepsilon, \varrho)$-capacity of the set M defined by (2.14).

Then for any $\kappa > 0$ there exists a positively invariant compact set $A_{q,\kappa} \subset M$ of finite fractal dimension satisfying
$$(2.35) \qquad \sup\{\mathrm{dist}(V^k u, A_{q,\kappa}) : u \in M\} \le q^k,\quad k = 1, 2, \ldots,$$

where $q = \eta + \delta < 1$, and

$$\dim_f A_{q,\kappa} \leq \left[\ln \frac{1}{\delta + \eta}\right]^{-1} \cdot (c_\varrho(M, \delta) + \kappa). \tag{2.36}$$

We note that we construct finite-dimensional exponentially attracting set $A_{q,\kappa}$ with the estimate of $\dim_f A_{q,\kappa}$ as close as we want to the estimate of the dimension of the negatively invariant set M in Theorem 2.14. It is also clear from the construction given below that the positively invariant sets $A_{q,\kappa}$ with properties (2.35) and (2.36) are not unique.

PROOF. We rely on some ideas presented in [**39**] (see also [**19**, Chap.1]).

By Lemma 2.18 V is an α-contraction on the set M in the sense of [**59**, p.14]. Therefore (see [**59**, Chap.2]) the set $M_0 = \cap_{n \geq 1} V^n M$ is a compact global attractor for the discrete dynamical system (M, V^k). By Theorem 2.14 it follows from (2.34) that $\dim_f M_0 < \infty$. We shall construct a set $A_{q,\kappa}$ as an extension of M_0.

It follows from (2.22) that

$$N(VM, q\varepsilon) \leq \exp\{\mathcal{C}_\varrho(M; \varepsilon, \delta\varepsilon)\} \cdot N(M, \varepsilon)$$

for any $\varepsilon > 0$, where $q = \eta + \delta$, $N(B, \varepsilon)$ is the cardinality of the minimal covering of B by its closed subsets with a diameter equal or less than 2ε and $\mathcal{C}_\varrho(M; \varepsilon, \delta\varepsilon)$ given by (2.14). Taking $V^{n-1}M$ instead of M in the previous formula we obtain that

$$N(V^n M, q\varepsilon) \leq \exp\{\mathcal{C}_\varrho(V^{n-1}M; \varepsilon, \delta\varepsilon)\} \cdot N(V^{n-1}M, \varepsilon), \quad n = 1, 2, \ldots$$

Since $V^{n-1}M \subset M$, it follows from (2.14) that

$$\mathcal{C}_\varrho(V^{n-1}M; \varepsilon, \delta\varepsilon) \leq \mathcal{C}_\varrho(M; \varepsilon, \delta\varepsilon), \quad n = 1, 2, \ldots$$

Thus we have that

$$N(V^n M, q\varepsilon) \leq \exp\{\mathcal{C}_\varrho(M; \varepsilon, \delta\varepsilon)\} \cdot N(V^{n-1}M, \varepsilon), \quad n = 1, 2, \ldots$$

Let $\kappa > 0$ be arbitrary and $\varepsilon_n = q^n \varepsilon_0$, where we choose $0 < \varepsilon_0 \leq 1/2$ such that

$$\mathcal{C}_\varrho(M; \varepsilon, \delta\varepsilon) \leq c_\varrho(M, \delta) + \kappa \quad \text{for all} \quad 0 < \varepsilon \leq \varepsilon_0.$$

In this case we have that

$$N(V^n M, \varepsilon_n) \leq \exp\{n(c_\varrho(M, \delta) + \kappa)\} \cdot N(M, \varepsilon_0), \quad n = 1, 2, \ldots \tag{2.37}$$

We denote by E_n the maximal $(2\varepsilon_n, d)$-distinguishable subset of $V^n M$. It follows from [**68**, Theorem 4] that

$$\text{Card} E_n = m_d(V^n M, 2\varepsilon_n) \leq N(V^n M, \varepsilon_n). \tag{2.38}$$

Thus by (2.37) E_n is a finite set for each $n = 1, 2, \ldots$ As in [**39**] (see also [**19**, Chap.1]) we set

$$A_{q,\kappa} = M_0 \cup \{V^k E_m : k, m = 0, 1, 2, \ldots\}.$$

This set depends on ε_0 and, hence, on κ.

Let us prove that the set $A_{q,\kappa}$ satisfies the conclusions of the theorem.

It is easy to to see that $A_{q,\kappa}$ is a compact positively invariant set. We also have that

$$\text{dist}_H(V^n y, A_{q,\kappa}) \leq \text{dist}_H(V^n y, E_n) \leq 2\varepsilon_n = 2q^n \varepsilon_0 \leq q^n, \quad n = 0, 1, 2, \ldots,$$

for every $y \in M$. This implies (2.35).

To estimate the fractal dimension of $A_{q,\kappa}$ we note that the property $E_m \subset V^m M$ implies that
$$A_{q,\kappa} \subset V^n M \cup \{V^k E_m : k+m \le n-1, k, m \ge 0\}$$
for every $n = 1, 2, \ldots$ Therefore
$$N(A_{q,\kappa}, \varepsilon_n) \le N(V^n M, \varepsilon_n) + \sum_{\substack{k,m \ge 0 \\ k+m \le n-1}} \mathrm{Card}\{V^k E_m\}$$

$$\le N(V^n M, \varepsilon_n) + \sum_{\substack{k,m \ge 0 \\ k+m \le n-1}} \mathrm{Card}\{E_m\}.$$

It follows from (2.37) and (2.38) that
$$N(A_{q,\kappa}, \varepsilon_n) \le \exp\{n(c_\varrho(M, \delta) + \kappa)\} \cdot N(M, \varepsilon_0) \cdot (1 + \Sigma_n),$$
where
$$\Sigma_n = \sum_{\substack{k,m \ge 0 \\ k+m \le n-1}} \exp\{-(n-m)(c_\varrho(M, \delta) + \kappa)\} \le \frac{n(n+1)}{2}.$$

Consequently
$$\ln N(A_{q,\kappa}, \varepsilon_n) \le n \cdot (c_\varrho(M, \delta) + \kappa) + \ln N(M, \varepsilon_0) + \ln\left(1 + \frac{n(n+1)}{2}\right).$$

Therefore, as in the proof of Theorem 2.14, we can conclude that
$$\ln N(A_{q,\kappa}, \varepsilon) \le n_\varepsilon \cdot (c_\varrho(M, \delta) + \kappa) + \ln N(M, q\varepsilon_0) + \ln\left(1 + \frac{n_\varepsilon(n_\varepsilon + 1)}{2}\right)$$
for any $\varepsilon \in (0, \varepsilon_0)$, where $n_\varepsilon \le [\ln(1/q)]^{-1} \cdot \ln(\varepsilon_0/\varepsilon)$. Thus estimate (2.36) follows from (2.13). □

Under the hypotheses of Theorem 2.22 the discrete dynamical system (M, V^k) possesses a compact global attractor M_0. This attractor uniformly attracts all the trajectory of the system (M, V^k) and by Theorem 2.14 $\dim_f M_0 < \infty$. Unfortunately, in general the speed of convergence to the attractor cannot be estimated. This speed can appear to be small. However Theorem 2.22 says that the global attractor is contained in finite-dimensional positively invariant set which attracts M uniformly and exponentially fast. Thus the dynamics of the system becomes finite-dimensional exponentially fast independent of the speed of convergence to the global attractor. Moreover, the reduction principle is applicable in this case. Thus finite-dimensional positively invariant exponentially attracting sets can be useful for description of qualitative behaviour of infinite-dimensional systems. These sets are frequently called *inertial sets* or *fractal exponential attractors*. In some cases they turn out to be surfaces in the phase space. For details we refer to [**39**] and to the references therein.

In conclusion of this section we give two corollaries of Theorem 2.22 which are important in subsequent considerations.

COROLLARY 2.23. *Let X be a Banach space and M be a bounded closed set in X. Assume that the mapping $V : M \mapsto M$ possesses the properties*

(i) *V is Lipschitz on M, i.e, there exists $L > 0$ such that (2.17) holds;*

(ii) there exist compact seminorms $n_1(x)$ and $n_2(x)$ on X such that (2.18) holds.

Then for any $\kappa > 0$ and $\delta \in (0, 1 - \eta)$ there exists a positively invariant compact set $A_{\kappa,\delta} \subset M$ of finite fractal dimension such that

$$\sup \left\{ \operatorname{dist}(V^k u, A_{\kappa,\delta}) \,:\, u \in M \right\} \leq q^k, \quad k = 1, 2, \ldots, \tag{2.39}$$

where $q = \eta + \delta < 1$, and

$$\dim_f A_{\kappa,\delta} \leq \left[\ln \frac{1}{\eta + \delta} \right]^{-1} \cdot \left[\kappa + \ln m_0 \left(\frac{2K(1+L^2)^{1/2}}{\delta} \right) \right],$$

where, as in Theorem 2.15, $m_0(R)$ is the maximal number of pairs (x_i, y_i) in $X \times X$ possessing the properties

$$\|x_i\|^2 + \|y_i\|^2 \leq R^2, \ n_1(x_i - x_j) + n_2(y_i - y_j) > 1, \ i \neq j.$$

PROOF. We apply Theorem 2.22 with the pseudometric

$$\varrho(x, y) = K \cdot (n_1(x - y) + n_2(Vx - Vy)).$$

By Lemma 2.19 ϱ is compact and we have that

$$c_\varrho(M; \delta) \leq \ln m_0 \left(\frac{2K(1+L^2)^{1/2}}{\delta} \right).$$

Therefore the statement of the corollary follows from Theorem 2.22. □

Theorem 2.22 also implies the following assertion.

COROLLARY 2.24. Let X be a Banach space and M be a bounded closed set in X. Assume that $V : M \mapsto M$ is a mapping such that it admits the splitting

$$V = S + K,$$

where S is Lipschitz and stable on M, i.e, there exists $0 < \eta < 1$ such that

$$\|Sv_1 - Sv_2\| \leq \eta \|v_1 - v_2\|, \ v_1, v_2 \in M,$$

and K is a Lipschitz mapping from M into Y, where Y is a Banach space which is compactly embedded in X, i.e.,

$$\|Kv_1 - Kv_2\|_Y \leq L_K \|v_1 - v_2\|, \ v_1, v_2 \in M.$$

Then for any $\kappa > 0$ and $\delta \in (0, 1 - \eta)$ there exists a positively invariant compact set $A_{\kappa,\delta} \subset M$ of finite fractal dimension such that (2.39) holds and

$$\dim_f A_{\kappa,\delta} \leq \left[\ln \frac{1}{\eta + \delta} \right]^{-1} \cdot \left[\kappa + \ln m_{Y,X} \left(\frac{2L_K}{\delta} \right) \right],$$

where $m_{Y,X}(R)$ is the maximal number of points x_i in the ball of the radius R in Y possessing the properties $\|x_i - x_j\| > 1$, $i \neq j$.

This corollary was proved earlier in [**43**] under the restriction that the contractivity constant η less than $1/2$.

PROOF. We apply Theorem 2.22 with

$$d(v_1, v_2) = \|v_1 - v_2\| \quad \text{and} \quad \varrho(v_1, v_2) = \|Kv_1 - Kv_2\|.$$

Relation (2.34) follows from (2.32). □

2.4. Gradient systems

The study of the structure of the global attractors is an important problem from point of view of applications. There are no universal approaches to this problem. Even in finite-dimensional case an attractor can possess extremely complicate structure. However for some cases one can give a geometrical description of the global attractor.

In this subsection for the sake of further references we collect several known results on the properties of dynamical systems possessing Lyapunov function. These results mainly concern the structure of global attractors.

Let $\mathcal{N}$ be the set of stationary points of the dynamical system (X, S_t), i.e.
$$\mathcal{N} = \{v \in X \,:\, S_t v = v \text{ for all } t \geq 0\},$$
and define the *unstable manifold* $\mathcal{M}^u(\mathcal{N})$ emanating from the set $\mathcal{N}$ as a set of all $y \in X$ such that there exists a full trajectory $\gamma = \{u(t) : t \in \mathbb{R}\}$ with the properties
$$u(0) = y \quad \text{and} \quad \lim_{t \to -\infty} \operatorname{dist}_X(u(t), \mathcal{N}) = 0.$$

It is clear that $\mathcal{M}^u(\mathcal{N})$ is an invariant set. It is also easy to prove the following assertion (see, e.g., [4], [19] or [**101**]).

PROPOSITION 2.25. *Let $\mathcal{N}$ be the set of stationary points of the dynamical system (X, S_t) possessing a global attractor A. Then $\mathcal{M}^u(\mathcal{N}) \subset A$.*

For gradient systems it is possible to prove that $\mathcal{M}^u(\mathcal{N}) = A$. We give the following definition.

DEFINITION 2.26. *Let $Y \subseteq X$ be a positively invariant set of a dynamical system (X, S_t).*

- *The continuous functional $\Phi(y)$ defined on Y is said to be the Lyapunov function for the dynamical system (X, S_t) on Y iff the function $t \mapsto \Phi(S_t y)$ is nonincreasing function for any $y \in Y$.*
- *The Lyapunov function $\Phi(y)$ is said to be strict on Y iff the equation $\Phi(S_t y) = \Phi(y)$ for all $t > 0$ and for some $y \in Y$ implies that $S_t y = y$ for all $t > 0$, i.e. y is a stationary point of (X, S_t).*
- *The dynamical system (X, S_t) is said to be gradient iff there exists a strict Lyapunov function for (X, S_t) on the whole phase space X.*

EXAMPLE 2.27. The system (H, S_t) generated in the space $H = H_0^1(\Omega) \times L_2(\Omega)$ by damped wave equation (1.8) with the Dirichlet boundary conditions is gradient provided (i) g and f satisfies the assumptions listed in Example 1.2, (ii) $g(s) = 0$ if and only if $s = 0$, and (iii) $s \mapsto sg(s)$ is a convex function on $\mathbb{R}$. Indeed, it follows from Proposition 1.10 and Remark 1.11 that the energy
$$\mathcal{E}(y) = \frac{1}{2} \int_\Omega \left(|u_1|^2 + |\nabla u_0|^2\right) dx + \int_\Omega \hat{f}(u_0) dx, \quad y = (u_0, u_1) \in H,$$
where $\hat{f}$ is the antiderivative of the nonlinear term f, satisfies the inequality
$$(2.40) \qquad \mathcal{E}(S_t y) + k \int_0^t u_t(\tau) g(u_t(\tau)) d\tau \leq \mathcal{E}(y),$$
for any $y \in H$. Here $(u(t), u_t(t)) = S_t y$. Therefore $t \mapsto \mathcal{E}(S_t y)$ is nonincreasing function for any $y \in H$. If $\mathcal{E}(S_t y) = \mathcal{E}(y)$ for some $t > 0$, then (2.40) implies that $\int_0^t u_t(\tau) g(u_t(\tau)) d\tau = 0$. Therefore $u_t(\tau) = 0$ for all $\tau \in [0, t]$. This is possible only

if $u(t) \equiv u_0$ is a stationary solution. Thus the system (H, S_t) possesses a strict Lyapunov function $\Phi(y) = \mathcal{E}(y)$.

2.4.1. Geometric structure of global attractors. We have the following result on the structure of a global attractor (for the proof we refer to any book from the list [4, 19, 59, 73, 101]).

THEOREM 2.28. *Let a dynamical system (X, S_t) possess a compact global attractor A. Assume that there exists a strict Lyapunov function on A. Then $A = \mathcal{M}^u(\mathcal{N})$. Moreover the global attractor A consists of full trajectories $\gamma = \{u(t) : t \in \mathbb{R}\}$ such that*
$$\lim_{t \to -\infty} \operatorname{dist}_X(u(t), \mathcal{N}) = 0 \text{ and } \lim_{t \to +\infty} \operatorname{dist}_X(u(t), \mathcal{N}) = 0.$$

The following assertion shows that in the case of gradient systems we can guarantee existence of a compact global attractor without assuming dissipativity in explicit form.

COROLLARY 2.29. *Assume that (X, S_t) is a gradient dynamical system which, moreover, is asymptotically smooth. Assume that Lyapunov function $\Phi(x)$ associated with the system is bounded from above on any bounded subset of X and the set $\Phi_R = \{x : \Phi(x) \leq R\}$ is bounded for every R. If the set $\mathcal{N}$ of stationary points of (X, S_t) is bounded, then (X, S_t) possesses a compact global attractor.*

PROOF. We choose R_0 such that $\mathcal{N} \subset \Phi_{R_0}$. For any $R > 0$ the set Φ_R is bounded and forward invariant. Therefore by Theorem 2.3 the dynamical system (Φ_R, S_t) possesses a compact global attractor A_R for every R. If $R \geq R_0$, then from Theorem 2.28 we have that $A_R = \mathcal{M}^u(\mathcal{N})$. Thus A_R does not depend on R for $R \geq R_0$ and it is global attractor for (X, S_t). □

If the system (X, S_t) is not gradient but it possesses a Lyapunov function, we cannot guarantee that $A = \mathcal{M}^u(\mathcal{N})$. However we can prove the following assertion.

THEOREM 2.30. *Let (X, S_t) be a asymptotically smooth dynamical system in a Banach space X. Assume that there exists a Lyapunov function $\Phi(x)$ for (X, S_t) on X such that $\Phi(x)$ is bounded from above on any bounded subset of X and the set $\Phi_R = \{x : \Phi(x) \leq R\}$ is bounded for every R. Let $\mathcal{B}$ be the set of elements $x \in X$ such that there exists a full trajectory $\{u(t) : t \in \mathbb{R}\}$ with the properties $u(0) = x$ and $\Phi(u(t)) = \Phi(x)$ for all $t \in \mathbb{R}$. If $\mathcal{B}$ is bounded, then (X, S_t) possesses a compact global attractor and $A = \mathcal{M}^u(\mathcal{B})$.*

PROOF. As in the proof of Corollary 2.29 we choose R_0 such that $\mathcal{B} \subset \Phi_{R_0}$. By Theorem 2.3 the dynamical system (Φ_R, S_t) possesses a compact global attractor A_R for every R. If $R \geq R_0$, then $\mathcal{B} \subset \Phi_R$ and therefore by Theorem 1.6.2 [19] $A_R = \mathcal{M}^u(\mathcal{B})$ for all $R \geq R_0$. Thus $A = \mathcal{M}^u(\mathcal{B})$ is global attractor for (X, S_t). □

The following assertion describes long-time behavior of individual trajectories (for the proof we refer to [19] or [101], for instance).

THEOREM 2.31. *Assume that a gradient dynamical system (X, S_t) possesses a compact global attractor A. Then for any $x \in X$ we have*
$$\lim_{t \to +\infty} \operatorname{dist}_X(S_t x, \mathcal{N}) = 0, \tag{2.41}$$

i.e. any trajectory stabilizes to the set $\mathcal{N}$ of stationary points.

Assume that $\mathcal{N} = \{z_1, \ldots, z_n\}$ is a finite set. In this case we have the relation
$$\mathcal{M}^u(\mathcal{N}) = \cup_{i=1}^n \mathcal{M}^u(z_i),$$
where $\mathcal{M}^u(z_i)$ is the unstable manifold of the stationary point z_i. This is to say, $M^u(z_i)$ consists of all $y \in X$ such that there exist a full trajectory
$$\gamma = \{u(t) : t \in \mathbb{R}\} \subset X$$
with the properties $u(0) = y$ and $u(t) \to z_i$ as $t \to -\infty$.

Theorems 2.28 and 2.31 imply the following assertion.

COROLLARY 2.32. *Assume that a gradient dynamical system (X, S_t) possesses a compact global attractor A and $\mathcal{N}$ is a finite set. Then*

(i) *the global attractor A consists of full trajectories $\gamma = \{u(t) : t \in \mathbb{R}\}$ connecting pairs of stationary points, i.e. any $u \in A$ belongs some full trajectory γ and for any $\gamma \subset A$ there exists a pair $\{z, z^*\} \subset \mathcal{N}$ such that*
$$u(t) \to z \text{ as } t \to -\infty \text{ and } u(t) \to z^* \text{ as } t \to +\infty;$$

(ii) *for any $v \in X$ there exists a stationary point z such that*
$$S_t v \to z \text{ as } t \to +\infty.$$

REMARK 2.33. Assume that the hypotheses of Corollary 2.32 hold. Introduce m_0 distinct values $\Phi_1 < \Phi_2 < \ldots < \Phi_{m_0}$ of the set $\{\Phi(x) : x \in \mathcal{N}\}$ and let
$$\mathcal{N}^j = \{x \in \mathcal{N} : \Phi(x) = \Phi_j\}, \quad j = 1, \ldots, m_0.$$
Then the sets $\mathcal{N}^1, \ldots, \mathcal{N}^{m_0}$ give us a *Morse decomposition* of the attractor A, i.e. (i) the subsets $\mathcal{N}^j$ are compact, invariant and disjoint and (ii) for any $x \in A \setminus \cup_j \mathcal{N}^j$ and every full trajectory $\gamma_x \subset A$ through x there exist $k > l$ such that $\alpha(\gamma_x) \in \mathcal{N}^k$ and $\omega(\gamma_x) \in \mathcal{N}^l$. Here $\alpha(\gamma_x)$ and $\omega(\gamma_x)$ are α-limiting and ω-limiting sets of the trajectory γ_x.

2.4.2. On the rate of convergence to global attractors. In many cases it is important to know how fast the trajectories starting from bounded sets converge to global attractors. For gradient systems this rate is related to the rates of convergence of individual trajectories to equilibria (see, e.g., the results of exponential convergence proved in [**4**] for the case of hyperbolic equilibria).

Our goal in this subsection is to present a result (see Theorem 2.35 below) which gives an estimate of the rate of convergence to the global attractor under the assumption that similar estimates are known in small vicinities of stationary points. To our best knowledge the first result in this direction has been obtained by Kostin [**69**] for the case of discrete dynamical systems. Inspired by technique developed by [**69**] we obtain a similar result for time continuous systems.

Our main assumption in this subsection is the following one.

ASSUMPTION 2.34. (X, S_t) *is a dynamical system on complete metric space X (endowed with the metric d) possessing the properties:*
- *There exists a compact global attractor $A = \mathcal{M}^u(\mathcal{N})$, where $\mathcal{N}$ is the set of all equilibria (we do not assume that $\mathcal{N}$ is finite).*
- *There exists a strict Lyapunov function on X such that (i) $\Phi(x)$ is bounded from above on any bounded subset of X, (ii) the set $\Phi_R = \{x : \Phi(x) < R\}$ is bounded for every R, and (iii) the set $\{\Phi(x) : x \in \mathcal{N}\}$ is finite and $\Phi_1 < \Phi_2 < \ldots < \Phi_m$ are its m distinct values.*

- There exists constants $c_R \geq 1$ and $L_R \geq 0$ such that
(2.42) $$d(S_t y_1, S_t y_2) \leq c_R e^{L_R t} d(y_1, y_2) \quad \text{for any} \quad y_1, y_2 \in \Phi_R.$$
- For every set
$$\mathcal{N}^j = \{x \in \mathcal{N} \,:\, \Phi(x) = \Phi_j\}, \quad j = 1, \ldots, m,$$
there exist a vicinity $\mathcal{O}_j$ of $\mathcal{N}^j$ and a decreasing continuous function $\psi_j : \mathbb{R}_+ \mapsto \mathbb{R}_+$ such that the property $S_t z \in \mathcal{O}_j$ for all $t \in [0, T]$ implies that
(2.43) $$\operatorname{dist}(S_t z, \mathcal{M}^u(\mathcal{N}^j)) \leq \psi_j(t), \quad t \in [0, T].$$

We note that under Assumption 2.34 the sets $\mathcal{N}^j$ comprise a Morse decomposition of the attractor A.

Let $\varepsilon > 0$ be chosen such that
(2.44) $$\varepsilon < \min_i \{\Phi_{i+1} - \Phi_i\} \quad \text{and} \quad \varepsilon < \frac{1}{2} \min_i \left\{ \Phi_i - \sup_{u \in A_i \setminus \mathcal{O}_i} \Phi(u) \right\},$$
where we have introduced the subsets
$$A_n = \bigcup_{j=1}^{n} \mathcal{M}^u(\mathcal{N}^j), \quad n = 1, 2, \ldots, m,$$
of the attractor A. It is clear that $A = A_m$. The choice of ε such that (2.44) holds is possible because $\Phi_i < \Phi_j$ for $i < j$ and $\sup_{u \in A_i \setminus \mathcal{O}_i} \Phi(u) < \Phi_i$. The latter can be easily obtained by a contradiction argument from monotonicity properties of $\Phi(u)$.

Under Assumption 2.34 in the space X we consider the following sets
$$\mathcal{B}_i = \{u \in X \,:\, \Phi(u) < \Phi_i - \varepsilon\}, \quad i = 2, \ldots, m,$$
and
$$\mathcal{B}_{m+1} = \{u \in X \,:\, \Phi(u) < R_*\},$$
where R_* is chosen such that the global attractor A belongs to $\mathcal{B}_{m+1}$ (and, thus, $\mathcal{B}_{m+1}$ is an absorbing set for (X, S_t)). It is clear that every set $\mathcal{B}_i$ is forward invariant. One can show (see, e.g., [4]) that for every $n = 1, \ldots, m$ the set A_n is a global attractor for the restriction $(\mathcal{B}_{n+1}, S_t)$ of (X, S_t) on $\mathcal{B}_{n+1}$.

THEOREM 2.35. *Under Assumption 2.34 there exist numbers*
$$0 < T_1 < T_2 < \ldots < T_m < \infty$$
such that
(2.45) $$\sup_{y \in \mathcal{B}_{k+1}} \operatorname{dist}(S_{T_k + t} y, A_k) \leq \Psi_k(t), \quad k = 1, 2, \ldots m, \ t \geq 0,$$
where $\Psi_1(t) = \psi_1(t)$ and
(2.46)
$$\Psi_{k+1}(t) = c_{R_*} e^{L_{R_*} T_k} \cdot \left[\psi_{k+1}(t) + \sup_{s \in [0,t]} \min\left\{ \psi_{k+1}(s) e^{L_{R_*}(t-s)} \,;\, \Psi_k(t-s) \right\} \right]$$
for every $k = 1, \ldots, m-1$. In particular, for any bounded set B in X there exists $t_B \geq 0$ such that
$$\sup_{y \in B} \operatorname{dist}(S_t y, A) \leq \Psi_m(t - t_B), \quad t \geq t_B,$$
where A is the global attractor for the system (X, S_t).

PROOF. It can be derived from the corresponding result for discrete dynamical systems (see [**69**, Theorem 2.2]). However, for the sake of completeness, we give independent arguments.

It is clear that $A_1 = \mathcal{M}^u(\mathcal{N}^1) = \mathcal{N}^1$. Therefore there exists $T_1 > 0$ such that
$$S_t \mathcal{B}_2 \subset \mathcal{O}_1 \quad \text{for all} \quad t \geq T_1.$$
Hence by assumption (2.43) we have that
$$\sup_{y \in \mathcal{B}_2} \operatorname{dist}(S_{T_1 + t} y, A_1) \leq \psi_1(t), \quad t \geq 0.$$
Thus (2.45) holds for $k = 1$.

Assume now that (2.45) holds for $k = n < m$ and prove it for $k = n + 1$. Let us first prove that
$$(2.47) \qquad A_{n+1} \subset \mathcal{B}_{n+1} \cup \mathcal{O}_{n+1}.$$
Indeed,
$$A_{n+1} = \bigcup_{j=1}^{n+1} \mathcal{M}^u(\mathcal{N}^j) = A_n \cup \mathcal{M}^u(\mathcal{N}^{n+1}).$$
Since $A_n \subset \mathcal{B}_{n+1}$, it is sufficient to prove that
$$\mathcal{M}^u(\mathcal{N}^{n+1}) \subset \mathcal{B}_{n+1} \cup \mathcal{O}_{n+1}.$$
If $y \in \mathcal{M}^u(\mathcal{N}^{n+1}) \setminus \mathcal{O}_{n+1} \subset A_{n+1} \setminus \mathcal{O}_{n+1}$, then by (2.44) we have
$$\Phi(y) \leq \sup_{u \in A_{n+1} \setminus \mathcal{O}_{n+1}} \Phi(u) \leq \Phi_{n+1} - 2\varepsilon.$$
Thus $y \in \mathcal{B}_{n+1}$ and therefore (2.47) holds.

Since A_{n+1} is a global attractor for $(\mathcal{B}_{n+2}, S_t)$, relation (2.47) implies that there exists T' such that
$$S_{T'+t} \mathcal{B}_{n+2} \subset \mathcal{B}_{n+1} \cup \mathcal{O}_{n+1} \quad \text{for all} \quad t \geq 0.$$
Furthermore, for any $v \in \mathcal{B}_{n+2}$ we have that either
$$S_{T'+t} v \in \mathcal{O}_{n+1} \quad \text{for all} \quad t \geq 0,$$
or there exists $\tau = \tau_v \in [0, +\infty)$ such that
$$S_{T'+t} v \in \mathcal{O}_{n+1} \quad \text{for all} \quad 0 \leq t < \tau$$
and
$$S_{T'+t} v \in \mathcal{B}_{n+1} \quad \text{for all} \quad t \geq \tau.$$
Therefore, we have that either
$$(2.48) \quad \operatorname{dist}(S_{T'+t} v, A_{n+1}) \leq \operatorname{dist}\left(S_{T'+t} v, \mathcal{M}^u(\mathcal{N}^{n+1})\right) \leq \psi_{n+1}(t) \quad \text{for all} \quad t \geq 0$$
or there exists $\tau = \tau_v \in [0, +\infty)$ such that
$$(2.49) \qquad \operatorname{dist}(S_{T'+t} v, A_{n+1}) \leq \psi_{n+1}(t) \quad \text{for} \quad 0 \leq t < \tau$$
and
$$(2.50) \qquad \operatorname{dist}(S_{T_n + T' + t} v, A_{n+1}) \leq \Psi_n(t - \tau) \quad \text{for} \quad t > \tau.$$
We also have from (2.42) and (2.49) that
$$\operatorname{dist}(S_{T'+t} v, A_{n+1}) \leq c_{R_*} e^{L_{R_*}(t-\tau)} \psi_{n+1}(\tau) \quad \text{for} \quad t > \tau$$

2.4. GRADIENT SYSTEMS

and, consequently, by (2.50)
$$\operatorname{dist}(S_{T_n+T'+t}v, A_{n+1}) \le \min\left\{c_{R_*}\psi_{n+1}(\tau)e^{L_{R_*}(T_n+t-\tau)}\,;\,\Psi_n(t-\tau)\right\}, \quad t > \tau.$$
Therefore, it follows from (2.46) that
$$(2.51) \qquad \operatorname{dist}(S_{T_n+T'+t}v, A_{n+1}) \le \Psi_{n+1}(t) \quad \text{for} \quad t > \tau.$$
In the case when $0 < t < \tau$ from (2.42) and (2.49) we have that
$$\begin{aligned}\operatorname{dist}(S_{T_n+T'+t}v, A_{n+1}) &\le c_{R_*}e^{L_{R_*}T_n} \cdot \operatorname{dist}(S_{T'+t}v, A_{n+1}) \\ &\le c_{R_*}e^{L_{R_*}T_n}\psi_{n+1}(t) \le \Psi_{n+1}(t).\end{aligned}$$
A similar estimate holds in the case of (2.48). Therefore from (2.51) we conclude that (2.45) holds for $k = n+1$ and $T_{n+1} = T_n + T'$. □

The following assertion give a condition for exponential rate of attraction of bounded sets.

COROLLARY 2.36. *Assume that Assumption 2.34 holds with* $\psi_j(t) = c_j e^{-\gamma_j t}$, *where c_j and γ_j are positive constants, $j = 1, \ldots, m$. Then there exists $\gamma_0 > 0$ such that for any bounded set B we can find positive constants C_B and t_B such that*
$$\sup_{y \in B} \operatorname{dist}(S_t y, A) \le C_B e^{-\gamma_0 t}, \quad t \ge t_B.$$

PROOF. It is sufficient to show that every function $\Psi_k(t)$ from the statement of Theorem 2.35 admits the estimate
$$(2.52) \qquad \Psi_k(t) \le C_k e^{-\beta_k t}, \quad k = 1, \ldots, m,$$
with positive C_k and β_k. Obviously this estimate holds for $k = 1$. If (2.52) holds for some $k = n < m$, then from (2.46) we obtain that
$$\Psi_{n+1}(t) \le C\left[e^{-\gamma_{n+1}t} + \sup_{0 \le s \le t} \min\left\{e^{-\gamma_{n+1}s + L_{R_*}(t-s)}\,;\,e^{-\beta_n(t-s)}\right\}\right]$$
with some positive constant C. Therefore, a simple calculation gives that (2.52) holds for $k = n+1$ with $\beta_{n+1} = \beta_n \gamma_{n+1}/(\gamma_{n+1} + \beta_n + L_{R_*})$. □

CHAPTER 3

Existence of compact global attractors for evolutions of the second order in time

The main goal in this chapter is to provide new results on the existence of global attractors governed by evolutions in (1.1). To accomplish this we shall employ two different approaches which will allow us to apply abstract results from Chapter 2.

The first approach, fairly general, is rooted in control theory and based on certain stabilizability-observability inequalities which have been recently proved for several PDE dynamics in the context of stabilization and controllability theory. We shall generalize this method and will show that a certain abstract variant of the said inequality is sufficient for proving asymptotic smoothness of the flow. The advantage of that approach is that it provides a unified treatment for dynamical systems with nonlinear dissipation even in the cases when the dissipation is very weak (the derivative of the damping operator D vanishes at the origin).

The second approach is based on a more traditional "splitting method" that will be combined with nonlinear interpolation theory. This is a more special method, problem dependent, which requires certain rather specific assumptions imposed on the nonlinear terms in the equation. However, in some cases, particularly when nonlinear terms in the equation are local (e.g. semilinear wave equation), this method appears a right tool for the problem.

The main results on existence of global attractors which were obtained by the first method are formulated in Theorem 3.26 and Theorem 3.58. A common tread to these results is "stabilizability" inequality derived for the system representing two different solutions. This stabilizability inequality is motivated by recent progress in control theory where this particular type of inequalities have attracted attention in the context of controllability and stabilization theory. In fact, this connection between control theory and dynamical system theory is exploited all along in the paper (see also the qualitative results presented in Chapter 4). A similar approach is applied in the proof of Theorem 3.34 which allows to avoid assumption of large damping parameter k in the critical case ($\widetilde{\eta} = 0$ in relation (3.61) below). However this theorem requires some compactness of the potential energy functional Π.

The second (splitting) method is presented in Theorem 3.44 and Theorem 3.47. The splitting method, in comparison with the first one, makes it possible to avoid both assumption of large damping parameter k in the critical case and compactness requirement for Π at the expense of additional hypotheses concerning the structure of the damping operator D and of the nonlinear term F. Since the main application that we have in mind is to wave equations, we consider the case $M = I$ in these two theorems.

Ultimate dissipativity is most often a preliminary step for proving existence of attractors -see Theorem 2.3. This property is fundamental to long time behaviour of solutions. We shall pay particular attention to the task of estimating absorbing radiuses with respect to the damping parameter. These estimates will be later critically used in asserting asymptotic smoothness of dynamics within critical region of parameters (such as critical exponents for nonlinear terms).

3.1. Ultimate dissipativity

There are several approaches to the proof of dissipativity (see, e.g., [4, 19, 20, 58, 59, 62, 73, 101] and the references therein). We shall detail some of these approaches, particularly these pertinent to our framework. However it should be noted that sometimes it is more convenient to study dissipativity for concrete examples directly, see e.g., [1, 4, 15, 20, 101].

The task of establishing ultimate dissipativity property is particularly difficult for dynamical systems whose energy is not decreasing. This may be due to presence of restoring forces which are not conservative. However, it is expected that the lack of monotonicity of the energy will be eventually overtaken by other dissipative mechanisms in the model. By this we mean that the system will be "ultimately" dissipative, i.e. the long time dynamics will be confined to a bounded set called absorbing set.

It is our aim in this section to demonstrate that a rather general class of abstract systems defined in (1.1) exhibits ultimate dissipativity property under some natural additional assumptions. It should also be noted that in applications to nonlinear PDE's not only dissipativity of the dynamical system is needed, but also control of the size of the absorbing set in terms of the parameters representing the damping. Accordingly, our emphasis is placed on carrying the estimates with a full control of the dependence on the damping parameter.

In order to prove ultimate dissipativity of the system (H, S_t) generated by (1.1) we need some additional assumptions concerning problem (1.1). Below we formulate these assumption treating separately the two case when non-conservative forces ($F^* \not\equiv 0$) are present and the case when these forces are absent ($F^* \equiv 0$). The treatment of the non-conservative case we split into two subcases which, roughly speaking, may be described as follows: (i) the damping operator D and the force F^* are linearly bounded, F^* is independent of u_t, and (ii) the damping operator D is allowed to be superlinear and the restoring force F^* may depend on u_t and satisfies some growth estimate which is stronger than in the subcase (i). We also note in the subcase (i) the mechanism which allows to prove ultimate dissipativity is some sort of hidden superlinearity of potential energy, in the subcase (ii) a similar role is played by superlinearity of the damping.

3.1.1. The model without non-conservative forces. In this subsection we consider problem (1.1) with $F^* \equiv 0$, i.e. we deal with the problem

$$
(3.1) \quad \begin{cases} Mu_{tt}(t) + \mathcal{A}u(t) + k \cdot D(u_t(t)) = F(u(t)), \\ u|_{t=0} = u_0 \in \mathcal{D}(\mathcal{A}^{1/2}), \; u_t|_{t=0} = u_1 \in V = \mathcal{D}(M^{1/2}), \end{cases}
$$

where $F(u) = -\Pi'(u)$. In addition to Assumption 1.1 we impose the following hypotheses concerning D and F.

ASSUMPTION 3.1. **(D)** • There exist constants $c_0 \geq 0$ and $c_1 > 0$ such that

$$(3.2) \quad (Mv, v) \leq c_0 + c_1(Dv, v) \quad \text{for any} \quad v \in \mathcal{D}(\mathcal{A}^{1/2}),$$

- for any $\delta > 0$ there exists a non-decreasing function $K_\delta(s) > 0$ such that

(3.3) $\quad |(Dv, u)| \leq K_\delta(E_0(u,v)) \cdot (Dv, v) + \delta \cdot (1 + E_0(u,v)), \quad u, v \in \mathcal{D}(\mathcal{A}^{1/2})$,

where $E_0(u,v) = \frac{1}{2}((Mv, v) + (\mathcal{A}u, u))$.

(F) There exist $0 \leq \eta < 1$ and a positive constant c_2 such that

(3.4) $\quad (u, F(u)) \leq \eta |\mathcal{A}^{1/2}u|^2 + c_2, \quad u \in \mathcal{D}(\mathcal{A}^{1/2})$.

REMARK 3.2. Assumption (3.3) holds true provided that for any $\delta > 0$ we can find a non-decreasing function $\widetilde{K}_\delta(s) > 0$ such that

(3.5) $\quad |Dv|^2_{[\mathcal{D}(\mathcal{A}^{1/2})]'} \leq \delta + \widetilde{K}_\delta(|v|_V) \cdot (Dv, v), \quad u, v \in \mathcal{D}(\mathcal{A}^{1/2})$.

Indeed, it follows from (3.5) that

$$\begin{aligned} |(Dv, u)| &\leq |Dv|_{[\mathcal{D}(\mathcal{A}^{1/2})]'} \cdot |\mathcal{A}^{1/2}u| \\ &\leq \sqrt{\delta} \cdot |\mathcal{A}^{1/2}u| + |\mathcal{A}^{1/2}u| \cdot \left(\widetilde{K}_\delta(|v|_V) \cdot (Dv, v)\right)^{1/2} \\ &\leq \sqrt{\delta} \cdot \left(1 + |\mathcal{A}^{1/2}u|^2\right) + K_\delta(E(u,v)) \cdot (Dv, v), \end{aligned}$$

where $K_\delta(s) = \frac{1}{2\sqrt{\delta}} \cdot \widetilde{K}_\delta(\sqrt{2s})$. After rescaling of δ, this implies (3.3).

REMARK 3.3. For sake of concreteness and future reference we note that in the case of wave equation (1.8) Assumption 3.1(D) holds provided $g \in C(\mathbb{R})$ is a monotone increasing function such that $g(0) = 0$ and there exist positive constants m_1 and m_2 such that

(3.6) $\quad m_1|s| \leq |g(s)| \leq m_2|s|^p \quad \text{for all } |s| \geq 1$

where $1 \leq p \leq 5$ when $n = 3$ and $1 \leq p < \infty$ for $n = 2$. Property (3.4) of nonlinear force $F(u)$ easily follows from (1.11). See Chapter 5 for details.

Our main result on dissipativity in the case $F^* \equiv 0$ is the following:

THEOREM 3.4. Under Assumptions 1.1 and 3.1 the system (H, S_t) generated by (3.1) in the space $H = \mathcal{D}(\mathcal{A}^{1/2}) \times V$ is ultimately dissipative, i.e. there exists $R > 0$ possessing the property: for any bounded set B from H there exists $t_0 = t_0(B)$ such that $\|S_t y\|_H \leq R$ for all $y \in B$ and $t \geq t_0$. We can choose a radius R of an absorbing ball such that R does not depend on the damping parameter k. Moreover, there exists a forward invariant absorbing set $\mathcal{B}_0$ with the size which does not depend on k.

PROOF. Our proof is based on Lyapunov's method. Though the method is standard, in our case more delicate arguments are needed. This is due to the necessity of tracing the size of absorbing set as a function of the parameter describing dissipation. Let
$$V(u, u_t) = \mathcal{E}(u, u_t) + \varepsilon(Mu_t, u),$$
where the energy $\mathcal{E}$ is given by (1.18). The parameter $\varepsilon > 0$ will be chosen later. Since $\mathcal{D}(\mathcal{A}^{1/2})$ continuously embedded into V, we have

(3.7) $\quad |(Mu_t, u)| \leq C|u_t|_V \cdot |u|_V \leq (Mu_t, u_t) + c|\mathcal{A}^{1/2}u|^2$.

Therefore by (1.30) there exists $\varepsilon_0 > 0$ such that

(3.8) $\quad \frac{1}{4} \cdot E(u_0, u_1) - c \leq V(u_0, u_1) \leq 4 \cdot E(u_0, u_1) + c, \quad u_0 \in \mathcal{D}(\mathcal{A}^{1/2}), u_1 \in V$, for all $0 < \varepsilon \leq \varepsilon_0$, where c does not depend on ε and $E(u_0, u_1)$ is defined by (1.19). This implies that there exist an increasing function $\varphi : \mathbb{R}_+ \mapsto \mathbb{R}_+$ and a constant c which do not depend on the damping parameter k such that

$$(3.9) \qquad \frac{1}{8} \cdot \|y\|_H^2 - c \leq V(y) \leq \varphi(\|y\|_H), \quad y \equiv (u_0, u_1) \in H = \mathcal{D}(\mathcal{A}^{1/2}) \times V.$$

Now we consider

$$\frac{d}{dt} V(u(t), u_t(t)) = \frac{d}{dt} \mathcal{E}(u, u_t) + \varepsilon \frac{d}{dt} (Mu_t, u),$$

where $u(t)$ is a strong solution to problem (1.1). Using (1.1) and energy relation (1.17), for $V(t) = V(u(t), u_t(t))$ we have that

$$(3.10) \qquad \frac{d}{dt} V(t) = -k \cdot (Du_t, u_t) + (F^*(u, u_t), u_t)$$
$$+ \varepsilon \left\{ (Mu_t, u_t) - (\mathcal{A}u, u) - k(Du_t, u) + (F(u), u) \right\}.$$

Since $F^* \equiv 0$, using (3.3) and (3.4) we obtain that

$$\frac{d}{dt} V(t) \leq -k \cdot [1 - \varepsilon K_\delta(E_0(u(t), u_t(t)))] \cdot (Du_t, u_t) + \varepsilon(2 - \eta)(Mu_t, u_t)$$
$$+ \varepsilon \left\{ -(2(1-\eta) - \delta k) E_0(u(t), u_t(t)) + \delta k + c_2 \right\}.$$

Thus from (3.2), choosing $\delta = (1-\eta)k^{-1}$, we get that

$$\frac{d}{dt} V(t) \leq -k \cdot \left\{ 1 - \varepsilon \left[\frac{2c_1}{k} + K_{(1-\eta)k^{-1}}(E_0(u(t), u_t(t))) \right] \right\} \cdot (Du_t, u_t)$$
$$+ \varepsilon \left\{ -(1-\eta) E_0(u(t), u_t(t)) + 2c_0 + c_2 + 1 \right\}.$$

Let $\mathcal{B}$ be a bounded set in $H = \mathcal{D}(\mathcal{A}^{1/2}) \times V$ and $(u_0, u_1) \in \mathcal{B}$. It follows from (1.30) and energy relation (1.17) with $F^* \equiv 0$ that

$$E_0(u(t), u_t(t)) \leq C(1 + \mathcal{E}(u(t), u_t(t))) \leq C(1 + \mathcal{E}(u_0, u_1)) \leq C_\mathcal{B}.$$

Therefore, we have that

$$\frac{d}{dt} V(t) \leq -k \cdot \left\{ 1 - \varepsilon \left[\frac{2c_1}{k} + C(\mathcal{B}, k) \right] \right\} \cdot (Du_t, u_t)$$
$$+ \varepsilon (1-\eta) \left\{ -E_0(u(t), u_t(t)) + C_0 \right\},$$

where $C_0 > 0$ does not depend on k. Consequently, there exists $\varepsilon_*(\mathcal{B}, k) \in (0, \varepsilon_0]$ such that

$$(3.11) \qquad \frac{d}{dt} V(t) \leq -\varepsilon (1-\eta) \left(E_0(u(t), u_t(t)) - C_0 \right), \quad \varepsilon \in (0, \varepsilon_*(\mathcal{B}, k)].$$

Since by the definition any generalized solution can be approximate by strong solutions, this formula (in the integral form) remains true for generalized solutions. Therefore, by the standard method (see, e.g., [**19**, Theorem 4.1, Chap.1]) we can conclude from (3.11) and (3.9) that there exists $R > 0$ independent of k such that the ball

$$B_R = \{ y \in H : \|y\|_H \leq R \}$$

is absorbing for (H, S_t). It follows from (1.30) that

$$B_R \subset \mathcal{B}_0 \equiv \left\{ (u_0, u_1) \in H : \mathcal{E}(u_0, u_1) \leq C_R^1 \right\} \subset \{ y \in H : \|y\|_H \leq C_R^2 \}$$

for some C_R^1 and C_R^2. Thus the set $\mathcal{B}_0$ is bounded and absorbing. Its invariance follows from energy relation (1.17) with $F^* \equiv 0$. $\square$

REMARK 3.5. It is clear from the argument given in the proof of Theorem 3.4 that, *without* controlling of the size of absorbing set, we can obtain dissipativity of (H, S_t) by changing assumption (3.3) into the following requirement: there exist a non-decreasing function $K(s) > 0$ and a constant C such that the relation

(3.12) $\quad |(Dv, u)| \leq K(E_0(u, v)) \cdot (Dv, v) + C + \delta \cdot E_0(u, v), \quad u, v \in \mathcal{D}(\mathcal{A}^{1/2}),$

holds for some $0 \leq \delta < (1 - \eta)k^{-1}$. As in Remark 3.2 one can show that (3.12) follows from the property

(3.13) $\quad |Dv|^2_{[\mathcal{D}(\mathcal{A}^{1/2})]'} \leq C + \widetilde{K}(|v|_V) \cdot (Dv, v), \quad u, v \in \mathcal{D}(\mathcal{A}^{1/2}),$

for some constant $C > 0$ and non-decreasing function $\widetilde{K}(s) > 0$.

REMARK 3.6. The fact that the energy $\mathcal{E}(u, u_t)$ decreases along trajectories in the case $F^* \equiv 0$ allows to consider superlinear dissipation. Indeed, the presence of the function $K_\delta(E(u, v))$ in condition (3.3) allows to handle potential superlinearity of the dissipation (see Remark 3.3).

3.1.2. The model with non-conservative forces: part I. In this subsection we assume that $F^* \not\equiv 0$. The main difference in comparison with the conservative case ($F^* \equiv 0$) is that by (1.17) the energy $\mathcal{E}$ does not necessarily decrease along trajectories. In fact, it may behave in an almost arbitrary fashion. This is a reason why dissipativity property requires more restrictive bounds imposed on the nonlinear terms when non-conservative forces are present. However, several equations in mathematical physics provide an ample justification and motivation for considering this class of models.

We shall begin with the case when the non-conservative term does not depend on velocity. Thus, we consider below the case when $F^* = F^*(u)$, i.e. the problem

(3.14) $\quad \begin{cases} Mu_{tt}(t) + \mathcal{A}u(t) + k \cdot D(u_t(t)) = -\Pi'(u(t)) + F^*(u(t)) \equiv F(u(t)), \\ u|_{t=0} = u_0 \in \mathcal{D}(\mathcal{A}^{1/2}), \ u_t|_{t=0} = u_1 \in V = \mathcal{D}(M^{1/2}). \end{cases}$

We impose the following

ASSUMPTION 3.7. **(D)** • Relation (3.2) holds, i.e. there exist constants $c_0 \geq 0$ and $c_1 > 0$ such that

(3.15) $\quad (Mv, v) \leq c_0 + c_1(Dv, v), \quad v \in \mathcal{D}(\mathcal{A}^{1/2}),$

• the operator D is linearly bounded in the following sense: there exist non-negative constant c_2 and c_3 such that

(3.16) $\quad |Dv|_{[\mathcal{D}(\mathcal{A}^{1/2})]'} \leq c_2 + c_3|v|_V, \quad v \in \mathcal{D}(\mathcal{A}^{1/2}).$

(F) • There exist positive constants $0 \leq \eta < 1$, c_4 and c_5 such that

(3.17) $\quad (u, F(u)) \leq \eta |\mathcal{A}^{1/2}u|^2 - c_4\Pi_0(u) + c_5, \quad u \in \mathcal{D}(\mathcal{A}^{1/2}),$

• for every $\eta > 0$ there exists $C_\eta > 0$ such that

(3.18) $\quad |u|^2 \leq C_\eta + \eta \left(|\mathcal{A}^{1/2}u|^2 + \Pi_0(u)\right), \quad u \in \mathcal{D}(\mathcal{A}^{1/2}),$

• the non-conservative term $F^*(u)$ satisfies the inequality

(3.19) $\quad |F^*(u)|^2_{V'} \leq c_6 + c_7|\mathcal{A}^{1/2-\delta}u|^2 \ \text{ for some } \ \delta > 0.$

REMARK 3.8. We note that inequality in (3.18) is often related to some sort of unique continuation properties. Indeed, the validity of this inequality is typically reduced (via compactness-contradiction argument) to the verification of the following implication $\Pi_0(u) = 0 \to u = 0$. The above implication, in many applications, can be interpreted as an uniqueness property of some nonlinear PDE's. This property is related to a "hidden" superlinearity of potential energy. In fact, in the case of von Karman evolutions this implication coincides with uniqueness of zero solutions to Monge–Ampere equations (see Chapter 6 and , e.g., [**26**] and the references therein). Relation (3.18) when combined with interpolation inequality imply that for any positive $\delta > 0$ and any positive number η there exists a constant C_η such that

$$(3.20) \qquad |\mathcal{A}^{1/2-\delta}u|^2 \leq \eta[|\mathcal{A}^{1/2}u|^2 + \Pi_0(u)] + C_\eta.$$

Moreover, we can suppose that $C_\eta = b(1/\eta)$, where $b(\xi)$ is a non-decreasing function of $\xi \in \mathbb{R}_+ \setminus \{0\}$.

REMARK 3.9. Von Karman model (1.12) and (1.13) considered in Example 1.3 with non-conservative term of the form $F^*(u) = L(u, u_t) = \partial_{x_1} u$ occurring in nonlinear elasticity with aeroelastic forces that contribute to non-conservative terms is a typical application of the setup considered in Assumption 3.7. Assumption 3.7(D) is valid for this model provided that the functions g_0 and $g = (g_1, g_2)$ possess the following additional property

$$(3.21) \qquad m_1 s^2 \leq s g_0(s) \leq m_2 s^2, \quad |s| \geq 1,$$

and

$$(3.22) \qquad m_1(s_1^2 + s_2^2) \leq \sum_{i=1,2} s_i g_i(s_1, s_2) \leq m_2(s_1^2 + s_2^2), \quad s_1^2 + s_2^2 \geq 1,$$

where m_1 and m_2 are positive constants. Assumption 3.7(D) is valid with $c_0 = c_2 = 0$, if relations (3.21) and (3.22) hold *for all* $s \in \mathbb{R}$ and $(s_1, s_2) \in \mathbb{R}^2$.

For this model Assumption 3.7(F) also holds (see Chapter 6 for details). The validity of this condition is based on the validity of uniqueness property $[w, w] = 0 \to w = 0$, a property linked with uniqueness of Monge–Ampere equations.

THEOREM 3.10. *Under Assumptions 1.1 and 3.7 the system* (H, S_t) *generated by (3.14) in the space* $H = \mathcal{D}(\mathcal{A}^{1/2}) \times V$ *is ultimately dissipative. It possesses a forward invariant bounded absorbing set (whose size may depend on the damping parameter k).*

Assume, in addition that Assumption 3.7(D) holds with $c_0 = c_2 = 0$ in (3.15) and (3.16), i.e. there exist constants $c_1 > 0$ and $c_3 > 0$ such that

$$(3.23) \qquad (Mv, v) \leq c_1(Dv, v), \quad v \in \mathcal{D}(\mathcal{A}^{1/2}),$$

and

$$(3.24) \qquad |Dv|_{[\mathcal{D}(\mathcal{A}^{1/2})]'} \leq c_3 |v|_V, \quad v \in \mathcal{D}(\mathcal{A}^{1/2}).$$

Then we can choose an absorbing ball with radius which is independent of the damping parameter $k \in [k_0, +\infty)$ for every fixed $k_0 > 0$. In this case a forward invariant bounded (uniformly with respect to $k \geq k_0$) absorbing set also exists.

PROOF. We first note that, without loss of generality, we can assume that $|v|_V = |M^{1/2}v|$ for any $v \in V$. It is also easy to see that Assumption 3.7(D) implies that

$$(3.25) \qquad |(Dv, u)|^2 \leq [1 + E(u, v)] \cdot [\widetilde{c}_2 + \widetilde{c}_3 (Dv, v)], \quad u, v \in \mathcal{D}(\mathcal{A}^{1/2}),$$

where $E(u, v)$ is given by (1.19) and

$$(3.26) \qquad \widetilde{c}_2 = 4(c_2^2 + c_0 c_3^2), \quad \widetilde{c}_3 = 4 c_1 c_3^2.$$

As in the proof of Theorem 3.4 we consider the functional

$$V(u, u_t) = \mathcal{E}(u, u_t) + \varepsilon (M u_t, u).$$

The parameter $\varepsilon > 0$ will be chosen later. As above, using (3.7) and (1.30) one can see that there exists $\varepsilon_0 > 0$ such that

$$(3.27) \quad \frac{1}{4} \cdot E(u_0, u_1) - c \leq V(u_0, u_1) \leq 4 \cdot E(u_0, u_1) + c, \quad u_0 \in \mathcal{D}(\mathcal{A}^{1/2}), \, u_1 \in V,$$

for all $0 < \varepsilon \leq \varepsilon_0$, where $E(u_0, u_1)$ is given by (1.19) and the constant $c > 0$ does not depend on ε and k.

From (3.15) we have that $|u_t|_V \leq \sqrt{c_0} + \sqrt{c_1} \cdot (D u_t, u_t)^{1/2}$. Therefore, we can write

$$(F^*(u), u_t) \leq |F^*(u)|_{V'} \cdot |u_t|_V \leq \frac{k}{4}(Du_t, u_t) + \sqrt{c_0}|F^*(u)|_{V'} + \frac{c_1}{k}|F^*(u)|^2_{V'}.$$

Consequently, for $V(t) = V(u(t), u_t(t))$ from (3.10) and (3.17) we obtain that

$$\frac{d}{dt} V(t) \leq -\frac{3}{4} k (Du_t, u_t) + \varepsilon(2 - \eta)(M u_t, u_t) - \varepsilon k(Du_t, u) + \sqrt{c_0}|F^*(u)|_{V'}$$

$$+ \frac{c_1}{k}|F^*(u)|^2_{V'} + \varepsilon \left\{ -2(1 - \eta) E_0(u, u_t) - c_4 \Pi_0(u) + c_5 \right\}.$$

Hence, it follows from property (3.15) that

$$\frac{d}{dt} V(t) \leq \left(-\frac{3}{4} k + 2\varepsilon c_1 \right) (Du_t, u_t) + \varepsilon k (Du_t, u)$$

$$(3.28) \qquad + \sqrt{c_0}|F^*(u)|_{V'} + \frac{c_1}{k}|F^*(u)|^2_{V'} + \varepsilon a_0 \left\{ -E(u, u_t) + a_1 \right\},$$

where $a_0 = \min\{2 - 2\eta, c_4\} > 0$ and $a_1 = a_0^{-1}(2c_0 + c_5) > 0$ do not depend on ε and k.

From (3.25) we have that

$$k|(Du_t, u)| \leq \frac{a_0}{2} (1 + E(u, u_t)) + \frac{k^2}{2a_0} (\widetilde{c}_2 + \widetilde{c}_3 \cdot (Du_t, u_t)),$$

where $\widetilde{c}_2$ and $\widetilde{c}_3$ are given by (3.26). Therefore (3.28) implies that

$$\frac{d}{dt} V(t) \leq \left(-\frac{3}{4} k + \varepsilon \left[2c_1 + \frac{\widetilde{c}_3 k^2}{2a_0} \right] \right) (Du_t, u_t)$$

$$(3.29) \qquad + \sqrt{c_0}|F^*(u)|_{V'} + \frac{c_1}{k}|F^*(u)|^2_{V'} + \frac{\varepsilon a_0}{2} \left\{ -E(u, u_t) + 1 + a_1 + \frac{\widetilde{c}_2 k^2}{2a_0^2} \right\}.$$

For every fixed $k > 0$ we can choose $\varepsilon > 0$ such that

$$\frac{d}{dt} V(t) \leq \sqrt{c_0}|F^*(u)|_{V'} + \frac{c_1}{k}|F^*(u)|^2_{V'} + a_2 \left\{ -E(u, u_t) + a_3 \right\},$$

where a_2 and a_3 are positive constants. Therefore, it follows from (3.27), (3.19) and (3.20) that
$$\frac{d}{dt} V(t) + \gamma V(t) \leq C$$
with positive constants γ and C (which may depend on k). After integration, this implies

(3.30) $\quad V(u(t); u_t(t)) \leq V(u_0; u_1) e^{-\gamma t} + C\gamma^{-1}(1 - e^{-\gamma t}), \ t > 0, \ (u_0; u_1) \in H$.

Therefore, for any $R > C\gamma^{-1}$ the set
$$\mathcal{V}_R = \{(u_0; u_1) \in H \ : \ V(u_0, u_1) \leq R\}$$
is a positively invariant absorbing set for the system considered. It is clear that $\mathcal{V}_R$ is bounded. Thus the first part of Theorem 3.10 is proved.

To prove the second part of this theorem we note that in this case $c_0 = 0$ in (3.15) and $c_2 = 0$ in (3.16). Therefore, by (3.26) $\widetilde{c}_2 = 0$ Hence, (3.29) can be rewritten in the form

$$\frac{d}{dt} V(t) \leq \left(-\frac{3}{4} k + \varepsilon \left[2c_1 + \frac{\widetilde{c}_3 k^2}{2a_0} \right] \right) (Du_t, u_t)$$

(3.31) $\qquad\qquad + \frac{c_1}{k} |F^*(u)|_{V'}^2 + \frac{\varepsilon a_0}{2} (-E(u, u_t) + 1 + a_1),$

where $a_0 > 0$ and $a_1 > 0$ does not depend on ε and k. We need to estimate the contribution of the term F^*. From (3.19) and (3.20) we have

$$\frac{c_1}{k} |F^*(u)|_{V'}^2 \leq \frac{C}{k} \left(1 + |\mathcal{A}^{1/2-\delta} u|^2 \right) \leq \frac{\eta}{k} \left((\mathcal{A}u, u) + \Pi_0(u) \right) + \frac{1}{k} \cdot b\left(\frac{1}{\eta} \right),$$

where $b(\xi) > 0$ is a non-decreasing function of $\xi \in \mathbb{R}_+ \setminus \{0\}$. Taking η proportional to εk we obtain

$$\frac{c_1}{k} |F^*(u)|_{V'}^2 \leq \frac{\varepsilon a_0}{8} \left((\mathcal{A}u, u) + \Pi_0(u) \right) + \frac{1}{k} \cdot \widetilde{b}\left(\frac{1}{\varepsilon k} \right).$$

Now we take
$$\varepsilon \equiv \varepsilon(k) = \frac{3k\delta_0}{8c_1 + 2a_0^{-1} \widetilde{c}_3 k^2},$$
where $0 < \delta_0 < 1$ is chosen such that $\varepsilon(k) \leq \varepsilon_0$ for all $k \geq 0$. In this case from (3.31) we obtain that

$$\frac{d}{dt} V(t) \leq -\frac{\varepsilon a_0}{4} E(u(t), u_t(t)) + \varepsilon \left[\frac{a_0}{2} (1 + a_1) + \frac{1}{\varepsilon k} \cdot \widetilde{b}\left(\frac{1}{\varepsilon k} \right) \right].$$

Since for any fixed $k_0 > 0$ there exists $\delta(k_0) > 0$ such that $k\varepsilon(k) \geq \delta(k_0)$ for all $k \geq k_0$, using (3.27) we conclude that

$$\frac{d}{dt} V(u(t), u_t(t)) + \alpha \varepsilon V(u(t), u_t(t)) \leq \varepsilon \cdot C,$$

where the constants $\alpha > 0$ and $C > 0$ do not depend on $k \in [k_0, +\infty)$ for every fixed $k_0 > 0$. After integration we obtain that

(3.32) $\qquad V(u(t), u_t(t)) \leq V(u_0, u_1) \cdot e^{-\alpha \varepsilon t} + C\alpha^{-1} \cdot \left(1 - e^{-\alpha \varepsilon t} \right).$

As before, the above inequality is obtained for strong solutions. Standard limit argument based on density allows to extend the inequality to all generalized solutions. Thus (3.32) implies ultimate dissipativity of (H, S_t) with absorbing set whose size

is independent of the damping parameter $k \in [k_0, +\infty)$. Moreover, as above one can see that for any $R > C\alpha^{-1}$ the set
$$\mathcal{V}_R = \{(u_0; u_1) \in H \ : \ V(u_0, u_1) \leq R\}$$
is a positively invariant absorbing set for the system considered with the size independent of $k \in [k_0, +\infty)$ for every fixed $k_0 > 0$. □

We note that Assumption 3.7 assumes implicity linear bound on the non-conservative term $F^*(u)$, which in addition needs to be compact. This is expressed by the last condition in Assumption 3.7. In addition, condition (3.16) in Assumption 3.7 forces linear bound on the damping. While these restrictions are to be expected in general when non-conservative potential terms are in the model, in some more special cases when the potential energy has strong coercivity properties, we will be able to relax these constraints. There are many models of physical interest (see Chapter 7 for examples) when non-conservative term is non-compact and may be superlinear. Similarly the damping may be superlinear. In such cases it is still possible to obtain ultimate dissipativity, however at the expenses of assuming large values of the damping parameter. The relevant result is formulated below.

THEOREM 3.11. *Let Assumption 1.1 be valid. Assume that Assumption 3.7 holds with the following two relations replacing respectively (3.16) and (3.19):*

$$(3.33) \quad k|(Dv, u)| \leq \eta \left[|\mathcal{A}^{1/2}u|^2 + \Pi_0(u)\right] + C_{\eta,k} + C_1 \cdot (Dv, v) \quad v, u \in \mathcal{D}(\mathcal{A}^{1/2}),$$

where $\eta > 0$ is arbitrary small and C_1 does not depend on k, and

$$(3.34) \quad |F^*(u)|_{V'}^2 \leq C_2 + C_3 \left(|\mathcal{A}^{1/2}u|^2 + \Pi_0(u)\right), \quad u \in \mathcal{D}(\mathcal{A}^{1/2}).$$

Then there exists $k_* > 0$ such that the system (H, S_t) generated by (3.14) in the space $H = \mathcal{D}(\mathcal{A}^{1/2}) \times V$ is ultimately dissipative for each $k \geq k_*$. It possesses a forward invariant bounded absorbing set (whose size may depend on the damping parameter k).

PROOF. The main idea of the proof is the same as in the first part of Theorem 3.10. However, rescaling of the parameters will be different. We start with relation (3.28). It follows from (3.33) with an appropriate choice of δ so that

$$(3.35) \quad k|(Du_t, u)| \leq \frac{a_0}{2} E(u, u_t) + d_1 \cdot (Du_t, u_t) + C_k,$$

where a_0 is the same as in (3.28) and the constant d_1 is independent of k. Therefore (3.28) implies that

$$\frac{d}{dt} V(t) \leq \left(-\frac{3}{4}k + \varepsilon \left[2c_1 + d_1\right]\right) (Du_t, u_t)$$

$$+ \sqrt{c_0} |F^*(u)|_{V'} + \frac{c_1}{k} |F^*(u)|_{V'}^2 - \frac{\varepsilon a_0}{2} E(u, u_t) + \varepsilon C_k.$$

Now using (3.34) we obtain that

$$\frac{d}{dt} V(t) \leq \left(-\frac{3}{4}k + \varepsilon d_2\right)(Du_t, u_t) - \frac{a_0}{4}\left(\varepsilon - \frac{d_3}{k}\right) E(u, u_t) + C_{k,\varepsilon}.$$

Therefore, taking $\varepsilon = \frac{3k}{4d_2}$ we find that from (3.27) that for each $k \geq 2\sqrt{d_2 d_3}$,

$$\frac{d}{dt} V(t) \leq -\omega_k V(t) + C_k,$$

where ω_k and C_k are positive constant depending on k. As above (cf. (3.30)), we complete the proof. $\square$

REMARK 3.12. Using (3.20) we can derive estimate (3.33) from the relation

(3.36) $$|Dv|_{[\mathcal{D}(\mathcal{A}^{1/2-\delta})]'} \leq c_1 + c_2|v|_V$$

for some $\delta > 0$. Indeed, by interpolation we have that

$$\begin{aligned} k|(Dv,u)| &\leq k \cdot (c_1 + c_2|v|_V)|\mathcal{A}^{1/2-\delta}u| \\ &\leq |v|_V^2 + C_{1,k} \cdot |\mathcal{A}^{1/2-\delta}u|^2 + C_{2,k}. \end{aligned}$$

Therefore, using (3.20) we obtain (3.33). Thus Theorem 3.11 remains true if we replace (3.33) by (3.36).

REMARK 3.13. As already indicated above, models that exhibit conservative forces complying with the setup of Theorem 3.11 are models with superlinear potential energy. A prototype for this framework is Kirchhoff limit model considered in Chapter 7. In that case conservative forces are superlinear: $F^*(u) = |\nabla u|^2$.

3.1.3. The model with non-conservative forces: part II. In this subsection we shall consider another type of non-conservative force that may depend on the velocity as well. In that case, the proof of ultimate dissipativity requires slightly different setup. The approach presented involves a well-known "barrier's" method which become a standard device in dealing with problems with non-monotone energy (see [**62**] and the references therein). This method allows to avoid linear type growth condition (3.16) imposed on the damping operator D. Instead, it requires a stronger (than (3.19)) hypothesis concerning growth of $F^*(u)$. However, in the case of superlinear damping the "barrier's" method allow us to include easily in the non-conservative force F^* an essential dependence on the velocity u_t.

Thus we consider the following problem

(3.37) $$\begin{cases} Mu_{tt}(t) + \mathcal{A}u(t) + k \cdot D(u_t(t)) = -\Pi'(u(t)) + F^*(u(t), u_t(t)), \\ u|_{t=0} = u_0 \in \mathcal{D}(\mathcal{A}^{1/2}), \quad u_t|_{t=0} = u_1 \in V = \mathcal{D}(M^{1/2}). \end{cases}$$

Here M and $\mathcal{A}$ are the same as above (see Section 1.2) and and D and Π satisfy the hypotheses listed in Assumption 1.1. In addition, we impose the following hypotheses concerning D, Π and F^*.

ASSUMPTION 3.14. • There exist constants $c_0 \geq 0$ and $c_1 > 0$ such that

(3.38) $$(Mv,v) \leq c_0 + c_1(Dv,v) \quad \text{for any} \quad v \in \mathcal{D}(\mathcal{A}^{1/2});$$

• there exist $0 \leq \eta < 1$, constants $c_2 > 0$ and $c_3, c_4 \geq 0$ such that

(3.39) $$\begin{aligned} -k(Dv,u) + (u, F^*(u,v)) &\leq (\Pi'(u),u) + \eta|\mathcal{A}^{1/2}u|^2 - c_2\Pi_0(u) \\ &\quad +c_3 + c_4 \cdot [1 + E(u,v)]^\gamma \cdot (Dv,v), \end{aligned}$$

for any $u,v \in \mathcal{D}(\mathcal{A}^{1/2})$, where $0 \leq \gamma < 1$, $E(u,v)$ is given by (1.19) and $\Pi_0(u)$ is the positive part of the potential energy Π (see Assumption 1.1);
• there exist a number $0 \leq \eta < 1$ and a non-decreasing continuous function $b : \mathbb{R}_+ \mapsto \mathbb{R}_+ \setminus \{0\}$ such that

(3.40) $$\lim_{s \to \infty} \left\{ s^{1-1/\gamma} \cdot b(s) \right\} = 0,$$

in the case $\gamma > 0$, where γ is the parameter from (3.39), and for any $0 < \delta \leq 1$ we have the relation

(3.41) $\quad (v, F^*(u,v)) \leq \eta k(Dv, v) + \delta E(u,v) + b\left(\dfrac{1}{\delta}\right), \quad u, v \in \mathcal{D}(\mathcal{A}^{1/2}),$

where $E(u,v)$ is given by (1.19).

We also assume that the conclusions of Theorem 1.5 hold for problem (3.37), that is (i) for every $(u_0, u_1) \in \mathcal{D}(\mathcal{A}^{1/2}) \times \mathcal{D}(\mathcal{A}^{1/2})$, such that $\mathcal{A}u_0 + kDu_1 \in V'$ there exists unique strong solution to problem (3.37) on any interval $[0,T]$ possessing property (1.16) and satisfying the energy relation
(3.42)
$$\mathcal{E}(u(t), u_t(t)) + k\int_0^t (Du_t(\tau), u_t(\tau))d\tau = \mathcal{E}(u_0, u_1) + \int_0^t (F^*(u(\tau), u_t(\tau)), u_t(\tau))d\tau,$$

where $\mathcal{E}(u_0, u_1)$ is given by (1.18), and (ii) for every $(u_0; u_1) \in \mathcal{D}(\mathcal{A}^{1/2}) \times V$ there exists unique generalized solution.

The above requirement is not an assumption in the case of locally Lipschitz restoring forces (see (1.3)), but in the generality considered (where $F^*(u,v)$ may be strongly nonlinear and non-locally Lipschitz) additional arguments for the well-posedness part may be needed (see, e.g., Remark 1.8).

Our main result on dissipativity in this subsection is the following:

THEOREM 3.15. *Under Assumption 3.14 the system (H, S_t) generated by (3.37) in the energy space H is ultimately dissipative for each $k > 0$, i.e. there exists $R_0 > 0$ (which may depend on k) possessing the property: for any bounded set B from H there exists $t_0 = t_0(B)$ such that $\|S_t y\|_H \leq R_0$ for all $y \in B$ and $t \geq t_0$. Moreover, there exists a forward invariant bounded absorbing set $\mathcal{B}_0$.*

PROOF. We use the usual Lyapunov's function along with a "barrier's" method (see [**62**] and the references therein). In fact, we use the same Lyapunov's function as in the previous cases. Let
$$V_c(u, u_t) = \mathcal{E}(u, u_t) + \varepsilon(Mu_t, u) + c, \quad \varepsilon \in (0, \varepsilon_0)$$
where the energy $\mathcal{E}$ is given by (1.18) and the constants $\varepsilon_0 > 0$ and $c > 0$ are chosen such that

(3.43) $\quad E(u_0, u_1) \leq 4V_c(u_0, u_1) \leq \varphi(\|y\|_H), \quad y \equiv (u_0, u_1) \in H = \mathcal{D}(\mathcal{A}^{1/2}) \times V,$

for all $0 < \varepsilon \leq \varepsilon_0$, where $\varphi : \mathbb{R}_+ \mapsto \mathbb{R}_+$ is an increasing function. This choice is possible due to relations (3.8) and (3.9). Computing the derivative of $V_c(t)$ on a strong solution $u(t)$, as above (cf. (3.10)) we obtain that

$$\begin{aligned}\dfrac{d}{dt} V_c(t) &= -k \cdot (Du_t, u_t) + (F^*(u, u_t), u_t) \\ &\quad + \varepsilon\left\{(Mu_t, u_t) - (\mathcal{A}u, u) - k(Du_t, u) - (\Pi'(u), u) + (F^*(u, u_t), u)\right\}.\end{aligned}$$

Therefore, from (3.39) and (3.41) we have that

$$\dfrac{d}{dt}V_c(t) = -k(1-\eta)\cdot(Du_t, u_t) + \delta E(u, u_t) + b\left(\dfrac{1}{\delta}\right) + \varepsilon \cdot (Mu_t, u_t)$$
$$+ \varepsilon\left\{-(1-\eta)|\mathcal{A}^{1/2}u|^2 - c_2\Pi_0(u) + c_3 + c_4 \cdot [1 + E(u, u_t)]^\gamma \cdot (Du_t, u_t)\right\}.$$

Therefore using (3.38), (3.43) and choosing δ proportional to ε, after rescaling of ε we find that there exist constants d_i independent on ε such that

$$\frac{d}{dt} V_c(t) + \varepsilon V_c(t) \leq d_0 \left\{ \varepsilon + b\left(\frac{d_1}{\varepsilon}\right) \right\} + d_2 \left\{ \varepsilon \left[1 + V_c(t)\right]^\gamma - d_3 \right\} \cdot (Du_t, u_t)$$

for any $\varepsilon \in (0, \varepsilon_0)$. Therefore

$$(3.44) \quad V_c(t) \leq e^{-\varepsilon(t-s)} V_c(s) + d_0 \left\{ 1 + \frac{1}{\varepsilon} b\left(\frac{d_1}{\varepsilon}\right) \right\}$$

$$+ d_2 \int_s^t e^{-\varepsilon(t-\tau)} \left\{ \varepsilon \left[1 + V_c(\tau)\right]^\gamma - d_3 \right\} \cdot (Du_t(\tau), u_t(\tau)) d\tau$$

for all $t \geq s \geq 0$. If $\gamma = 0$, then by choosing $\varepsilon < d_3$ one can easily see from the relation above that the system dissipative.

Consider the case $\gamma > 0$ and show that the integral term in (3.44) can also be eliminated. Let $\sigma(V)$ be a solution to the algebraic equation

$$(3.45) \qquad \{1 + V + d_0(1 + \sigma \cdot b(d_1 \sigma))\}^\gamma = \frac{d_3 \sigma}{2}$$

for each $V \in \mathbb{R}_+$. Since $\sigma \cdot b(d_1 \sigma)$ is a strictly increasing continuous function, (3.40) holds and $\gamma < 1$, this equation has unique solution and the function $V \mapsto \sigma(V)$ is continuous, positive and strictly increasing on $\mathbb{R}_+$. Moreover, $\sigma(V) \to +\infty$ as $V \to +\infty$.

Now we take $\varepsilon = [\sigma(V_c(s))]^{-1}$ in (3.44) and apply "barrier's" method. We claim that

$$(3.46) \qquad \varepsilon \left[1 + V_c(t)\right]^\gamma - d_3 < 0 \quad \text{for all} \quad t \geq s$$

for this choice of ε. Indeed, it is clear from (3.45) that

$$\varepsilon \left[1 + V_c(s)\right]^\gamma - d_3 < -\frac{d_3}{2} < 0.$$

Thus by continuity of $V_c(t)$ we have (3.46) for t from some interval $[s, s+T)$. If $T < +\infty$, then there exists $T^* > 0$ such that

$$(3.47) \quad \varepsilon \left[1 + V_c(t)\right]^\gamma - d_3 < 0, \ t \in [s, s+T^*), \quad \text{and} \quad \varepsilon \left[1 + V_c(s+T^*)\right]^\gamma - d_3 = 0.$$

Under this condition we obtain from (3.44) that

$$V_c(t) \leq V_c(s) + d_0 \left\{ 1 + \frac{1}{\varepsilon} b\left(\frac{d_1}{\varepsilon}\right) \right\}, \quad t \in [s, s+T^*).$$

Thus by (3.45) we have that

$$\varepsilon \left[1 + V_c(t)\right]^\gamma \leq \varepsilon \left[1 + V_c(s) + d_0 \left\{ 1 + \frac{1}{\varepsilon} b\left(\frac{d_1}{\varepsilon}\right) \right\} \right]^\gamma = \frac{d_3}{2}, \quad t \in [s, s+T^*).$$

By continuity of $V_c(t)$ this relation contradicts to the second relation in (3.47). Thus

$$(3.48) \qquad V_c(t) \leq e^{-\varepsilon(V_c(s))(t-s)} V_c(s) + f(V_c(s)) \quad \text{for all} \quad t \geq s \geq 0,$$

where

$$\varepsilon(V) = [\sigma(V)]^{-1}, \quad f(V) = d_0 \{1 + \sigma(V) b\left(d_1 \sigma(V)\right)\}$$

and $\sigma = \sigma(V)$ solves (3.45). In particular, from (3.48) we have that

$$(3.49) \qquad V_c(t) \leq V_c(0) + f(V_c(0)) \quad \text{for all} \quad t \geq 0.$$

Let us assume that $V_c(0) \leq R$ for some $R > 0$. Since $\sigma(V)$ is increasing function, relation (3.49) implies that there exists $\varepsilon_R > 0$ such that
$$\varepsilon(V_c(s)) \geq \varepsilon_R > 0 \quad \text{for all} \quad s \geq 0.$$
Therefore
$$V_c(t) \leq e^{-\varepsilon_R(t-s)} V_c(s) + f(V_c(s)) \quad \text{for all} \quad t \geq s \geq 0,$$
provided $V_c(0) \leq R$. Since $f(V)$ is an increasing function, we obtain that
$$(3.50) \quad W_R(t) \leq e^{-\varepsilon_R(t-s)} W_R(s) + f(W_R(s)) \quad \text{for all} \quad t \geq s \geq 0,$$
where $W_R(t) = \sup_{V_c(0) \leq R} V_c(t)$. This implies that
$$W_R^\infty \equiv \limsup_{t \to \infty} W_R(t) \leq f(W_R(s)) < +\infty \quad \text{for all} \quad s \geq 0,$$
Therefore the continuity of the function $f(V)$ gives that $W_R^\infty \leq f(W_R^\infty)$. However, it is easy to see from the definition of the function f and from relation (3.40) that $\lim_{V \to \infty} \{V^{-1} f(V)\} = 0$. This implies that there exists $V_0 > 0$ (independent of R) such that $W_R^\infty \leq V_0$. Therefore dissipativity of the system (H, S_t) generated by (3.37) follows from (3.43).

If the set $\mathcal{B} = \{y \in H : \|y\|_H^2 \leq R\}$ is absorbing for (H, S_t), then $S_t \mathcal{B} \subset \mathcal{B}$ for all $t \geq t_\mathcal{B}$. This implies that $\mathcal{B}_0 = \cup_{t \geq t_\mathcal{B}} S_t \mathcal{B}$ is a forward invariant absorbing set and $\mathcal{B}_0 \subseteq \mathcal{B}$. This completes the proof of Theorem 3.15. $\square$

We note the fact that the proof collapses when condition (3.40) for $\gamma > 0$ does not hold. On the other hand this condition is very natural in applications, as will be seen in Example 3.19 below.

Now we consider a special case when the non-conservative F^* has the form $F^*(u, u_t) = D^*(u_t) + G^*(u)$. In this case we impose the following hypotheses.

ASSUMPTION 3.16. **(D)** Operators D and D^* possess the properties:
- relation (3.38) holds;
- for any $\delta > 0$ there exist constants $a_1^\delta > 0$ and $a_2^\delta > 0$ such that

$$(3.51) \quad (-kDv + D^*v, u) \leq \delta E(u,v) + a_1^\delta + a_2^\delta \cdot [1 + E(u,v)]^\gamma \cdot (Dv, v),$$

  for any $u, v \in \mathcal{D}(\mathcal{A}^{1/2})$, where $\gamma \in [0,1)$ and $E(u,v)$ is given by (1.19);
- there exists a number $0 \leq \eta < 1$ and a constant a_3 such that

$$(3.52) \quad (v, D^*v) \leq \eta k(Dv, v) + a_3, \quad v \in \mathcal{D}(\mathcal{A}^{1/2}).$$

(F) The nonlinear operator $F(u) \equiv -\Pi'(u) + G^*(u)$ satisfies relation (3.17) and there exists a non-decreasing continuous function $b : \mathbb{R}_+ \mapsto \mathbb{R}_+ \setminus \{0\}$ such that (i) if $\gamma > 0$ in (3.51), then (3.40) holds, and (ii) for any $0 < \delta \leq 1$ we have the relation

$$(3.53) \quad |G^*(u)|_{V'}^2 \leq \delta \left(|\mathcal{A}^{1/2} u|^2 + \Pi_0(u)\right) + b\left(\frac{1}{\delta}\right), \quad u, v \in \mathcal{D}(\mathcal{A}^{1/2}),$$

where $\Pi_0(u)$ is the positive part of the potential energy Π (see Assumption 1.1).

A direct application of Theorem 3.15 gives the following assertion.

COROLLARY 3.17. *Let* $F^*(u, u_t) = D^*(u_t) + G^*(u)$. *Then under Assumptions 3.16 the system* (H, S_t) *generated by (3.37) is ultimately dissipative for each* $k > 0$ *and possesses a forward invariant bounded absorbing set.*

PROOF. We check that Assumption 3.16 implies Assumption 3.14 for the case considered.

It follows from (3.51) and (3.17) that

$$(-kDv, u) + (F^*(u, v), u) - (\Pi'(u), u) \leq \eta |\mathcal{A}^{1/2} u|^2 - c_4 \Pi_0(u)$$
$$+ c_5 + \delta E(u, v) + a_1^\delta + a_2^\delta \cdot [1 + E(u, v)]^\gamma \cdot (Dv, v).$$

Therefore using (3.38) and choosing an appropriate δ we obtain (3.39).

From (3.53) and (3.38) we have that

$$\begin{aligned}(v, G^*(u)) &\leq |v|_V |G^*(u)|_{V'} \leq \varepsilon |v|_V^2 + \frac{1}{4\varepsilon} |G^*(u)|_{V'}^2 \\ &\leq \varepsilon (c_0 + c_1(Dv, v)) + \frac{1}{4\varepsilon} \left[\delta \left(|\mathcal{A}^{1/2} u|^2 + \Pi_0(u) \right) + b\left(\frac{1}{\delta}\right) \right] \end{aligned}$$

for any $\varepsilon, \delta \in (0, 1)$. Therefore using (3.52), choosing $\varepsilon = \varepsilon_0$ small enough and rescaling δ we obtain (3.41) with the function $b(s)$ of the form $d_1 + d_2 b(d_3 s)$. Thus Assumption 3.14 holds for the case considered. □

REMARK 3.18. We note that in the case when the restoring force F^* is only velocity dependent (i.e. $G^*(u) \equiv 0$), the assumptions required are much simpler, since there is no need for constructing a function $b(s)$. The conclusion of Corollary 3.17 holds under the (D)-part of Assumption 3.16 and standard requirement (3.17) imposed on $F(u) = -\Pi'(u)$.

In order to illustrate the abstract setup and to provide motivation for the assumptions imposed, we shall consider two canonical examples of wave equation and von Karman equation. Through these examples we shall see that the hypotheses imposed are natural and satisfied for many classical models.

EXAMPLE 3.19 (**Wave equation with non-monotone damping**). In a bounded domain $\Omega \subset \mathbb{R}^n$, $n = 2, 3$, with a smooth boundary Γ we consider the following equation

(3.54) $$w_{tt} - \Delta w + kg(w_t) + g^*(w_t) + f(w) = 0 \text{ in } Q = [0, \infty) \times \Omega$$

subject to the boundary condition either Dirichlet $w = 0$ on $\Sigma \equiv [0, \infty) \times \Gamma$ or else Robin type $\partial_\nu w + w = 0$ in Σ and the initial conditions $w(0) = w_0$ and $w_t(0) = w_1$. We assume that k is a positive parameter and the functions $g(s)$ and $f(s)$ possess properties (1.9)–(1.11) (see Example 1.2). In addition we assume that $g(s)$ satisfies (3.6) and $g^*(s)$ is a continuous function such that

(3.55) $$\limsup_{|s| \to \infty} \left| \frac{g^*(s)}{g(s)} \right| < k.$$

One can see that Assumption 3.16 holds with $G^* \equiv 0$. Indeed, condition (3.55) implies at once relation (3.52) in Assumption 3.16(D). As for (3.51), on the strength of (3.55) it suffices to verify the inequality (3.51) for (Dv, u) term only. Let us take

$n = 3$ (the analysis for $n = 2$ is simpler). In this case the growth condition (3.6) imposed on $g(s)$ along with Sobolev's embeddings yield:

$$\begin{aligned}
-(D(v), u) &= -\int_\Omega g(v) u\, dx \leq |u|_{L_6(\Omega)} \left[\int_\Omega |g(v)|^{\frac{6}{5}} dx\right]^{\frac{5}{6}} \\
&\leq \|u\|_{L_6(\Omega)} \left[C + \int_\Omega |g(v) v|\right]^{\frac{5}{6}} \\
&\leq \delta \|u\|_{H^1(\Omega)}^2 + \frac{C_1}{\delta} + C_2 \|u\|_{H^1(\Omega)} \int_\Omega g(v) v\, dx,
\end{aligned}$$

which implies inequality (3.51) with $\gamma = 1/2$. Thus, if problem (3.54) is well-posed (in the sense of Theorem 1.5), then the corresponding dynamical system is dissipative. On the other hand, problem (3.54) is certainly well-posed if we assume that $g(s)$ and f are subject to the usual conditions specified in (1.9) - (1.11) (see Example 1.2) and the operator $v \to g^*(v)$ is locally Lipschitz on $L_2(\Omega)$. A particular case is, of course, a linear antidamping $g^*(s) = -const^2 \cdot s$. Thus, the result presented above extends significantly that found in [**104**], where $g(s) = s^3$ and $g^*(s) = -s$ are considered.

The same result on dissipativity remains true if we put in the wave system a strong (nonlinear) damping, i.e. if instead of (3.54) we consider the equation

(3.56) $\qquad w_{tt} - \Delta w + \nabla \widetilde{g}(\nabla w_t) + k g(w_t) + g^*(w_t) + f(w) = 0$

subject to the Dirichlet boundary condition, where the vector function $\widetilde{g} : \mathbb{R}^n \mapsto \mathbb{R}^n$ has the form $\widetilde{g}(s_1, \ldots, s_n) = (g_1(s_1), \ldots, g_n(s_n))$. If in addition to the hypotheses above we assume that $g_i(s)$ are nondecreasing functions such that $g_i(0) = 0$ and $|g_i(s)| \leq C|s|$ for all $|s| \geq 1$, then we can establish dissipativity of the dynamical system generated by (3.56) in the space $H_0^1(\Omega) \times L_2(\Omega)$. Indeed, to check Assumption 3.16 we need only to estimate the contribution of the term

$$-\int_\Omega \nabla \widetilde{g}(\nabla v) u\, dx = \int_\Omega \widetilde{g}(\nabla v) \nabla u\, dx \equiv \sum_{i=1}^n \int_\Omega g_i(v_{x_i}) u_{x_i} dx$$

in relation (3.51). This can be done in the following way.

Let $\Omega_1 = \{x \in \Omega : |v_{x_i}| \geq 1\}$ and $\Omega_2 = \Omega \setminus \Omega_1$. Then

$$\int_\Omega g_i(v_{x_i}) u_{x_i} dx \leq \int_{\Omega_1} |g_i(v_{x_i})| |u_{x_i}| dx + C \|u_{x_i}\|_{L_2(\Omega)}.$$

Using the obvious relation $|u_{x_i}| \leq \eta |u_{x_i}|^2 |v_{x_i}|^{-1} + (4\eta)^{-1} |v_{x_i}|$ with arbitrary $\eta > 0$, we obtain that

$$\int_\Omega g_i(v_{x_i}) u_{x_i} dx \leq \frac{1}{4\eta} \int_\Omega g_i(v_{x_i}) v_{x_i} dx + \eta \int_{\Omega_1} \frac{|g_i(v_{x_i})|}{|v_{x_i}|} |u_{x_i}|^2 dx + C \|u_{x_i}\|_{L_2(\Omega)}.$$

Since $|g_i(s)| \leq C|s|$ for $|s| \geq 1$, this implies that

$$\int_\Omega \widetilde{g}(\nabla v) \nabla u\, dx \leq \delta \|\nabla u\|_{L_2(\Omega)}^2 + C_\delta \left(1 + \int_\Omega \widetilde{g}(\nabla v) \nabla v\, dx\right)$$

for any $\delta > 0$. Consequently we can establish (3.51) for the case considered and hence the system generated by (3.56) is dissipative. We also refer to Section 7.4 for other results on long-time dynamics of wave equations with a strong damping.

EXAMPLE 3.20 (**Von Karman evolution equations**). We can apply Corollary 3.17 to von Karman problem (1.12) and (1.13) with $\alpha > 0$, $g(s_1, s_2) = (g_1(s_1), g_2(s_2))$, and with non-conservative term $L(u, u_t) = D^*(u_t) + G^*(u)$ with

$$D^*(u_t) = g^*(u_t, \nabla u_t), \quad G^*(u_t) = f(u, \nabla u),$$

where $g^*, f : \mathbb{R}^3 \mapsto \mathbb{R}$ are continuous functions with the properties

(3.57) $$\limsup_{\sum_i |s_i| \to \infty} \frac{|s_0 g^*(s_0, s_1, s_2)|}{s_0 g_0(s_0) + \alpha(s_1 g_1(s_1) + s_2 g_2(s_2))} < k,$$

$$|g^*(s_0, s_1, s_2)| \leq C \left(1 + s_0 g_0(s_0) + \alpha[s_1 g_1(s_1) + s_2 g_2(s_2)]\right),$$

where C is a positive constant, and

(3.58) $$\limsup_{\sum_i |s_i| \to \infty} \frac{|f(s_0, s_1, s_2)|}{(|s_0| + |s_1| + |s_2|)^r} < \infty$$

with some $r < 1/2$.

Indeed, a similar argument as for wave equations, allow us to derive (3.51) with $\gamma = 1/2$ and (3.52) from (3.57).

As for Assumption 3.16(F) we note that (3.58) implies

$$|f(s_0, s_1, s_2)| \leq C \cdot (1 + |s_0| + |s_1| + |s_2|)^r.$$

Therefore

$$\begin{aligned}|G^*(u)|_{V'}^2 &= \|G^*(u)\|_{H^{-1}(\Omega)}^2 \leq C_r \|G^*(u)\|_{L_{2/r}(\Omega)}^2 \\ &\leq C\left(1 + \|u\|_{H^1(\Omega)}^{2r}\right) \leq C\left(1 + \delta\|u\|_{H^1(\Omega)}^2 + \delta^{-r/(1-r)}\right).\end{aligned}$$

This implies (3.53) with $b(s) = C_1 + C_2 s^{r/(1-r)}$. To satisfy (3.40) with $\gamma = 1/2$ we need $r/(1-r) < 1$, i.e. $0 < r < 1/2$.

In the case $\alpha = 0$ with $f \equiv 0$ instead of (3.57) we should assume that $g^*(s_0, s_1, s_2) \equiv g^*(s_0)$ depends on s_0 only and

$$\limsup_{|s| \to \infty} \left|\frac{g^*(s)}{g_0(s)}\right| < k,$$

as in the case of wave equations.

We note that in the present example non-conservative forces involve displacement and velocity. In such case, an appropriate scaling between superlinearity of the damping and sublinearity of the non-potential force seems necessary. This is expressed by the fact that $r < 1/2$ is required in (3.58).

3.2. Asymptotic smoothness: the main assumption

Starting from this section we assume that the dynamical system generated by (1.1) is dissipative. Sufficient conditions for this are given in the previous section. Therefore, in order to prove an existence of global attractor for evolution (1.1), by appealing to Theorem 2.3 one needs to establish an asymptotic smoothness. For this, some continuity type of assumptions are necessary. These are formulated below.

ASSUMPTION 3.21. **(D)** There exist a strictly increasing, concave function $H_0 \in C(\mathbb{R}^+)$ with the property $H_0(0) = 0$, such that

(3.59) $$H_0((D(u+v) - D(u), v)) \geq (Mv, v) \quad \text{for any} \quad u, v \in \mathcal{D}(\mathcal{A}^{1/2}).$$

Moreover, there exist parameters $\kappa \in (0, 2]$ and $0 < \delta < 1/2$ such that for every $\epsilon > 0$ we have inequality

$$|(D(u+v) - D(u), w)|$$

(3.60)
$$\leq C_1^{\epsilon}(r) \cdot (D(u+v) - D(u), v)$$

$$+ C_2(r)\left(1 + (D(u), u) + (D(u+v), u+v)\right)\left[|\mathcal{A}^{1/2-\delta}w|^{\kappa} + \epsilon|\mathcal{A}^{1/2}w|^2\right]$$

for any $u, v, w \in \mathcal{D}(\mathcal{A}^{1/2})$ such that $|\mathcal{A}^{1/2}w| + |M^{1/2}u| + |M^{1/2}v| \leq r$ with arbitrary $r > 0$, where $C_1^{\epsilon}(r)$ and $C_2(r)$ are non-decreasing functions of r, $C_2(r)$ does not depend on ϵ.

(F) There exist $0 \leq \widetilde{\eta} \leq 1/2$ and a compact seminorm on V denoted by n_V such that

(3.61)
$$|(F(u, v) - F(\hat{u}, \hat{v})|_{V'}^2 \leq C(r)\left[|\mathcal{A}^{1/2-\widetilde{\eta}}(u - \hat{u})|^2 + n_V^2(v - \hat{v})\right]$$

for any (u, v) and $(\hat{u}, \hat{v})$ from $\mathcal{D}(\mathcal{A}^{1/2}) \times V$ such that $|\mathcal{A}^{1/2}u| + |v|_V \leq r$ and $|\mathcal{A}^{1/2}\hat{u}| + |\hat{v}|_V \leq r$. Here $r > 0$ is arbitrary, $C(r)$ is a non-decreasing function of r.

REMARK 3.22. We recall that a seminorm n_V is compact on V, if $n_V(v_n) \to 0$ for any sequence $\{v_n\} \subset V$ such that $v_n \to 0$ weakly in V as $n \to \infty$. A typical example of compact semi-norm can be given by $n_V(v) \equiv |\mathcal{A}^{-\delta}M^{1/2}v|_V$ for some positive δ.

We also note that in a special case when $H_0(s) = k_0^{-1}s$ condition (3.59) is nothing else but strong monotonicity requirement imposed on the damping operator D. However, in a more general case when H_0 is nonlinear, condition (3.59) allows to obtain estimates with a damping whose behaviour at the origin is not quantified (e.g., superlinear). In practical applications where $D(v) = g(v)$ and g is a monotone (scalar) function, we take $H_0(s) = s, |s| \geq 1$, and the behaviour of $H_0(s)$ for $|s| \leq 1$ is described by a suitable concave function depending on g [**80**]. Condition (3.60) is, instead, continuity type of requirement. We also note that coercivity condition (3.59) implies property (3.2) in Assumption 3.1 (see also (3.15)). Concerning (3.60) we note that the substitution $u = 0$ in (3.60) leads to a less restrictive condition in comparison with (3.3). Finally, condition (3.61) states local Lipschitz continuity of the nonlinear term F (cf. (1.3)).

We conclude this section with some preliminary assertions which are important tool in further considerations.

LEMMA 3.23. *Under Assumption 1.1 there exists $T_0 > 0$ and a constant $c > 0$ independent of T such that for any pair w and v of strong solutions to (1.1) we have the following relation*

$$TE_z(T) + \int_0^T E_z(t)dt \leq c\left\{\int_0^T |M^{1/2}z_t(s)|^2 ds + k\int_0^T (D(t, z_t), z_t)dt\right.$$

(3.62)
$$\left. + k\int_0^T |(D(t, z_t), z)|\,dt + \Psi_T(w, v)\right\}$$

3.2. ASYMPTOTIC SMOOTHNESS: THE MAIN ASSUMPTION

for every $T \geq T_0$, where $z(t) = w(t) - v(t)$ and we use the following notations:

$$E_z(t) = E_0(z(t), z_t(t)) = \frac{1}{2}\left((Mz_t(t), z_t(t)) + (\mathcal{A}z(t), z(t))\right),$$

$$D(t, z_t) = D(v_t(t) + z_t(t)) - D(v_t(t)),$$

$$\Psi_T(w, v) = \left|\int_0^T (G_{w,v}(\tau), z_t(\tau))d\tau\right| + \left|\int_0^T (G_{w,v}(t), z(t))dt\right|$$

(3.63)
$$+ \left|\int_0^T dt \int_t^T (G_{w,v}(\tau), z_t(\tau))d\tau\right|$$

with $G_{w,v}(t)$ given by

(3.64) $$G_{w,v}(t) = F(w(t), w_t(t)) - F(v(t), v_t(t))$$

PROOF. The variable z satisfies the equation

(3.65) $$Mz_{tt} + \mathcal{A}z + kD(t, z_t) = G_{w,v}(t)$$

and hence we have the following energy relation

(3.66) $$E_z(T) + k\int_t^T (D(\tau, z_t), z_t)d\tau = E_z(t) + \int_t^T (G_{w,v}(\tau), z_t(\tau))d\tau.$$

Multiplying (3.65) by z after integration we obtain that

(3.67) $$\int_0^T E_z(t)dt \leq c_0\left(E_z(T) + E_z(0)\right) + \int_0^T |M^{1/2}z_t(s)|^2 ds$$
$$+ \frac{k}{2}\int_0^T |(D(s, z_t), z)|ds + \frac{1}{2}\int_0^T (G_{w,v}(s), z(s))ds.$$

From (3.66) we have

$$E_z(0) = E_z(T) + k\int_0^T (D(\tau, z_t), z_t)d\tau - \int_0^T (G_{w,v}(\tau), z_t(\tau))d\tau$$

and integrating (3.66) from 0 to T

$$TE_z(T) \leq \int_0^T E_z(t)dt + \int_0^T dt \int_t^T (G_{w,v}(\tau), z_t(\tau))d\tau.$$

Therefore (3.62) follows from (3.67). □

REMARK 3.24. The inequality stated in Lemma 3.23 will constitute a common first step in the proofs of several assertions in this and the next chapters on the existence and finite dimensionality of global attractors. The inequality in (3.62) represents equipartition of the energy. The potential energy is reconstructed from the kinetic energy and the nonlinear quantities entering the equation. Eventually, these quantities will need to be absorbed ('modulo' lower order terms) by the damping. The realization of this step depends heavily on the assumptions imposed on the model.

For the estimates involving the damping operator the following inequality will be frequently used:

LEMMA 3.25. *Let Assumption 1.1 be in force. Assume that w and v are strong solutions to (1.1) possessing properties*

$$\max_{s\in[0,T]} \left\{ |\mathcal{A}^{1/2}w(s)|^2 + |M^{1/2}w_t(s)|^2 + |\mathcal{A}^{1/2}v(s)|^2 + |M^{1/2}v_t(s)|^2 \right\} \leq R^2 \quad (3.68)$$

for some $R > 0$. Then, with the notations from the previous lemma, we have that

$$\frac{1}{2}\max_{t\in[0,T]} |\mathcal{A}^{1/2}z(t)|^2 \leq E_z(T) + k\int_0^T (D(t,z_t), z_t)dt + c_R \int_0^T E_z(t)dt, \quad (3.69)$$

hence, trivially,

$$\int_0^T [(D(w_t), w_t) + (D(v_t), v_t)] |\mathcal{A}^{1/2}z|^2 dt$$
$$\leq 2D_0^T \cdot \left[E_z(T) + k\int_0^T (D(t,z_t), z_t)dt + c_R \int_0^T E_z(t)dt \right], \quad (3.70)$$

where $c_R > 0$ does not depend on k and T,

$$D_0^T = D_0^T(w,v) \equiv \int_0^T [(D(w_t), w_t) + (D(v_t), v_t)] dt. \quad (3.71)$$

PROOF. With the same notations as in Lemma 3.23 we have that

$$|(G_{w,v}(t), z_t)| \leq C_R \left(|\mathcal{A}^{1/2}z| + |M^{1/2}z_t| \right) |M^{1/2}z_t| \leq C_R E_z(t)$$

under condition (3.68). Therefore, it follows from (3.66) that

$$\max_{t\in[0,T]} E_z(t) \leq E_z(T) + k\int_0^T (D(\tau, z_t), z_t)d\tau + C_R \int_0^T E_z(\tau)d\tau.$$

Since $|\mathcal{A}^{1/2}z(t)| \leq 2E_z(t)$, this implies (3.69) and (3.70). □

Our main results on existence of global attractors for evolutions in (1.1) are formulated below. We shall distinguish two separate cases: subcritical ($\widetilde{\eta} > 0$ in (3.61)) and critical ($\widetilde{\eta} = 0$) nonlinearity.

3.3. Global attractors in subcritical case

Our main results read as follows.

THEOREM 3.26. *We assume Assumptions 1.1 and 3.21 with $\widetilde{\eta} > 0$. In addition, we also assume that $\mathcal{D}(\mathcal{A}^{1/2})$ is compactly embedded into V and there exists a closed bounded forward invariant set $\mathcal{B} \subset H \equiv \mathcal{D}(\mathcal{A}^{1/2}) \times V$ for the dynamical system (H, S_t) generated by problem (1.1). Then the restriction $(\mathcal{B}, S_t)$ of the dynamical system (H, S_t) on $\mathcal{B}$ possesses a compact global attractor A.*

The proof of this theorem is given later.
The following assertion is a direct corollary of the Theorem 3.26.

COROLLARY 3.27. *Let Assumptions 1.1 and Assumption 3.21 with $\widetilde{\eta} > 0$ be satisfied. If $\mathcal{D}(\mathcal{A}^{1/2})$ is compactly embedded into V, then any bounded semitrajectory $\gamma = \{S_t y : t \geq 0\}$ of the dynamical system (H, S_t) is a relatively compact set in H.*

PROOF. We apply Theorem 3.26 with $\mathcal{B} = \overline{\{S_t y : t \geq 0\}}$. □

3.3. GLOBAL ATTRACTORS IN SUBCRITICAL CASE

COROLLARY 3.28. *We assume Assumptions 1.1 and 3.21 with $\widetilde{\eta} > 0$. In addition, we also assume that $\mathcal{D}(\mathcal{A}^{1/2})$ is compactly embedded into V. If the dynamical system (H, S_t) generated by problem (1.1) is dissipative, then (H, S_t) possesses compact global attractor A.*

PROOF. Let B_0 be a bounded absorbing set for (H, S_t). By the definition there exists $t_0 \geq 0$ such that $S_t B_0 \subset B_0$ for all $t \geq t_0$. Let $\widetilde{B} = \overline{\cup_{t \geq t_0} S_t B_0}$. It is clear that $\widetilde{B}$ is a closed bounded forward invariant set for this system. Since for any bounded set B we have that $S_t B \subset B_0$ for all $t \geq t(B)$, we obtain that $S_t B \subset \widetilde{B}$ for all $t \geq t_0 + t(B)$. Hence $\widetilde{B}$ is also an absorbing set for this system. Thus, by Theorem 3.26 $(\widetilde{B}, S_t)$ possesses compact global attractor A. By absorption property of $\widetilde{B}$ the set A is also a global attractor for (H, S_t). □

The following assertion shows that in some cases one does not need to know a priori the ultimate dissipativity of (H, S_t). In fact, this property follows a-posteriori from asymptotic smoothness and Lyapunov's structure of the dynamics considered.

COROLLARY 3.29. *Assume that $F^* \equiv 0$. Let Assumptions 1.1 and 3.21 with $\widetilde{\eta} > 0$ hold and $\mathcal{D}(\mathcal{A}^{1/2})$ is compactly embedded into V. If the set of the stationary solutions to (1.1) is bounded in $\mathcal{D}(\mathcal{A}^{1/2})$, then the dynamical system (H, S_t) generated by problem (1.1) possesses compact global attractor A.*

PROOF. We can derive this assertion from Theorem 2.30. However for the sake of self-containess we give independent argument.

Let $\mathcal{N}$ be the set of the stationary points for (H, S_t). Every element of $\mathcal{N}$ has the form $(w; 0)$, where w is a stationary solution to (1.1). Thus $\mathcal{N}$ is a bounded set in H.

By Theorem 1.5 the energy $\mathcal{E}(u, u_t)$ given by (1.18) satisfies the relation
$$\mathcal{E}(u(t), u_t(t)) + k \int_0^t (Du_t(\tau), u_t(\tau))d\tau = \mathcal{E}(u_0, u_1)$$
on strong solutions. By (3.59) with $u = 0$ this implies that for all strong solutions we have

(3.72) $$\mathcal{E}(u(t), u_t(t)) + k \int_0^t H_0^{-1}\left(|M^{1/2}u_t(\tau)|^2\right) d\tau \leq \mathcal{E}(u_0, u_1).$$

Thus after passing with a limit on strong solutions (3.72) remains valid for any generalized solution.

We denote $\mathcal{E}_c = \{y = (u_0; u_1) \in H : \mathcal{E}(u_0, u_1) \leq c\}$, where $c \geq c_0$ and c_0 is chosen such that $\mathcal{N} \subset \mathcal{E}_{c_0}$. It is clear from (3.72) that $\mathcal{E}_c$ is a closed bounded forward invariant set for (H, S_t). Therefore by Theorem 3.26 the system $(\mathcal{E}_c, S_t)$ possesses compact global attractor A_c for each $c \geq c_0$. Property (3.72) implies that the functional $V(y) \equiv \mathcal{E}(u_0, u_1)$, $y = (u_0; u_1) \in \mathcal{H}$, is a strict Lyapunov function for $(\mathcal{E}_c, S_t)$, i.e. (i) $V(S_t y)$ is not increasing with respect to $t > 0$ and (ii) the equality $V(S_t y) = V(y)$ for some $t > 0$ implies that y is a stationary point, $S_t y = y$ for all $t \geq 0$. Thus $(\mathcal{E}_c, S_t)$ is a gradient system (see Definition 2.26). This implies (see Theorem 2.28) that the global attractor A for $(\mathcal{E}_c, S_t)$ has the form $A = \mathcal{M}^u(\mathcal{N})$, where $\mathcal{M}^u(\mathcal{N})$ is the unstable manifold emanating from the set $\mathcal{N}$ of equilibria for (1.1), $\mathcal{N} = \{(w; 0) : \mathcal{A}w = F(w)\}$. In particular, A_c does not depend on $c \geq c_0$, $A_c = A_{c_0}$ for all $c \geq c_0$. Since any bounded set belongs to $\mathcal{E}_c$ for some $c \geq c_0$, the set A_{c_0} is a compact global attractor for (H, S_t). □

Proof of Theorem 3.26. In order to establish existence of a compact attractor, it suffices to prove that the semiflow is asymptotically smooth. This property, roughly speaking, amounts to the statement that the difference of two solutions is uniformly stable modulo metric with zero noncompactness measure (cf. Corollary 2.7). This involves studying the flow corresponding to the difference of two solutions.

Step 0: preliminaries. We shall apply Theorem 2.4 on the forward invariant set $\mathcal{B}$ which we consider as a metric space with the distance $d(y_1, y_2) = \|y_1 - y_2\|_H$, $y_1, y_2 \in \mathcal{B}$. We use the same idea as in [**27**], where the case $M = I$ is considered.

Let $w(t)$ and $v(t)$ be two generalized solutions to (1.1) corresponding to initial data in the invariant set $\mathcal{B}$:

$$(3.73) \qquad (w(t), w_t(t)) \equiv S_t y_0, \ (v(t), v_t(t)) \equiv S_t y_1, \ y_0, y_1 \in \mathcal{B}.$$

We shall prove that the requirements imposed by Theorem 2.4 are satisfied. Since all terms in (2.3) are continuous with respect to the distance d given by the energy norm $\|\cdot\|_H$, we can assume that $w(t)$ and $v(t)$ are two strong solutions. We shall also use the following shortcut notations described in (3.63).

Let $T > 0$. Since $\mathcal{B}$ is a bounded forward invariant set, from energy equality (1.17) and continuity of F we always have:

$$(3.74) \qquad D_0^T \equiv \int_0^T (D(v_t), v_t) dt + \int_0^T (D(w_t), w_t) dt \leq C_\mathcal{B}(1 + T).$$

We denote $z(t) = w(t) - v(t)$. The variable z satisfies

$$(3.75) \qquad M z_{tt} + \mathcal{A} z + k D(t, z_t) = G_{w,v}(t), \quad (z(0), z_t(0)) = y_0 - y_1,$$

where as in (3.63) and (3.64) we use the notations

$$D(t, z_t) = D(z_t + v_t) - D(v_t)$$

and

$$G_{w,v}(t) = F(z + v, z_t + v_t) - F(v, v_t).$$

The standard energy method gives with any $t \in 0, T]$

$$(3.76) \qquad E_z(T) + k \int_t^T (D(t, z_t), z_t) dt = E_z(t) + \int_t^T (G_{w,v}(t), z_t) dt.$$

Step 1: energy reconstruction. It follow from the definition of Ψ_T given in (3.63) and from (3.61) that

$$(3.77) \qquad \Psi_T(w, v) \leq C_{\mathcal{B},\epsilon}(T) \int_0^T \left[|\mathcal{A}^{1/2 - \tilde{\eta}} z|^2 + n_V^2(z_t) \right] dt + \epsilon \int_0^T E_z(t) dt$$

for every $\epsilon > 0$. From coercivity assumption (3.59) and Jensen's inequality we also obtain

$$(3.78) \qquad \int_0^T |M^{1/2} z_t|^2 dt \leq \mathcal{H}_0 \left(\int_0^T (D(t, z_t), z_t) dt \right),$$

where $\mathcal{H}_0(s) = TH_0(s/T)$. Therefore applying Lemma 3.23 and using (3.77) with $\epsilon > 0$ small enough, we obtain that

$$\begin{aligned}
(3.79) \quad & TE_z(T) + \frac{1}{2}\int_0^T E_z(t)dt \\
& \leq c\left\{(\mathcal{H}_0 + k)\left(\int_0^T (D(t,z_t), z_t)dt\right) + k\int_0^T |(D(t,z_t), z)|\,ds\right\} \\
& \quad + C_{\mathcal{B}}(T)\int_0^T \left[|\mathcal{A}^{1/2-\tilde{\eta}}z|^2 + n_V^2(z_t)\right]dt
\end{aligned}$$

for every $T \geq T_0$. Thus using (3.60) more we have that

$$\begin{aligned}
& TE_z(T) + \int_0^T E_z(t)dt \leq C_{\mathcal{B},\epsilon} \cdot (\mathcal{H}_0 + I)\left(\int_0^T (D(t,z_t), z_t)dt\right) \\
& \quad + C_{\mathcal{B}}\int_0^T [1 + (D(w_t), w_t) + (D(v_t), v_t)]\left[|\mathcal{A}^{1/2-\delta}z|^\kappa + \epsilon|\mathcal{A}^{1/2}z|^2\right]dt \\
(3.80) \quad & \quad + C_{\mathcal{B}}(T)\int_0^T \left[|\mathcal{A}^{1/2-\tilde{\eta}}z|^2 + n_V^2(z_t)\right]dt.
\end{aligned}$$

Step 2: handling of the damping. Using (3.74) and Lemma 3.25 we obtain

$$\begin{aligned}
& \int_0^T [1 + (D(w_t), w_t) + (D(v_t), v_t)]|\mathcal{A}^{1/2}z|^2 dt \\
(3.81) \quad & \leq C_{\mathcal{B},T} \cdot \left(E_z(T) + k\int_0^T (D(t,z_t), z_t)dt + \int_0^T E_z(t)dt\right).
\end{aligned}$$

For the further use we note that this relation holds for *all* $0 \leq \tilde{\eta} \leq 1/2$.

Selecting $\epsilon \equiv \epsilon(\mathcal{B}, T)$ in (3.80) suitably small and using (3.81) yields

$$\begin{aligned}
& TE_z(T) + \int_0^T E_z(t)dt \leq C_{\mathcal{B},T} \cdot (\mathcal{H}_0 + I)\left(\int_0^T (D(t,z_t), z_t)dt\right) \\
& \quad + C_{\mathcal{B}}\int_0^T [1 + (D(w_t), w_t) + (D(v_t), v_t)]|\mathcal{A}^{1/2-\delta}z|^\kappa dt \\
(3.82) \quad & \quad + C_{\mathcal{B},T}\int_0^T \left[|\mathcal{A}^{1/2-\tilde{\eta}}z|^2 + n_V^2(z_t)\right]dt
\end{aligned}$$

for all $T \geq T_0$. By (3.74) we also have that

$$\begin{aligned}
& \int_0^T [1 + (D(w_t), w_t) + (D(v_t), v_t)]|\mathcal{A}^{1/2-\delta}z|^\kappa dt \\
(3.83) \quad & \leq C_{\mathcal{B},T} \sup_{t \in [0,T]} |\mathcal{A}^{1/2-\delta}z(t)|^\kappa.
\end{aligned}$$

Combining with (3.82) we obtain that

$$E_z(T) \leq C_{\mathcal{B},T}(\mathcal{H}_0 + I)\left(\int_0^T (D(t,z_t), z_t)dt\right)$$

$$(3.84) \qquad + C_{\mathcal{B},T}\left\{\sup_{t\in[0,T]} |\mathcal{A}^{1/2-\widetilde{\delta}}z(t)|^\kappa + \int_0^T n_V^2(z_t(t))dt\right\},$$

where $\widetilde{\delta} = \min\{\widetilde{\eta}, \delta\}$. Let $G_0(s) = (\mathcal{H}_0 + I)^{-1}\left(\frac{s}{2C_{\mathcal{B},T}}\right)$. Since $G_0(s)$ is a convex function and

$$(\mathcal{H}_0 + I)^{-1}(s) \leq s, \quad \text{for} \quad s \geq 0,$$

from (3.84) we obtain that

$$G_0(E_z(T)) \leq \frac{1}{2}\int_0^T (D(t,z_t), z_t)dt$$

$$(3.85) \qquad + \frac{1}{2}\left\{\sup_{t\in[0,T]} |\mathcal{A}^{1/2-\widetilde{\delta}}z(t)|^\kappa + \int_0^T n_V^2(z_t(t))dt\right\}.$$

From energy relation (3.76) we have

$$k\int_0^T (D(t,z_t), z_t)dt \leq E_z(0) - E_z(T)$$

$$+ C_{\mathcal{B},T}\sup_{t\in[0,T]} |\mathcal{A}^{1/2-\widetilde{\delta}}z(t)| + C_{\mathcal{B}}\int_0^T n_V(z_t(t))dt.$$

Thus from (3.85) we obtain

$$E_z(T) + 2kG_0(E_z(T))$$

$$\leq E_z(0) + C_{\mathcal{B},T}\left\{\sup_{t\in[0,T]} |\mathcal{A}^{1/2-\widetilde{\delta}}z(t)|^{\min\{1,\kappa\}} + \int_0^T n_V(z_t(t))dt\right\}.$$

Since $z(t)$ is uniformly bounded in $\mathcal{D}(\mathcal{A}^{1/2})$, by interpolation we have that

$$E_z(T) + 2kG_0(E_z(T))$$

$$\leq E_z(0) + C_{\mathcal{B},T}\left\{\sup_{t\in[0,T]} |z(t)|^{\widetilde{\kappa}} + \int_0^T n_V(z_t(t))dt\right\}$$

for some $\widetilde{\kappa} \in (0,1]$. This implies that

$$\|S_T y_1 - S_T y_2\|_H \leq q\left(\|y_1 - y_2\|_H + \rho_{\mathcal{B}}^T(\{S_\tau y_1\}, \{S_\tau y_2\})\right).$$

Here $q(s) = \sqrt{2}\left([I + 2kG_0]^{-1}(s^2/2)\right)^{1/2}$ and

$$\rho_{\mathcal{B}}^T(\{S_t y_1\}, \{S_t y_2\}) = C_{\mathcal{B},T}\left\{\sup_{t\in[0,T]} |v(t) - w(t)|^{\kappa^*} + \int_0^T n_V(v_t(t) - w_t(t))dt\right\}$$

for some $\kappa^* \in (0, 1/2]$, where $S_t y_1$ and $S_t y_2$ has the form (3.73).

Clearly, the function q satisfies all the requirements of Theorem 2.4.

Now we prove that the pseudometric $\rho_{\mathcal{B}}^T$ is precompact on the set $\mathcal{L}_{\mathcal{B},T}$ all solutions to (1.1) on $[0,T]$ with initial data from $\mathcal{B}$.

Let V_1 be the completion of V with respect to the norm $|\cdot|_{V_1}$ given by

$$|v|_{V_1} = |\mathcal{A}^{-1/2}Mv| + n_V(v)$$

and W be the completion of V with respect to the norm $|\cdot|_W = |\mathcal{A}^{-1/2}M\cdot|$. It is clear that $V \subset V_1 \subset W$ and the embedding $V \subset V_1$ is compact.

It follows from (3.60) with $u = 0$ that
$$|(D(v), w)| \le C \cdot [1 + (D(v), v)] \quad \text{for any} \quad w \in \mathcal{D}(\mathcal{A}^{1/2}), \ |\mathcal{A}^{1/2}w| \le 1.$$
Therefore
$$|\mathcal{A}^{-1/2}D(v)| \le C \cdot [1 + (D(v), v)].$$
Hence, using (3.74) and equation (1.1) we can conclude that

(3.86) $$\int_0^T |\mathcal{A}^{-1/2}Mv_{tt}(t)|dt \le C_{\mathcal{B},T} \quad \text{for all} \quad v \in \mathcal{L}_{\mathcal{B},T},$$

i.e. the set $\{v_{tt} : v \in \mathcal{L}_{\mathcal{B},T}\}$ is bounded in $L_1(0, T; W)$. Since
$$\sup_{t \in [0,T]} |v_t(t)|_V \le C_\mathcal{B} \quad \text{for every} \quad v \in \mathcal{L}_{\mathcal{B},T},$$
using [96, Corollary 4] we can conclude that the set $\{w \equiv v_t : v \in \mathcal{L}_{\mathcal{B},T}\}$ is precompact in $L_p(0, T; V_1)$ for every $p \ge 1$. This means that $\int_0^T n_V(v_t(t) - w_t(t))dt$ is a precompact pseudometric on $\mathcal{L}_{\mathcal{B},T}$. By the compactness of embedding
$$C([0,T], \mathcal{D}(\mathcal{A}^{1/2})) \cap C^1([0,T], V) \subset C([0,T], \mathcal{H}))$$
we also have precompactness of the pseudometric $\sup_{t \in [0,T]} |v(t) - w(t)|^{\kappa^*}$. Therefore the pseudometric $\rho_\mathcal{B}^T$ is also precompact on $\mathcal{L}_{\mathcal{B},T}$.

Thus we can apply Theorem 2.4 to obtain the asymptotic smoothness of (H, S_t) and, hence, the statement of Theorem 3.26. □

The following remark is important in the proof of finite-dimensionality of the global attractor. We use it in Chapter 4.

REMARK 3.30. In the case when $H_0(s) = k_0^{-1} \cdot s$ under the conditions of Theorem 3.26 we can obtain the following stabilizability estimate

(3.87) $$\begin{aligned}E_0(z(t), z_t(t)) &\le Ce^{-\omega_\mathcal{B} t} E_0(z(0), z_t(0)) \\ &\quad + C(\mathcal{B}) \left\{ \sup_{0 \le s \le t} |\mathcal{A}^{1/2-\widetilde{\delta}}z(s)|^\kappa + \int_0^t e^{-\omega_\mathcal{B}(t-\tau)} n_V^2(z_t(\tau))dt \right\}\end{aligned}$$

for the difference z of two solutions v and w with initial data from $\mathcal{B}$, where $\omega_\mathcal{B} > 0$, $\widetilde{\delta} = \min\{\widetilde{\eta}, \delta\}$ and $\widetilde{\eta}, \delta, \kappa$ are the constants from Assumption 3.21. Indeed, it follows from (3.84) that

$$\begin{aligned}E_z(T) &\le C_1 \int_0^T (D(z_t + v_t) - D(v_t), z_t)dt \\ &\quad + C_2 \left\{ \sup_{t \in [0,T]} |\mathcal{A}^{1/2-\widetilde{\delta}}z(s)|^\kappa + \int_0^T n_V^2(z_t(\tau))dt \right\},\end{aligned}$$

where C_1 and C_2 may depend on $\mathcal{B}$ and $T > 1$. Since $H_0(s) = k_0^{-1} \cdot s$, from energy relation (3.76) and estimate (3.61) it is easy to see that

$$\begin{aligned}c_0(k, k_0) \int_0^T (D(z_t + v_t) - D(v_t), z_t)dt &\le E_z(0) - E_z(T) \\ &\quad + C_\mathcal{B} \left\{ \sup_{t \in [0,T]} |\mathcal{A}^{1/2-\widetilde{\delta}}z(s)|^\kappa + \int_0^T n_V^2(z_t(\tau))dt \right\}.\end{aligned}$$

Therefore
(3.88)
$$E_z(T) \leq C_{\mathcal{B}}(E_z(0) - E_z(T)) + C_{\mathcal{B}} \left\{ \sup_{t \in [0,T]} |\mathcal{A}^{1/2-\widetilde{\delta}}z(s)|^{\kappa} + \int_0^T n_V^2(z_t(\tau))dt \right\}.$$

The above implies
$$E_z(T) \leq \gamma_{\mathcal{B}} E_z(0) + \gamma_{\mathcal{B}} \left\{ \sup_{t \in [0,T]} |\mathcal{A}^{1/2-\widetilde{\delta}}z(s)|^{\kappa} + \int_0^T n_V^2(z_t(\tau))dt \right\},$$

where $\gamma_{\mathcal{B}} \equiv \frac{C_{\mathcal{B}}}{C_{\mathcal{B}}+1} < 1$ and $T > 1$. Therefore we obtain that
$$E_z((m+1)T) \leq \gamma_{\mathcal{B}} E_z(mT) + \gamma_{\mathcal{B}} c_m, \quad m = 0, 1, 2, \ldots,$$

where
$$c_m \equiv \sup_{t \in [mT,(m+1)T]} |\mathcal{A}^{1/2-\widetilde{\delta}}z(t)|^{\kappa} + \int_{mT}^{(m+1)T} n_V^2(z_t(\tau))dt.$$

This yields
$$E_z(mT) \leq \gamma_{\mathcal{B}}^m E_z(0) + \sum_{l=1}^m \gamma_{\mathcal{B}}^{m+1-l} c_{l-1}.$$

Since $\gamma_{\mathcal{B}} < 1$, there exists $\omega_{\mathcal{B}} > 0$ such that
$$\begin{aligned}E_z(mT) &\leq Ce^{-\omega_{\mathcal{B}} mT} E_z(0) \\ &\quad + C_{\mathcal{B}} \left\{ \sup_{0 \leq s \leq mT} |\mathcal{A}^{1/2-\widetilde{\delta}}z(s)|^{\kappa} + \int_0^{mT} e^{-\omega_{\mathcal{B}}(mT-\tau)} n_V^2(z_t(\tau))dt \right\}.\end{aligned}$$

Using Proposition 1.15 we obtain (3.87).

REMARK 3.31. As we will see in Chapter 4 the stabilizability estimate in (3.87) (with $\kappa = 2$) is invoked in the proof of finiteness of fractal dimension of global attractors. However we need this estimate *only* on the trajectories which belong to the global attractor (see, e.g., Theorem 4.4 in Chapter 4). Moreover, as we will see in Sect. 4.2, stabilizability estimates are useful in the study of smoothness properties of elements from the attractor. We will use this observation in the study of global attractors for wave equations in Chapter 5 and for von Karman model with $\alpha = 0$ in Chapter 6.

The observation made in the following remark will be invoked in Chapter 7 in the study of strongly damped systems.

REMARK 3.32. The statement of Theorem 3.26 remains true if we replace Assumption 3.21(F) by the following requirement:
- There exist $0 < \widetilde{\eta} < 1/2$ and $k_0 < k$ such that
  (i) $\mathcal{D}(\mathcal{A}^{1/2-\widetilde{\eta}})$ is continuously embedded into V;
  (ii) for $u, \hat{u} \in \mathcal{D}(\mathcal{A}^{1/2})$, $v, \hat{v} \in V$ satisfying the relations
  $$|\mathcal{A}^{1/2}u|^2 + |v|_V^2 \leq r \quad \text{and} \quad |\mathcal{A}^{1/2}\hat{u}|^2 + |\hat{v}|_V^2 \leq r$$
  the estimate
  $$\begin{aligned}|(F(u,v) - F(\hat{u},\hat{v}), v - \hat{v})| &\leq k_0(D(v) - D(\hat{v}), v - \hat{v}) \\ &\quad + C(r)\left[|\mathcal{A}^{1/2-\widetilde{\eta}}(u-\hat{u})|^2 + n_V^2(v-\hat{v})\right]\end{aligned}$$
  (3.89)

holds, where $C(r) > 0$ is non-decreasing function of r and n_V is a compact seminorm on V.

Indeed, the property $\mathcal{D}(\mathcal{A}^{1/2-\tilde{\eta}}) \subset V$ and relation (1.3) imply that

$$(F(u,v) - F(\hat{u},\hat{v}), u - \hat{u}) \leq |u - \hat{u}|_V \cdot |(F(u,v) - F(\hat{u},\hat{v})|_{V'}$$
$$\leq C(r)|\mathcal{A}^{1/2-\tilde{\eta}}(u-\hat{u})| \cdot \left[|\mathcal{A}^{1/2}(u-\hat{u})| + |v-\hat{v}|_V\right]$$
(3.90)
$$\leq \varepsilon \left[|\mathcal{A}^{1/2}(u-\hat{u})|^2 + |v-\hat{v}|_V^2\right] + C_\varepsilon(r)|\mathcal{A}^{1/2-\tilde{\eta}}(u-\hat{u})|^2$$

for any $\varepsilon > 0$. Using (3.89) and (3.90), with the notations of the proof of Theorem 3.26, instead of (3.77) we obtain that

$$\Psi_T(w,v) \leq C_T \int_0^T (D(t,z_t), z_t) dt$$
$$+ C_{\mathcal{B},\epsilon}(T) \int_0^T \left[|\mathcal{A}^{1/2-\tilde{\eta}}z|^2 + n_V^2(z_t)\right] dt + \epsilon \int_0^T E_z(t) dt.$$

This relation allow us to apply Lemma 3.23 and obtain (3.80) for the case considered. It also follows from energy relation (3.76) and from (3.89) that

$$(k - k_0) \int_0^T (D(z_t + v_t) - D(v_t), z_t) dt \leq E_z(0) - E_z(T)$$
(3.91)
$$+ C_{\mathcal{B},T} \sup_{t \in [0,T]} |\mathcal{A}^{1/2-\tilde{\eta}}z(t)|^2 + C_\mathcal{B} \int_0^T n_V^2(z_t(t)) dt.$$

Therefore we can complete the proof in the same way as in Theorem 3.26.

We also note that inequality (3.91) and the same argument as in Remark 3.30 make it possible to establish estimate (3.87) in the case considered.

3.4. Global attractors in critical case

When the contribution of nonlinear term F is no longer of a lower order with respect to the energy (i.e., $\tilde{\eta} = 0$ in (3.61)) we still obtain an existence of a compact global attractor under additional assumptions requiring that either the damping parameter k be sufficiently large, or else, the nonlinear terms F and D satisfy additional structural conditions. We begin with the latter case splitting it into two subcases.

3.4.1. The case of a compact potential energy functional. Our key assumption in this subsection is a compactness requirement concerning the nonlinear part of potential energy functional Π. As we shall see later, when discussing various applications, the postulated compactness property of the energy is a very natural requirement satisfied by most models arising in mathematical physics. A use of this property in the context of a simple von Karman equation was illustrated in [**66**]. More precisely, in addition to our basic Assumption 1.1 we impose the following hypotheses on D, Π and F^*.

ASSUMPTION 3.33. **(D)** • (i) For any $\eta > 0$ there exist $C_\eta > 0$ such that
(3.92) $\quad (Mv, v) \leq \eta + C_\eta \cdot (D(u+v) - D(u), v) \quad$ for any $\quad u, v \in \mathcal{D}(\mathcal{A}^{1/2})$.

• (ii) Relation (3.60) holds true;

(F)
- (i) The potential energy functional $\Pi(u)$ is continuous on $\mathcal{D}(\mathcal{A}^{1/2-\delta})$ for some $\delta > 0$;
- (ii) the mapping $u \mapsto \mathcal{A}^{-l}\Pi'(u)$ is continuous from $\mathcal{D}(\mathcal{A}^{1/2-\delta})$ into $\mathcal{H}$ for some $l, \delta > 0$;
- (iii) there exist $0 < \widetilde{\eta} \leq 1/2$ and a compact seminorm on V denoted by n_V such that

$$(3.93) \quad |(F^*(u,v) - F^*(\hat{u}, \hat{v})|_{V'}^2 \leq C(r) \left[|\mathcal{A}^{1/2-\widetilde{\eta}}(u-\hat{u})|^2 + n_V^2(v-\hat{v}) \right]$$

for any (u,v) and $(\hat{u}, \hat{v})$ from $\mathcal{D}(\mathcal{A}^{1/2}) \times V$ such that $|\mathcal{A}^{1/2}u| + |v|_V \leq r$ and $|\mathcal{A}^{1/2}\hat{u}| + |\hat{v}|_V \leq r$. Here $r > 0$ is arbitrary, $C(r)$ is a non-decreasing function of r.

Instead of condition (F)(iii) we can assume that $F^*(u,v)$ is compact in the following sense:

- for any interval $[0,T]$ and for any sequence $\{u^n(t)\}$ of solutions to (1.1) for initial data from invariant (bounded) absorbing set such that

$$(u^n(t); u_t^n(t)) \to (u(t); u_t(t)) \quad \text{weakly* in} \quad L_\infty(0,T; \mathcal{D}(\mathcal{A}^{1/2}) \times V)$$

we have that

$$\lim_{n \to \infty} \int_0^T |(F^*(u^n(t); u_t^n(t)) - F^*(u(t); u_t(t))|_{V'} dt = 0.$$

Condition (F)(ii) can be altered in a similar way.

Under the above assumptions the following result holds.

THEOREM 3.34. *Let Assumptions 1.1 and 3.33 hold. Assume that $\mathcal{D}(\mathcal{A}^{1/2-\beta})$ is compactly embedded in V for some $\beta > 0$ and the system (H, S_t) generated by problem (1.1) in $H = \mathcal{D}(\mathcal{A}^{1/2}) \times V$ is dissipative. Then, the semiflow generated by problem (1.1) possesses a compact global attractor A.*

REMARK 3.35. As we shall see in Chapter 5, Theorem 3.34 which in applications allows to relax compactness condition imposed on F by replacing this by a weaker notion of compactness of the corresponding part of potential energy, leads to new set of results for the second order systems such as wave equation and plate equations. Indeed, for semilinear wave equations with critical nonlinearity (both in the source and the damping), Theorem 3.34 allows to establish existence of attractors with a damping which may be strongly nonlinear and of an arbitrary size. This result appears to be new in the literature, where previous results require either the damping to be sufficiently large, or else globally Lipschitz [**48, 49, 50, 51, 90**]. Similarly, in the case of plate equations with critical forces or sources Theorem 3.34 allows to dispense with the previous requirement that the damping should be sufficiently large. First example of this was given in [**66**] where a simple (with zero in-plane force) von Karman plate was considered.

Once we take into considerations the assumption imposed on potential energy (the idea motivated by [**66**]), the proof of Theorem 3.34 is essentially embedded in the proof of Theorem 3.26. For reader's convenience we shall repeat the details. It suffices to prove

PROPOSITION 3.36. *Assume that Assumptions 1.1 and 3.33 hold true and the embedding $\mathcal{D}(\mathcal{A}^{1/2-\beta}) \subset V$ is compact for some $\beta > 0$. Then, the semiflow S_t generated by problem (1.1) is asymptotically smooth.*

PROOF. The proof is essentially embedded in the proof of Theorem 3.26 but it employs a new idea that is credited to [66]. We shall apply Proposition 2.10 on the forward invariant set $\mathcal{B}$ which we consider as a metric space with the distance $d(y_1, y_2) = \|y_1 - y_2\|_H$, $y_1, y_2 \in \mathcal{B}$.

Let $w(t)$ and $v(t)$ be two solutions to (1.1) corresponding to initial data in the invariant set $\mathcal{B}$:

$$(w(t), w_t(t)) \equiv S_t y_0, \ (v(t), v_t(t)) \equiv S_t y_1, \ y_0, y_1 \in \mathcal{B}. \tag{3.94}$$

Let $T > 0$. Since $\mathcal{B}$ is a bounded forward invariant set, from energy inequality and continuity of F we always have that

$$D_0^T \equiv \int_0^T (D(w_t), w_t) dt + \int_0^T (D(v_t), v_t) dt \leq C_\mathcal{B}(1 + T). \tag{3.95}$$

We also note that, as in proof of Theorem 3.26 in further calculations we can assume that $w(t)$ and $v(t)$ are strong solutions to (1.1). Thus $z(t) \equiv w(t) - v(t)$ satisfies the equation

$$Mz_{tt} + \mathcal{A}z + kD(t, z_t) = G_{w,v}(t), \quad (z(0), z_t(0)) = y_0 - y_1, \tag{3.96}$$

where we use the same notations as in (3.63), i.e.
(3.97)
$$D(t, z_t) = D(z_t(t) + v_t(t)) - D(v_t(t)), \quad G_{w,v}(t) = F(w(t), w_t(t)) - F(v(t), v_t(t)).$$

The standard energy method gives with any $t \in 0, T]$

$$E_z(T) + k \int_t^T (D(t, z_t), z_t) dt = E_z(t) + \int_t^T (G_{w,v}(\tau), z_t(\tau)) d\tau, \tag{3.98}$$

where, as above we use the following shortcut notation

$$E_z(t) \equiv E_0(z(t), z_t(t)) = \frac{1}{2} \left((Mz_t(t), z_t(t)) + (\mathcal{A}z(t), z(t)) \right).$$

We start with the following consequence of Lemma 3.23.

LEMMA 3.37. *For any $\eta > 0$ and $T > T_0$ there exist positive constants $C^{(1)}_{\mathcal{B},\eta}$ and $C^{(2)}_{\mathcal{B},T}$ such that*

$$E_z(T) \leq \eta + \frac{C^{(1)}_{\mathcal{B},\eta}}{T} \cdot [1 + \Psi_T(w, v)] + C^{(2)}_{\mathcal{B},T} \sup_{t \in [0,T]} |\mathcal{A}^{1/2-\delta} z(t)|^\kappa, \tag{3.99}$$

where δ and κ are the constants from (3.60) and $\Psi_T(w, v)$ is given by (3.63), i.e.

$$\Psi_T(w, v) = \left| \int_0^T (G_{w,v}(\tau), z_t(\tau)) d\tau \right| + \left| \int_0^T (G_{w,v}(t), z(t)) dt \right|$$
$$+ \left| \int_0^T dt \int_t^T (G_{w,v}(\tau), z_t(\tau)) d\tau \right| \tag{3.100}$$

with $G_{w,v}(t)$ given in (3.97).

PROOF. Using (3.60) from Lemma 3.23 we have that

$$TE_z(T) \leq c\left\{\int_0^T |M^{1/2}z_t|^2 dt + \Psi_T(w,v)\right\} + C_{\mathcal{B},\epsilon}\int_0^T (D(t,z_t), z_t) dt$$
$$+ C_{\mathcal{B}}\int_0^T [1 + (D(w_t), w_t) + (D(v_t), v_t)]\left[|\mathcal{A}^{1/2-\delta}z|^\kappa + \epsilon|\mathcal{A}^{1/2}z|^2\right] dt.$$

Since $|\mathcal{A}^{1/2}z(t)|^2 \leq C_\mathcal{B}$ for all $t \geq 0$, by (3.95) we have that

$$\int_0^T [1 + (D(w_t), w_t) + (D(v_t), v_t)]\left[|\mathcal{A}^{1/2-\delta}z|^\kappa + \epsilon|\mathcal{A}^{1/2}z|^2\right] dt$$
$$\leq \epsilon C_\mathcal{B}(1+T) + C_{\mathcal{B},T}\sup_{t\in[0,T]}|\mathcal{A}^{1/2-\delta}z(t)|^\kappa.$$

Therefore relation (3.92) after rescaling $\epsilon C_\mathcal{B} := \epsilon$ yields that

(3.101)
$$TE_z(T) \leq \epsilon + (\eta + \epsilon)\cdot T + C_{\mathcal{B},\eta,\epsilon}\int_0^T (D(t,z_t), z_t) dt + c\Psi_T(w,v)$$
$$+ C_{\mathcal{B},T}\sup_{t\in[0,T]}|\mathcal{A}^{1/2-\delta}z(t)|^\kappa$$

for any $\eta > 0$ and $\epsilon > 0$. Since $E_z(0) \leq C_\mathcal{B}$, using energy relation (3.98) with $t=0$ we obtain

$$k\int_0^T (D(t,z_t), z_t) dt \leq C_\mathcal{B} + \int_0^T (G_{w,v}(\tau), z_t(\tau)) d\tau \leq C_\mathcal{B} + \Psi_T(w,v).$$

Therefore substituting $\eta + \epsilon := \eta$ in (3.101) we obtain that

$$TE_z(T) \leq \eta\cdot T + C_{\mathcal{B},\eta} + C_{\mathcal{B},\eta}\Psi_T(w,v) + C_{\mathcal{B},T}\sup_{t\in[0,T]}|\mathcal{A}^{1/2-\delta}z(t)|^\kappa.$$

This implies (3.99). □

Using Lemma 3.37 we obtain the following assertion.

LEMMA 3.38. *For any $\eta > 0$ and $T > T_0$ there exist positive constants $C_{\mathcal{B},\eta}^{(1)}$ and $C_{\mathcal{B},\eta,T}^{(3)}$ such that*

(3.102)
$$E_z(T) \leq \eta + \frac{C_{\mathcal{B},\eta}^{(1)}}{T}\cdot[1 + \Psi_*(w,v;T)],$$

where $z = w - v$ and

(3.103)
$$\Psi_*(w,v;T) = \left|\int_0^T (G_0(\tau), z_t(\tau)) d\tau\right| + \left|\int_0^T dt \int_t^T (G_0(\tau), z_t(\tau)) d\tau\right|$$
$$+ C_{\mathcal{B},\eta,T}^{(3)}\left[\sup_{t\in[0,T]}|\mathcal{A}^{1/2-\beta_*}z(t)|^{\kappa_*} + \int_0^T n_V(z_t(t)) dt\right]$$

with positive β_* and κ_* and with $G_0(t)$ given by

(3.104)
$$G_0(t) \equiv G_0(w(t), v(t)) = -\Pi'(w(t)) + \Pi'(v(t)).$$

3.4. GLOBAL ATTRACTORS IN CRITICAL CASE

PROOF. It follows from (3.97) and (3.93) that $G_{w,v}(t) = G_0(t) + G_1(t)$, where
$$|G_1(t)| \leq C_\mathcal{B} \left[|\mathcal{A}^{1/2-\tilde{\eta}} z(t)| + n_V(z_t(t)) \right].$$
We also have that
$$\int_0^T (G_0(t), z(t)) dt \leq C_\mathcal{B} \int_0^T |z(t)|_V dt.$$
Therefore, since the function $\Psi_T(w, v)$ given by (3.100) is sub-additive with respect to $G_{w,v}$, relation (3.102) follows from (3.99). This completes the proof of Lemma 3.38. □

To apply Proposition 2.10 it is sufficient to prove that
$$(3.105) \qquad \liminf_{m \to \infty} \liminf_{n \to \infty} \Psi_*(w_n, w_m; T) = 0$$
for every $T > 0$, where $\Psi_*(w, v; T)$ is given by (3.103) and $(w^n(t); w_t^n(t)) \equiv S_t y_0^n$ are solutions to (1.1) with initial data $\{y_0^n\}_{n=1}^\infty$ from $\mathcal{B}$. We can assume that
$$(3.106) \quad (w^n(t); w_t^n(t)) \to (w(t); w_t(t)) \quad \text{*-weakly in} \quad L_\infty(0, T; \mathcal{D}(\mathcal{A}^{1/2}) \times V).$$

Step 1: The following claim results from the assumed compactness.
(3.107)
$$\lim_{m,n \to \infty} \left[\sup_{t \in [0,T]} |\mathcal{A}^{1/2 - \beta_*}(w^n(t) - w^m(t))|^{\kappa_*} + \int_0^T n_V(w_t^n(t) - w_t^m(t)) dt \right] = 0.$$
Indeed, let V_1 be the completion of V with respect to the norm $|\cdot|_{V_1}$ given by
$$|v|_{V_1} = |\mathcal{A}^{-1/2} M v| + n_V(v)$$
and W be the completion of V with respect to the norm $|\cdot|_W = |\mathcal{A}^{-1/2} M \cdot|$. It is clear that $V \subset V_1 \subset W$ and the embedding $V \subset V_1$ is compact.

It follows from (3.60) with $u = 0$ that
$$|(D(v), w)| \leq C \cdot [1 + (D(v), v)] \quad \text{for any} \quad w \in \mathcal{D}(\mathcal{A}^{1/2}), \, |\mathcal{A}^{1/2} w| \leq 1.$$
Therefore
$$|\mathcal{A}^{-1/2} D(v)| \leq C \cdot [1 + (D(v), v)].$$
Hence, using (3.95) and equation (1.1) we can conclude that
$$\int_0^T |\mathcal{A}^{-1/2} M w_{tt}^n(t)| dt \leq C_{\mathcal{B},T} \quad \text{for all} \quad n = 1, 2, \ldots,$$
i.e. the sequence $\{w_{tt}^n\}$ is bounded in $L_1(0, T; W)$. Since
$$\sup_{t \in [0,T]} |w_t^n(t)|_V \leq C_\mathcal{B} \quad \text{for all} \quad n = 1, 2, \ldots,$$
using [96, Corollary 4] we can conclude that the sequence $\{w_t^n\}$ is precompact in $L_1(0, T; V_1)$. This implies that
$$\int_0^T n_V(w_t^n(t) - w_t^m(t)) dt \to 0 \quad \text{as} \quad m, n \to \infty.$$
Therefore, by the compactness of embedding
$$(3.108) \qquad C([0,T], \mathcal{D}(\mathcal{A}^{1/2})) \cap C^1([0,T], V) \subset C([0,T], \mathcal{D}(\mathcal{A}^{1/2 - \beta_*}))$$
we obtain (3.107).

Step 2: We claim that

$$\lim_{m\to\infty}\lim_{n\to\infty}\int_t^T (G_0^{n,m}(\tau), w_t^n(\tau) - w_t^m(\tau))d\tau = 0 \tag{3.109}$$

for any $t < T$, where $G_0^{n,m}(t) = G_0(w^n(t), w^m(t))$ and $G_0(w,v)$ is defined by (3.104). Indeed, since

$$\int_t^T (G_0^{n,m}(\tau), w_t^n(\tau) - w_t^m(\tau))d\tau$$
$$= -\Pi(w^n(T)) + \Pi(w^n(t)) - \Pi(w^m(T)) + \Pi(w^m(t)) + \mathcal{I}_t^T(n,m),$$

where

$$\mathcal{I}_t^T(n,m) = \int_t^T [(\Pi'(w^n(\tau)), w_t^m(\tau)) + (\Pi'(w^m(\tau)), w_t^n(\tau))]\, d\tau,$$

by (F)(i) and (3.108) we have that

$$\lim_{m\to\infty}\lim_{n\to\infty}\int_t^T (G_0^{n,m}(\tau), w_t^n(\tau) - w_t^m(\tau))d\tau$$
$$= -2\Pi(w(T)) + 2\Pi(w(t)) + \lim_{m\to\infty}\lim_{n\to\infty}\mathcal{I}_t^T(n,m). \tag{3.110}$$

It follows from (F)(ii), (3.106) and (3.108) that

$$\Pi'(w^n(t)) \to \Pi'(w(t)) \quad \text{*-weakly in} \quad L_\infty(0,T;V').$$

Therefore using (3.106) we have

$$\lim_{m\to\infty}\lim_{n\to\infty}\int_t^T (\Pi'(w^n(\tau)), w_t^m(\tau))d\tau = \lim_{m\to\infty}\int_t^T (\Pi'(w(\tau)), w_t^m(\tau))d\tau$$
$$= \int_t^T (\Pi'(w(\tau)), w_t(\tau))d\tau$$

In a similar way we also have that

$$\lim_{m\to\infty}\lim_{n\to\infty}\int_t^T (\Pi'(w^m(\tau)), w_t^n(\tau))d\tau = \lim_{m\to\infty}\int_t^T (\Pi'(w^m(\tau)), w_t(\tau))d\tau$$
$$= \int_t^T (\Pi'(w(\tau)), w_t(\tau))d\tau.$$

Therefore

$$\lim_{m\to\infty}\lim_{n\to\infty}\mathcal{I}_t^T(n,m) = 2\int_t^T (\Pi'(w(\tau)), w_t(\tau))d\tau. \tag{3.111}$$

Hence after substituting (3.111) in (3.110) we get (3.109).

Concluding Step: It follows from (3.103), (3.107) and (3.109) and also from the Lebesgue Dominated Convergence Theorem that (3.105) holds. Thus we can apply Proposition 2.10 to obtain the asymptotic smoothness of S_t and, hence, the statement of Proposition 3.36. □

In the case when the nonlinear forces $F(u, u_t) \equiv F(u)$ in (1.1) are conservative, the conditions imposed on the damping operator D in Assumption 3.33 can be relaxed. More precisely, let us introduce the following hypotheses.

ASSUMPTION 3.39. **(D)**
- (i) For any $\eta > 0$ there exist $C_\eta > 0$ such that
$$(3.112) \quad (Mv, v) \leq \eta + C_\eta \cdot (D(v), v) \quad \text{for any} \quad v \in \mathcal{D}(\mathcal{A}^{1/2}).$$
- (ii) Relation (3.3) holds true, i.e. for any $\delta > 0$ there exists a nondecreasing function $K_\delta(s) > 0$ such that

$$(3.113) \quad |(Dv, u)| \leq K_\delta(E_0(u,v)) \cdot (Dv, v) + \delta \cdot (1 + E_0(u,v)), \quad u, v \in \mathcal{D}(\mathcal{A}^{1/2}),$$

where $E_0(u, v) = \frac{1}{2}((Mv, v) + (\mathcal{A}u, u))$.

(F)
- (i) The potential energy functional $\Pi(u)$ is continuous on $\mathcal{D}(\mathcal{A}^{1/2-\delta})$ for some $\delta > 0$;
- (ii) the mapping $u \mapsto \mathcal{A}^{-l}\Pi'(u)$ is continuous from $\mathcal{D}(\mathcal{A}^{1/2-\delta})$ into $\mathcal{H}$ for some $l, \delta > 0$;
- (iii) the nonlinear forces $F(u, v)$ are conservative, i.e. $F(u, v) \equiv F(u) = -\Pi'(u)$.

Under the above assumptions the following result holds.

THEOREM 3.40. *Let Assumptions 1.1 and 3.39 hold. Assume that $\mathcal{D}(\mathcal{A}^{1/2-\beta})$ is compactly embedded in V for some $\beta > 0$ and the nonlinear operator $F(u)$ possesses the property: there exist $0 \leq \eta < 1$ and a positive constant c_2 such that*

$$(3.114) \quad (u, F(u)) \leq \eta |\mathcal{A}^{1/2}u|^2 + c_2, \quad u \in \mathcal{D}(\mathcal{A}^{1/2}).$$

Then, the system (H, S_t) generated by problem (1.1) in $H = \mathcal{D}(\mathcal{A}^{1/2}) \times V$ possesses a compact global attractor A.

PROOF. Relation (3.114) and Assumption 3.39(D) imply Assumption 3.1 and hence by Theorem 3.4 the system (H, S_t) is dissipative. Thus to prove the existence of a compact global attractor it is sufficient to establish the following assertion.

PROPOSITION 3.41. *Assume that Assumptions 1.1 and 3.39 hold true and the embedding $\mathcal{D}(\mathcal{A}^{1/2-\beta}) \subset V$ is compact for some $\beta > 0$. Then, the semiflow S_t generated by problem (1.1) is asymptotically smooth.*

PROOF. As above we shall apply Proposition 2.10 on the forward invariant set $\mathcal{B}$. For this we will check that in the case of conservative forces it is possible to establish Lemma 3.37 and Lemma 3.38 under Assumption 3.39.

With the same notations as in the proof of Proposition 3.36, for any pair of strong solutions $(w(t), w_t(t))$ and $(v(t), v_t(t))$ with initial data from $\mathcal{B}$ from energy inequality we always have that

$$(3.115) \quad D_0^T \equiv \int_0^T (D(w_t), w_t) dt + \int_0^T (D(v_t), v_t) dt \leq C_\mathcal{B},$$

where, in contrast with (3.95), the constant $C_\mathcal{B}$ does not depend on T. This fact is crucial in our argument.

We shall apply Lemma 3.23. For this we need to estimate the terms involving the damping operator. As above we can assume that $w(t)$ and $v(t)$ are strong solutions to (1.1).

From (3.113) we have that

$$\begin{aligned} |(D(t, z_t), z)| &\leq |(D(w_t), z)| + |(D(v_t), z)| \\ &\leq C_\mathcal{B}^\delta \left[(D(w_t), w_t) | + (D(v_t), v_t) \right] + \delta C_\mathcal{B} \end{aligned}$$

for any $\delta > 0$. Therefore by (3.115) after rescaling of δ we obtain that
$$\int_0^T |(D(t,z_t),z)|dt \leq \delta T + C_\mathcal{B}^\delta$$
for any $\delta > 0$. In a similar way relation (3.112) implies that
$$\int_0^T |z_t|^2 dt \leq \int_0^T \left(|w_t|^2 + |v_t|^2\right) dt \leq \eta T + C_\mathcal{B}^\eta$$
for any $\eta > 0$. Therefore from Lemma 3.23 we get the estimate
$$TE_z(T) \leq (\delta + \eta) \cdot T + C_{\mathcal{B},\delta,\eta} + c\Psi_T(w,v), \quad T \geq T_0,$$
for any positive δ and η, where $\Psi_T(w,v)$ is given by (3.100). This implies relation (3.99) in Lemma 3.37. Thus we can obtain the conclusion of Lemma 3.38 with $\Psi_*(w,v;T)$ given by (3.103) with $n_V \equiv 0$. Now we can apply Proposition 2.10. The argument is the same as in the proof of Proposition 3.36. □

To complete the proof of Theorem 3.40 we use Proposition 3.41 and Theorem 2.3. □

3.4.2. The case of a non-compact potential energy functional.
In this subsection we do not assume the compactness property of Assumption 3.33(F)(i). This requires more structural hypotheses concerning the nonlinear operators F and D.

We impose the following structural conditions.

ASSUMPTION 3.42. **(F)**
- (i) There exists a constant $\gamma > 0$ such that
$$\Pi(u) - \Pi(0) - (\Pi'(0), u) + \frac{\gamma}{2}|u|^2 \geq 0.$$
Moreover, we assume that the following uniqueness property holds: $\mathcal{A}u + \Pi'(u) - \Pi'(0) + \gamma u = 0 \Rightarrow u \equiv 0$ for $u \in \mathcal{D}(\mathcal{A}^{1/2})$.
- (ii) there exists $\epsilon > 0$ and a continuous function $C(r)$ such that $\Pi'(\cdot)$ maps $\mathcal{D}(\mathcal{A}^{1/2+\varepsilon})$ into $\mathcal{D}(\mathcal{A}^\varepsilon)$ and
$$|\mathcal{A}^\epsilon(\Pi'(u+z) - \Pi'(u))| \leq C(r)|\mathcal{A}^{1/2+\epsilon}z|$$
for all $u, z \in \mathcal{D}(\mathcal{A}^{1/2})$ such that $|\mathcal{A}^{1/2}u|, |\mathcal{A}^{1/2}z| \leq r$.
- (iii) $|\mathcal{A}^\epsilon F_*(w,v)| \leq C(|\mathcal{A}^{1/2}w|, |v|_V)$ for $(w,v) \in \mathcal{D}(\mathcal{A}^{1/2}) \times V$ and $\widetilde{\eta} = 0$ in (3.61).
- (iv) there exists $\epsilon > 0$ and a continuous function $C(r)$ such that
(3.116) $$|\mathcal{A}^{-1}(\Pi'(u+z) - \Pi'(u))| \leq C(r)|\mathcal{A}^{1/2-\epsilon}z|$$
for all $u, z \in \mathcal{D}(\mathcal{A}^{1/2})$, $v, \hat{v} \in V$ such that $|\mathcal{A}^{1/2}u|, |\mathcal{A}^{1/2}z| \leq r$.

(D) We assume that the operator D satisfies the conditions:
- (i) There exists the Frechet derivative $D'(u)$ satisfying the inequality
$$|\mathcal{A}^{-1/2}D'(v)u| \leq C(r)(D'(v)u, u)^{1/2} \quad \text{for} \quad |v| \leq r, \, u \in \mathcal{D}(\mathcal{A}^{1/2}).$$
- (ii) Either (a) $|D(v)| \leq C(r)|\mathcal{A}^{1/2}v|$, $|v| \leq r$ or else, (b) for any $0 < T \leq 1$ and $v \in C([0,T], \mathcal{D}(\mathcal{A}^{1/2}))$ we have
$$\int_0^T |D(v)|^2 dt \leq C \left(\int_0^T (D(v),v)ds\right) \sup_{t \in [0,T]} |\mathcal{A}^{1/2}v(t)|^2$$

and

(3.117) $$|\mathcal{A}^{-1}(F^*(u+\hat{u}, v+\hat{v}) - F^*(u,v))| \leq C(r)\left[|\mathcal{A}^{1/2-\epsilon}\hat{u}| + |\mathcal{A}^{-\epsilon}\hat{v}|\right]$$

for all $u, \hat{u} \in \mathcal{D}(\mathcal{A}^{1/2})$, $v, \hat{v} \in V$ with the property

(3.118) $$|\mathcal{A}^{1/2}u|, |\mathcal{A}^{1/2}\hat{u}|, |v|_V, |\hat{v}|_V \leq r.$$

Here above $C(\cdot)$ is a nondecreasing function.
- (iii) For any $\epsilon > 0$ there exists a constant C_ϵ such that

$$|(D(u+v) - D(u), w)|$$

(3.119) $$\leq C_\epsilon C_1(r) \cdot (D(u+v) - D(u), v)$$

$$+ \epsilon C_2(r)\left(1 + (D(u), u) + (D(u+v), u+v)\right)$$

for any $u, v \in \mathcal{D}(\mathcal{A}^{1/2})$, $w \in \mathcal{D}(\mathcal{A}^l)$ such that $|\mathcal{A}^l w| + |M^{1/2}u| + |M^{1/2}v| \leq r$ with arbitrary $r > 0$, where $C_1(r)$ and $C_2(r)$ are nondecreasing functions of r and l is a positive integer.

REMARK 3.43. We note that Assumption 3.42(D)(iii) follows from (3.60) with $\kappa > 1$ in Assumption 3.21. Indeed, it is clear from (3.60) that

$$|(D(u+v) - D(u), w)| \leq C_1(r) \cdot (D(u+v) - D(u), v)$$

$$+ C_2(r)\left(1 + (D(u), u) + (D(u+v), u+v)\right)|\mathcal{A}^{1/2}w|^\kappa.$$

Thus after the substitution $w := w \cdot \epsilon^{1/(\kappa-1)}$ we obtain (3.119) with $C_\epsilon = \epsilon^{-1/(\kappa-1)}$ and $l = 1$.

THEOREM 3.44. *Let Assumptions 1.1 and 3.21 with $M = I$, $\tilde{\eta} = 0$ in (3.61) hold and the system (H, S_t) generated by problem (1.1) in $H = \mathcal{D}(\mathcal{A}^{1/2}) \times \mathcal{H}$ is dissipative and that the resolvent of $\mathcal{A}$ is compact in $\mathcal{H}$. Moreover we assume that the structural conditions imposed on F and D in Assumption 3.42 are satisfied. Then, the semiflow generated by problem (1.1) possesses a compact global attractor $\mathbf{A}$.*

REMARK 3.45. The assumption that $M = I$ is not essential. If M is not an identity, the assumption requiring local Lipschitz continuity of F needs to be appropriately modified. Since this particular setup finds most interesting applications in wave dynamics, where $M = I$ (see Chapter 5), we did not strive for a fully general abstract formulation accounting also for the operator M.

A situation when structural assumption imposed on the critical nonlinearity F in Assumption 3.42 may be dispensed with is when the damping provides some regularizing effect on the dynamics. This case is typical for semilinear waves with superlinear coercive damping. The theorem stated below pertains to this situation under the following hypotheses:

ASSUMPTION 3.46.
- (i) There exists a Banach space $V_0 \subset \mathcal{H} \subset V_0'$, such that $C_\delta \cdot (Dv, v) + \delta \geq |v|_{V_0}^2$, for some $\delta > 0$ and $C_\delta > 0$.
- (ii) Assumption 3.42(D)(iii) holds.
- (iii) The mapping $(u, v) \mapsto F(u, v)$ is compact from the space

$$\left[H^1(0, T; \mathcal{H}) \cap L_r(0, T; \mathcal{D}(\mathcal{A}^{1/2}))\right] \times L_r(0, T; \mathcal{H}) \quad \text{into} \quad L_2(0, T; V_0')$$

for some $r \geq 2$. Here the space $H^1(0,T;\mathcal{H}) \equiv W_2^1(0,T;\mathcal{H})$ is given by (1.14) and
$$F(u,v)(t) \equiv F(u(t),v(t)) \equiv -\Pi'(u(t)) + F^*(u(t),v(t))$$
is the substitution operator.

THEOREM 3.47. *Let Assumptions 1.1 and 3.21 with $M = I$, $\tilde{\eta} = 0$ in (3.61) hold. Assume that the system (H, S_t) generated by problem (1.1) is dissipative and that the resolvent of $\mathcal{A}$ is compact in $\mathcal{H}$. Then under Assumption 3.46 the semiflow S_t possesses a compact global attractor A in $H = \mathcal{D}(\mathcal{A}^{1/2}) \times \mathcal{H}$.*

In the special case, when $F^* = 0$ one can dispense with the a priori assumption on ultimate dissipativity of the system. In fact, an analogous result to that of Corollary 3.29 holds true.

COROLLARY 3.48. *Assume that $F^* = 0$, and the set of stationary solutions is bounded in $\mathcal{D}(\mathcal{A}^{1/2})$. Then, the results of Theorems 3.44 and Theorem 3.47 hold without assuming that the system (H, S_t) is dissipative.*

The proof of this Corollary is identical to that of Corollary 3.29.

Proof of Theorem 3.44. The proof of this result relies on a classical idea of decomposition of the flow into uniformly stable part and compact part (see Proposition 2.11). The technical details follow closely arguments given in [**20**] which paper deals with wave equation subject to boundary dissipation and critical exponent semilinearity.

We begin by decomposing the flow $S_t x = (w(t), w_t(t))$, $x = (u_0, u_1)$, into two parts, $S_t = S(t) + K(t)$, where the "stable" part $S(t)x \equiv (v(t), v_t(t))$ and compact part $K(t)x \equiv (z(t), z_t(t))$ are defined as follows:

$$v_{tt} + \mathcal{A}v + k(D(z_t + v_t) - D(z_t)) + \Pi'(v) - \Pi'(0) + \gamma v = 0,$$
(3.120)
$$(v(0), v_t(0)) = (u_0, u_1),$$

and

$$z_{tt} + \mathcal{A}z + kD(z_t) + \Pi'(w) - \Pi'(w-z) - \gamma(w-z) + \Pi'(0) - F^*(w, w_t) = 0,$$
(3.121)
$$(z(0), z_t(0)) = (0, 0).$$

It is a simple matter to show that both equations have global, finite energy solutions. Moreover, these solutions are strong, provided the initial conditions are taken in $\mathcal{D}(A)$, where A is defined by (1.23) with $M = I$. In fact, we argue as follows. We begin with local well-posedness of the second equation (3.121). We already know that for all initial conditions $(u_0, u_1) \in H$ we have weak solution $(w, w_t) \in C([0,\infty), H)$ which, moreover, is uniformly bounded in $t \geq 0$. Since Π' is locally Lipschitz from $\mathcal{D}(\mathcal{A}^{1/2})$ into $\mathcal{H}$, the map
$$z \to \Pi'(w) - \Pi'(w-z) - \gamma(w-z),$$
is locally Lipschitz $C([0,T], \mathcal{D}(\mathcal{A}^{1/2})) \to C([0,T])\mathcal{H})$. Thus equation (3.121) can be seen as a locally Lipschitz perturbation of monotone problem driven by the forcing term $-\Pi'(0) + F^*(w, w_t) \in C([0,\infty), \mathcal{H})$. Standard monotone operator theory argument gives local existence and uniqueness of generalized solutions to (3.121), i.e: $(z, z_t) \in C([0, T_m, \mathcal{H})$ for some $T_m \leq \infty$ - maximal time of existence. Moreover,

by Theorem 1.5 for $(u_0, u_1) \in \mathcal{D}(A)$ where A is given in (1.23) with $M = I$, solutions (w, w_t) are strong and possess properties (1.16) with $V = V' = \mathcal{H}$, i.e.

(3.122)
$$(w_t, w_{tt}) \in L_\infty(0, T; \mathcal{D}(\mathcal{A}^{1/2}) \times \mathcal{H}), \quad w_t \in C_r([0, T]; \mathcal{D}(\mathcal{A}^{1/2})),$$
$$w_{tt} \in C_r([0, T]; \mathcal{H}) \quad \text{and} \quad \mathcal{A}w(t) + kDw_t(t) \in C_r([0, T]; \mathcal{H}).$$

As a consequence, solutions (z, z_t) corresponding to "smooth" initial data (u_0, u_1) from $\mathcal{D}(A)$ are also strong (a priori only local) solutions and enjoy the properties (3.122) with $T = T_{max}$. Thus the function $(v, v_t) = (w - z, w_t - z_t) \in C(0, T_m, H)$ satisfies (3.120) in a weak sense. Moreover, for initial data $(u_0, u_1) \in \mathcal{D}(A)$ we have that (3.122) holds with $T = T_{max}$. From the structure of equation (3.120) we can conclude that the solution v is global in time. Indeed, by noting that

$$(D(z_t + v_t) - D(z_t), v_t) \geq 0$$

for strong solutions, applying energy identity first to strong solutions, and then extending to weak solutions we obtain

(3.123)
$$\mathcal{E}_v(t) \leq \mathcal{E}_v(0), \quad t \geq 0,$$

where the energy is given by

$$\mathcal{E}_v(t) \equiv \frac{1}{2}\left(|\mathcal{A}^{1/2}v(t)|^2 + |v_t(t)|^2\right) + \Pi(v(t)) - \Pi(0) - (\Pi'(0), v) + \frac{\gamma}{2}|v(t)|^2.$$

From Assumption 3.42 we know that for a suitably large γ

$$\mathcal{E}_v(t) \geq C|\mathcal{A}^{1/2}v(t)|^2 + |v_t(t)|^2 \quad \text{and} \quad \mathcal{E}_v(0) \leq C(|\mathcal{A}^{1/2}v(0)|, |v_t(0)|).$$

Thus, selecting $\gamma > 0$ suitably large we obtain global bound for all $t \geq 0$

$$|\mathcal{A}^{1/2}v(t)|^2 + |v_t(t)|^2 \leq C(|\mathcal{A}^{1/2}u_0|, |v_1|)$$

and consequently, since $(w(t), w_t(t))$ enjoy similar bound we conclude as well

$$|\mathcal{A}^{1/2}z(t)|^2 + |z_t(t)|^2 \leq C(|\mathcal{A}^{1/2}u_0|, |v_1|).$$

Thus, local solutions to (3.120) and (3.121) become global solutions and uniformly bounded for positive times. This shows that $T_{max} = \infty$. Finally, since (u_0, u_1) belong to absorbing set we can restrict our analysis to the functions with values in a fixed bounded set. Taking initial conditions $(u_0, u_1) \in \mathcal{D}(A)$ implies the improved regularity of solutions as well.

In order to prove Theorem 3.44 it suffices to establish that $S(t)$ is uniformly stable and the operator $K(t)$ is compact for positive t (see Proposition 2.11). This will be done below with the help of supporting Lemmas.

LEMMA 3.49. *The evolution operator $S(t)$ is uniformly stable. This is to say, there exists a continuous function $k : [0, \infty) \times [0, \infty) \to \mathbb{R}_+$ such that $k(t, r) \to 0$ for every $r > 0$ as $t \to \infty$ and $|S(t)y|_H \leq k(t, r)$ for all $y \in B_H(0, r)$.*

PROOF. We choose $\gamma > 0$ such that Assumption 3.42F(i) holds. Due to the dissipative nature of the term $\gamma v + \Pi'(v)$, and monotonicity of $D(z_t + v_t) - D(z_t)$, the proof of uniform stability is very much similar to the proof of Theorem 3.26 (see also Theorem 3.58 below). The lack of linear growth condition for D from below is responsible for potential lack of exponential stability. For sake if completeness we shall outline the details.

As in the proof of Theorem 3.26 it suffices to consider strong solutions. We define
$$\mathcal{E}_v(t) \equiv \frac{1}{2}\left(|\mathcal{A}^{1/2}v(t)|^2 + |v_t(t)|^2\right) + \Pi(v(t)) - \Pi(0) - (\Pi'(0), v) + \frac{\gamma}{2}|v(t)|^2.$$

On the strength of the first condition in Assumption 3.42(F) we have that $\mathcal{E}_v(t) \geq 0$. Moreover

$$(3.124) \qquad \mathcal{E}_v(t) + k \int_s^t (D(z_t + v_t) - D(z_t))v_t d\tau = \mathcal{E}_v(s), \quad t \geq s.$$

Step 1: We notice first that Lipschitz continuity of Π' from $\mathcal{D}(\mathcal{A}^{1/2}) \to V' = \mathcal{H}$ yields the estimate

$$(3.125) \qquad |(\Pi'(v) - \Pi'(0) + \gamma v, v)| \leq C_{\mathcal{B},\epsilon}|v|^2 + \epsilon|\mathcal{A}^{1/2}v|^2$$

for every $\epsilon > 0$. Multiplying equation (3.120) by v, integrating by parts and exploiting (3.124) and (3.125) along with interpolation inequalities yields

$$\int_\tau^{\tau+T} \mathcal{E}_v(t)dt \leq C\left[\mathcal{E}_v(\tau+T) + \int_\tau^{\tau+T} |v_t|^2 ds\right]$$
$$+ C_k^1 \int \int_\tau^{\tau+T} (D(z_t + v_t) - D(z_t))v_t ds$$
$$+ C_k^2 \int_\tau^{\tau+T} |(D(z_t + v_t) - D(z_t), v)| ds + C_{\mathcal{B}} \int_\tau^{\tau+T} |v(s)|^2 ds$$

for every $\tau \geq 0$, where the constants in the inequality above do not depend on τ. Noting from (3.124) that

$$(3.126) \int_\tau^{\tau+T} (D(z_t + v_t), z_t + v_t) + (D(z_t), z_t))ds \leq C_{\mathcal{B}} \quad \text{for all} \quad \tau \geq 0$$

and recalling Assumption 3.21(D) (see (3.59) and (3.60)) imposed on D gives (after taking the parameter ϵ sufficiently small in (3.60) and using the corresponding analog of relation (3.81)):

$$\int_\tau^{\tau+T} \mathcal{E}_v(t)dt \leq C_1 \mathcal{E}_v(\tau+T) + C_2(\mathcal{H}_0 + I)\left(\int_\tau^{\tau+T} (D(z_t + v_t) - D(z_t), v_t)ds\right)$$
$$+ C_3 \sup_{s \in (0,T)} |\mathcal{A}^{1/2-\delta}v(\tau+s)|^\kappa,$$

where $\mathcal{H}_0(s) = T H_0(s/T)$ (a similar argument was used in the proof of Theorem 3.26). Recalling monotonicity of $\mathcal{E}_v(t)$ which follows from (3.124), we have, with sufficiently large T that

$$\mathcal{E}_v(\tau+T) \leq C_1(\mathcal{H}_0 + I)\left(\int_\tau^{\tau+T} (D(z_t + v_t) - D(z_t), v_t)ds\right)$$
$$(3.127) \qquad + C_2 \sup_{s \in (0,T)} |\mathcal{A}^{1/2-\delta}v(\tau+s)|^\kappa$$

for every $\tau \geq 0$, where the constants C_1 and C_2 do not depend on τ.

3.4. GLOBAL ATTRACTORS IN CRITICAL CASE

Step 2: In order to establish the uniform stability of $S(t)$ it suffices to prove the following claim: *For every $M > 0$ there exists a constant $p(M) > 0$ such that we have:*

$$(3.128) \qquad k \int_\tau^{\tau+T} (D(z_t + v_t) - D(z_t), v_t) dt \geq p(M)$$

for every $\tau \geq 0$ and for all solutions $(v(t), v_t(t))$ of (3.120) with the property that $\mathcal{E}_v(\tau + T) \geq M$. Indeed, notice first that in view of energy identity (3.124) the above inequality is equivalent to

$$(3.129) \qquad \mathcal{E}_v(\tau + T) + p(M) \leq \mathcal{E}_v(\tau)$$

Thus, assuming (3.129) holds and taking $\tau = mT$ yielding

$$(3.130) \qquad \mathcal{E}_v((m+1)T) + p(M) \leq \mathcal{E}_v(mT), \quad m = 1, 2 \ldots,$$

where $p(M) > 0$, as long as $\mathcal{E}_v((m+1)T) \geq M$. We claim that inequality (3.130) implies the uniform stability of $S(t)$. If not, there exists $\epsilon_0 > 0$ and a sequence of solutions $(v_n(t), v_n(t))$ in the absorbing ball, along with a sequence of points $t_n \to \infty$ such that

$$\epsilon_0 < \mathcal{E}_{v_n}(t_n) \leq R_0, \ t_n \to \infty.$$

Since $\mathcal{E}_v(t) \leq \mathcal{E}_v(s)$ for $t \geq s$, we have that

$$\epsilon_0 < \mathcal{E}_{v_n}(t) \leq R_0 \quad \text{for all} \quad t \in [0, t_n].$$

Therefore, setting $M = \epsilon_0$ in (3.130) we obtain

$$\mathcal{E}_{v_n}((m+1)T) + p(\epsilon_0) \leq \mathcal{E}_{v_n}(mT)$$

provided $(m+1)T \leq t_n$. Hence

$$\epsilon_0 + mp(\epsilon_0) \leq \mathcal{E}_{v_n}((m+1)T) + mp(\epsilon_0) \leq \mathcal{E}_{v_n}(0) \leq R_0$$

under the condition $(m+1)T \leq t_n$. Since $t_n \to \infty$, we have that

$$\epsilon_0 + mp(\epsilon_0) \leq R_0, \quad \text{for all} \quad m = 1, 2 \ldots$$

which is impossible because $p(\epsilon_0) > 0$.

Thus, it suffices to prove (3.128). To this end we argue, again, by contradiction. If not, there exist a positive constant $M > 0$, a sequence of numbers τ_n and a sequence of solutions $v_n(t)$ such that

$$0 < M \leq \mathcal{E}_{v_n}(\tau_n + t) \leq R_0 \quad \text{for all} \quad t \in [0, T],$$

$$(3.131) \qquad \int_{\tau_n}^{\tau_n + T} (D(z_t + v_{n,t}) - D(z_t), v_{n,t}) ds \to 0, \quad n \to \infty.$$

Let $u^n(t) = v_n(\tau_n + t)$. From (3.131) we have

$$0 < M \leq \mathcal{E}_{u^n}(t) \leq R_0 \quad \text{for all} \quad t \in [0, T],$$

$$(3.132) \qquad \int_0^T (D(z_t^n + u_t^n) - D(z_t^n), u_t^n) ds \to 0, \quad n \to \infty,$$

where $z^n(t) = z(\tau_n + t)$. Thus, after renumeration we have that there exists $u(t)$ lying in $L_\infty(0, T, \mathcal{D}(\mathcal{A}^{1/2}))$ such that $u_t \in L_\infty(0, T, \mathcal{H})$ and

$$(3.133) \qquad \begin{aligned} u^n &\to u \quad \text{*-weakly in} \quad L_\infty(0, T, \mathcal{D}(\mathcal{A}^{1/2})), \\ u_t^n &\to u_t \quad \text{*-weakly in} \quad L_\infty(0, T, \mathcal{H}). \end{aligned}$$

From (3.132) and (3.59) we infer that $\int_0^T H_0^{-1}(|u_t^n|^2)dt \to 0$. Hence we have that $u_t^n \to 0$ *-weakly in $L_\infty(0,T,\mathcal{H})$. Due to compactness of the resolvent $\mathcal{A}^{-1}$, (3.133) and Aubin's Lemma (see Corollary 4 in [**96**]) we have

(3.134) $$\sup_{s \in [0,T]} |\mathcal{A}^{1/2-\delta}(u^n(s) - u(s))| \to 0.$$

Therefore Assumption 3.42(F)(iv) implies that $\Pi'(u^n(t)) \to \Pi'(u(t))$ in $[\mathcal{D}(\mathcal{A})]'$ for every $t \geq 0$.

Under Assumption 3.42(D)(iii) we have that for every $\epsilon > 0$ there exists C_ϵ such that

$$\int_0^T |(D(z_t^n + u_t^n) - D(z_t^n), \varphi)| ds \leq C_\epsilon C_1 \cdot \int_0^T (D(z_t^n + u_t^n) - D(z_t^n), u_t^n) ds$$

$$+ \epsilon C_2 \int_0^T [1 + (D(z^n + u_t^n), z^n + u_t^n) + (D(z_t^n), z_t^n)] ds$$

for any $\varphi \in L_\infty(0,T,\mathcal{D}(\mathcal{A}^l))$ such that $\sup_{s \in [0,T]} |\mathcal{A}^l \varphi(s)| < 1$. By (3.132) and (3.126) we have that

$$\limsup_{n \to \infty} \int_0^T |(D(z^n + u_t^n) - D(z_t^n), \varphi)| ds \leq C_\mathcal{B} \cdot \epsilon$$

for any $\epsilon > 0$. This implies that

$$\lim_{n \to \infty} \int_0^T |(D(z^n + u_t^n) - D(z_t^n), \varphi)| ds = 0 \quad \text{for any} \quad \varphi \in L_\infty(0,T,\mathcal{D}(\mathcal{A}^l)).$$

Now we can pass with the limit in the equation

$$u_{tt}^n + \mathcal{A}u^n + k(D(z_t^n + u_t^n) - D(z_t^n)) + \Pi'(u^n) - \Pi'(0) + \gamma u^n = 0,$$

which follows from (3.120), on the interval $[0,T]$ to obtain that u solves the equation

$$\mathcal{A}u + \Pi'(u) - \Pi'(0) + \gamma u = 0.$$

Hence, by the assumed uniqueness property in Assumption 3.42(F)(i) we must have $u \equiv 0$. Thus (3.134) holds with $u \equiv 0$ and hence, by (3.127) $\mathcal{E}_{u^n}(T) \equiv \mathcal{E}_{v_n}(\tau_n + T) \to 0$, which leads to contradiction. $\square$

LEMMA 3.50. *For any $T > 0$, the operator $K(T)$ is compact on the space $H = \mathcal{D}(\mathcal{A}^{1/2}) \times \mathcal{H}$.*

PROOF. The argument is different for the two cases considered in Assumption 3.42(D): (i) Condition (ii)(a) holds, (ii) Condition (ii)(b) holds. We shall begin with

Case 1: Condition (ii)(a) in Assumption 3.42(D) holds. The statement of the Lemma follows from the following regularity result.

PROPOSITION 3.51. *Under the Assumptions of Theorem 3.44 there exists $\epsilon > 0$ such that for any solution $(w(t), w_t(t))$ from the absorbing ball $\mathcal{B}$ and for all $T > 0$ solution $z(t)$ to equation (3.121) satisfies the conditions*

$$z \in C([0,T]; \mathcal{D}(\mathcal{A}^{1/2+\epsilon})) \quad \text{and} \quad z_t \in C([0,T]; \mathcal{D}(\mathcal{A}^\epsilon)).$$

Moreover $|\mathcal{A}^{1/2+\epsilon} z(t)| + |\mathcal{A}^\epsilon z_t(t)| \leq C(\mathcal{B}, T)$ for all $t \in [0,T]$.

3.4. GLOBAL ATTRACTORS IN CRITICAL CASE

PROOF. The proof is based on the study of regularity of a monotone problem

(3.135) $$z_{tt} + \mathcal{A}z + kD(z_t) = f, \quad z(0) = z_t(0) = 0,$$

along with nonlinear interpolation theorem applied to the map

$$\mathcal{T} : L_1([0,T]; \mathcal{H}) \to H \text{ defined by } \mathcal{T}f \equiv (z(T), z_t(T)),$$

where z is given by (3.135).

In order to apply Tartar's interpolation theorem [**100**] we need to establish supporting results:

PROPOSITION 3.52. *For any solution z to (3.135) and all $0 \leq \theta \leq 1/2$ the following estimate holds for $t \geq 0$:*

(3.136) $$|\mathcal{A}^{1/2+\theta}z(t)|^2 + |\mathcal{A}^\theta z_t(t)|^2 \leq C\left(|f|_{L_1(0,t,\mathcal{H})}\right)\left[\int_0^t |\mathcal{A}^\theta f|^2 dt + |f|^2_{L_\infty(0,t,\mathcal{H})}\right].$$

PROOF. The proof of this Proposition is based on interpolation of the regularity posted in (3.136) with indexes $\theta = 0$ and $\theta = 1/2$.

Step 1: For $\theta = 0$ we apply standard energy inequality along with positivity of $(D(z_t), z_t)$ to obtain the relation

(3.137) $$|\mathcal{A}^{1/2}z(t)|^2 + |z_t(t)|^2 \leq C\left[\int_0^t |f(s)|ds\right]^2,$$

Step 2: For $\theta = 1/2$ the argument is more involved. In fact, the resulting estimate takes the form:
(3.138)
$$|\mathcal{A}z(t)|^2 + |\mathcal{A}^{1/2}z_t(t)|^2 + |z_{tt}(t)|^2 \leq C\left(|f|_{L_1(0,t,\mathcal{H})}\right)\left[\int_0^t |\mathcal{A}^{1/2}f|^2 dt + |f|^2_{L_\infty(0,t,\mathcal{H})}\right].$$

In order to prove (3.138) we consider the equation in the variable $\bar{z} \equiv z_t$:

(3.139) $$\begin{aligned}\bar{z}_{tt} + \mathcal{A}\bar{z} + kD'(z_t)\bar{z}_t &= f_t \\ \bar{z}(0) = 0, \bar{z}_t(0) &= f(0)\end{aligned}$$

For a fixed z_t equation (3.139) is linear. Multiplying (3.139) by $\bar{z}_t$ and integrating in time yields:

$$|\bar{z}_t(t)|^2 + |\mathcal{A}^{1/2}\bar{z}(t)|^2 + 2k\int_0^t (D'(z_t)\bar{z}_t, \bar{z}_t)ds = |f(0)|^2 + 2\int_0^t (f_t, \bar{z}_t)ds$$

(3.140) $$= |f(0)|^2 + 2(f(t), \bar{z}_t(t)) - 2|f(0)|^2 - 2\int_0^t (f, \bar{z}_{tt})ds.$$

Computing the last term above with the use of equation (3.139) gives

(3.141) $$\int_0^t (f, \bar{z}_{tt})ds = \int_0^t (f, -\mathcal{A}\bar{z} - kD'(z_t)\bar{z}_t + f_t)ds$$

$$= \frac{1}{2}\left(|f(t)|^2 - |f(0)|^2\right) - \int_0^t (\mathcal{A}^{1/2}f, \mathcal{A}^{1/2}\bar{z})ds$$

$$- k\int_0^t (\mathcal{A}^{1/2}f, \mathcal{A}^{-1/2}D'(z_t)\bar{z}_t)ds.$$

Combining (3.140) with (3.141) gives

$$|\bar{z}_t(t)|^2 + |\mathcal{A}^{1/2}\bar{z}(t)|^2 + 2k \int_0^t (D'(z_t)\bar{z}_t, \bar{z}_t)ds$$
$$\leq C\left[|f|^2_{L_\infty(0,t,\mathcal{H})} + \left(1 + \frac{k}{\epsilon}\right)\int_0^t |\mathcal{A}^{1/2}f|^2 ds\right]$$
$$+ \epsilon k \int_0^t |\mathcal{A}^{-1/2}D'(z_t)\bar{z}_t|^2 ds + C\int_0^t |\mathcal{A}^{1/2}\bar{z}|^2 ds.$$

Therefore using Assumption 3.42(D)(i) imposed on D' we have that

$$|\bar{z}_t(t)|^2 + |\mathcal{A}^{1/2}\bar{z}(t)|^2 + 2k \int_0^t (D'(z_t)\bar{z}_t, \bar{z}_t)ds$$
$$\leq C\left[|f|^2_{L_\infty(0,t,\mathcal{H})} + \left(1 + \frac{k}{\epsilon}\right)\int_0^t |\mathcal{A}^{1/2}f|^2 ds\right]$$
$$+ \epsilon k C(|z_t|_{L_\infty(0,t;\mathcal{H})}) \int_0^t (D'(z_t)\bar{z}_t, \bar{z}_t)ds + C\int_0^t |\mathcal{A}^{1/2}\bar{z}|^2 ds.$$

From (3.137) and the fact that the function $C(s)$ is increasing we conclude that

$$|\bar{z}_t(t)|^2 + |\mathcal{A}^{1/2}\bar{z}(t)|^2 + 2k \int_0^t (D'(z_t)\bar{z}_t, \bar{z}_t)ds$$
$$\leq C\left[|f|^2_{L_\infty(0,t,\mathcal{H})} + \left(1 + \frac{k}{\epsilon}\right)\int_0^t |\mathcal{A}^{1/2}f|^2 ds\right]$$
(3.142)
$$+ \epsilon k C(|f|_{L_1(0,t;\mathcal{H})}) \int_0^t (D'(z_t)\bar{z}_t, \bar{z}_t)ds + C\int_0^t |\mathcal{A}^{1/2}\bar{z}|^2 ds.$$

Selecting $\epsilon C(|f|_{L_1(0,t;\mathcal{H})}) < 1$ and applying Gronwall's lemma give

$$|\bar{z}_t(t)|^2 + |\mathcal{A}^{1/2}\bar{z}(t)|^2$$
(3.143)
$$\leq C\left[|f|^2_{L_\infty(0,t,\mathcal{H})} + \left(1 + C_k\left(|f|_{L_1(0,t;\mathcal{H})}\right)\right)\int_0^t |\mathcal{A}^{1/2}f|^2 ds\right].$$

The above inequality gives the estimates for $|z_{tt}|$ and $|\mathcal{A}^{1/2}z_t|$. In order to obtain the estimate for $\mathcal{A}z$, we rewrite equation (3.135) in the form

$$\mathcal{A}z = -\bar{z}_t - kD(z_t) + f.$$

Using Assumption 3.42(D)(ii)(a) on $D(z_t)$ we have that

$$|\mathcal{A}z(t)| \leq |\bar{z}_t(t)| + |f(t)| + k|D(z_t)|$$
$$\leq |\bar{z}_t(t)| + |f(t)| + kC(|z_t(t)|)|\mathcal{A}^{1/2}\bar{z}(t)|.$$

Combining with (3.143) and (3.137) we obtain

(3.144) $$|\mathcal{A}z(t)|^2 \leq C\left(|f|_{L_1(0,t;\mathcal{H})}\right)\left[|f|^2_{L_\infty(0,t,\mathcal{H})} + + \int_0^t |\mathcal{A}^{1/2}f|^2 ds\right].$$

Now relations (3.143) and (3.144) imply the estimate in (3.138).

Step 3: interpolation. We shall use the following interpolation result due to [100]. For reader's convenience, we quote this theorem in a form suitable for our applications.

THEOREM 3.53 ([**100**], **Theorem 2**). *Let $A_0 \subset A_1, B_0 \subset B_1$ be given Banach spaces. Let $\mathcal{T}: A_1 \to B_1$ be given nonlinear operator with the following properties:*
(1) $\|\mathcal{T}a - \mathcal{T}b\|_{B_1} \leq f_1(\|a\|_{A_1}, \|b\|_{A_1})\|a - b\|_{A_1}$.
(2) $\|\mathcal{T}a\|_{B_0} \leq f_2(\|a\|_{A_1})\|a\|_{A_0}$,

where f_1 and f_2 are continuous functions. Then
$$\|\mathcal{T}a\|_{[B_0, B_1]_\theta} \leq Ch(\|a\|_{A_1})\|a\|_{[A_0, A_1]_\theta} \quad \text{for all} \quad 0 < \theta < 1,$$
where $h(t) = f_2(2t)^{1-\theta} f_1(t, 2t)^\theta$.

We shall apply this theorem with the following choices of spaces:
- $A_1 \equiv L_1(0, t; \mathcal{H})$,
- $A_0 \equiv L_2(0, t; \mathcal{D}(\mathcal{A}^{1/2})) \cap L_\infty(0, t; \mathcal{H})$,
- $B_1 \equiv H = \mathcal{D}(\mathcal{A}^{1/2}) \times \mathcal{H}$,
- $B_0 \equiv \mathcal{D}(\mathcal{A}) \times \mathcal{D}(\mathcal{A}^{1/2})$,

where $t > 0$ is fixed. The operator $\mathcal{T}$ we define by the formula
$$\mathcal{T}(f) \equiv (z(t), z_t(t)), \text{ where } z \text{ satisfies (3.135).}$$

Since (3.135) is monotone, by monotone operator theory we infer that for any $f \in L_1(0, t; \mathcal{H})$, there exist unique generalized solution $(z, z_t) \in C([0, t]; H)$ [**95**, Theorem 4.1A, Chapter IV]. Thus, the nonlinear operator $\mathcal{T}$ is a well defined bounded operator from $L_1(0, t; \mathcal{H})$ into H. Moreover, the following Lipschitz estimate

(3.145) $$\|\mathcal{T}f_1 - \mathcal{T}f_2\|_H \leq \int_0^t |f_1(s) - f_2(s)|_\mathcal{H} dt$$

holds [**95**, Theorem 4.1A, Chapter IV]. This gives the first inequality (with $f_1 \equiv 1$) assumed in Theorem 3.53. Inequality in (3.138) validates the second inequality in Theorem 3.53 satisfied with $f_1(s) = C(s)$. Thus, nonlinear interpolation theorem applied to estimates in (3.138) and (3.145) and the relation
$$L_2(0, t; \mathcal{D}(\mathcal{A}^\theta)) \cap L_\infty(0, t; \mathcal{H}) \subset [A_0, A_1]_\theta$$
complete the proof of the Proposition 3.52 (see also [**20**, Section 4.2] for a similar argument in the case of boundary damping). □

We apply the result of Proposition 3.52 with
$$f \equiv \Pi'(w) - \Pi'(w - z) + \Pi'(0) + \gamma(w - z) + F^*(w, w_t), \quad \theta = \epsilon.$$
Recalling the hypotheses imposed on Π and F^* in Assumption 3.42(F)(ii,iii) yield the following inequality valid for solutions to equation (3.121)
(3.146)
$$|\mathcal{A}^\epsilon f| \leq C(r) \left[|\mathcal{A}^{1/2+\epsilon} z| + \gamma |\mathcal{A}^\epsilon z| + |\mathcal{A}^{1/2-\delta} w| + 1 \right] \leq C(\mathcal{B}) \left[1 + |\mathcal{A}^{1/2+\epsilon} z| \right].$$
Moreover $|f|_{L_\infty(0,T;\mathcal{H})} \leq C(\mathcal{B})$. Hence

(3.147) $$|\mathcal{A}^{1/2+\epsilon} z(t)|^2 + |\mathcal{A}^\epsilon z_t(t)|^2 \leq C_\mathcal{B} \left[1 + \int_0^t |\mathcal{A}^{1/2+\epsilon} z|^2 dt \right],$$

where the constant $C_\mathcal{B}$ denotes the dependence on the absorbing set. Applying Cronwall's inequality to inequality in (3.147) yields the desired result in Proposition 3.51. □

Proposition 3.51 along with the compactness of the resolvent of $\mathcal{A}$ in $\mathcal{H}$ completes the proof of Lemma 3.50 in the first case.

We proceed next with the second case in Assumption 3.42(D)(ii).

Case 2: condition (ii)(b) is assumed. In this case some modifications of the previous argument are needed. Instead of Proposition 3.51 we now have a weaker regularity for the variable z.

PROPOSITION 3.54. *Under the Assumptions of Theorem 3.44 there exist $\epsilon > 0$ and $T > 0$ such that for any solution $(w(t), w_t(t))$ from the absorbing ball $\mathcal{B}$ the solution $z(t)$ to equation (3.121) satisfies:*

$$z \in L_2(0, T; \mathcal{D}(\mathcal{A}^{1/2+\epsilon})) \quad \text{and} \quad z_t \in C([0, T; \mathcal{D}(\mathcal{A}^\epsilon)).$$

Moreover,

$$|z|_{L_2(0,T;\mathcal{D}(\mathcal{A}^{1/2+\epsilon}))} + |z_t|_{C(0,T;\mathcal{D}(\mathcal{A}^\epsilon))} \leq C_\mathcal{B}.$$

PROOF. In analogy with the previous case, we shall consider a nonlinear map

$$\mathcal{T}: L_1([0,T]; \mathcal{H}) \to L_2(0, T; \mathcal{D}(\mathcal{A}^{1/2})) \times \mathcal{H} \text{ defined by } \mathcal{T}f \equiv (z(\cdot), z_t(T)),$$

where z is given by (3.135). We have:

PROPOSITION 3.55. *For any solution z to (3.135) and all $0 \leq \theta \leq 1/2$ the following estimate holds for $T \geq 0$:*
(3.148)
$$\int_0^T |\mathcal{A}^{1/2+\theta} z(t)|^2 dt + |\mathcal{A}^\theta z_t(T)|^2 \leq C\left(|f|_{L_1(0,T,\mathcal{H})}\right) \left[\int_0^T |\mathcal{A}^\theta f|^2 dt + |f|^2_{L_\infty(0,T,\mathcal{H})}\right]$$

and with $\theta < 1/2$

$$(3.149) \quad \int_0^T |\mathcal{A}^{1/2+\theta} z(t)|^2 dt \leq C_T \cdot C\left(|f|_{L_1(0,T,\mathcal{H})}\right) \left[\int_0^T |\mathcal{A}^\theta f|^2 dt + |f|^2_{L_\infty(0,T,\mathcal{H})}\right],$$

where $C_T \to 0$ as $T \to 0$.

PROOF. As before, the proof of this Proposition is based on interpolation of the regularity posted in (3.148) with indexes $\theta = 0$ and $\theta = 1/2$.

Step 1: For $\theta = 0$ we apply the standard energy inequality which gives

$$(3.150) \quad |\mathcal{A}^{1/2} z(t)|^2 + |z_t(t)|^2 + \int_0^t (D(z_t), z_t) ds \leq C \left[\int_0^t |f| dt\right]^2.$$

Hence

$$(3.151) \quad \int_0^T |\mathcal{A}^{1/2} z(t)|^2 dt \leq CT \left[\int_0^T |f| dt\right]^2.$$

Step 2: For $\theta = 1/2$ we have

$$(3.152) \quad |\mathcal{A}^{1/2} z_t(t)|^2 + |z_{tt}(t)|^2 \leq C\left(|f|_{L_1(0,t,\mathcal{H})}\right) \left[\int_0^t |\mathcal{A}^{1/2} f|^2 dt + |f|^2_{L_\infty(0,t,\mathcal{H})}\right].$$

The above inequality (cf. (3.143)) is obtained by arguments identical to the ones used in the previous case (the proof does not depend on conditions (D)(ii) in the Assumption 3.42). In order to obtain higher regularity of the displacement we take

advantage of the condition (b) in Assumption 3.42(D)(ii) applied to the elliptic problem:
$$\mathcal{A}z = -z_{tt} - kD(z_t) + f$$
along with the estimate in (3.152). These give:

$$\int_0^T |\mathcal{A}z(t)|^2 dt \leq 3\int_0^T \left[|z_{tt}(t)|^2 + |f(t)|^2 + k^2|D(z_t)|^2\right] dt$$

using the second assumption (ii)(b) on $D(z_t)$

$$\leq 3T \sup_{t\in[0,T]} \left[|z_{tt}(t)|^2 + |f(t)|^2\right]$$

$$+C\left(\int_0^T (Dz_t, z_t)dt\right) \sup_{t\in[0,T]} |\mathcal{A}^{1/2}z_t(t)|^2.$$

Combining the above inequality with (3.150) and (3.152) yields

(3.153) $$\int_0^T |\mathcal{A}z(t)|^2 dt \leq C\left(|f|_{L_1(0,T;\mathcal{H})}\right) \left[|f|^2_{L_\infty(0,T,\mathcal{H})} + \int_0^T |\mathcal{A}^{1/2}f|^2 dt\right],$$

which leads to the estimate (3.148) with $\theta = 1/2$.

Step 3: interpolation. Here the argument is similar to that used before - though applied twice. Indeed, we first apply Interpolation Theorem 3.53 with the following choice of spaces:

- $A_1 \equiv L_1(0,T;\mathcal{H})$,
- $A_0 \equiv L_2(0,T;\mathcal{D}(\mathcal{A}^{1/2})) \cap L_\infty(0,T;\mathcal{H})$,
- $B_1 \equiv L_2(0,T;\mathcal{D}(\mathcal{A}^{1/2})) \times \mathcal{H}$,
- $B_0 \equiv L_2(0,T;\mathcal{D}(\mathcal{A})) \times \mathcal{D}(\mathcal{A}^{1/2})$,

and with the operator $\mathcal{T}$ given by:

$$\mathcal{T}(f) \equiv (z(\cdot), z_t(T)), \text{ where } z \text{ satisfies (3.135).}$$

Inequalities (3.150) and (3.153) along with Interpolation Theorem imply the first estimate in Proposition 3.55.

For the second estimate, we need control of the size of the constant C_T. This can be accomplished by applying the same interpolation Theorem in a slightly different setting:

- $A_1 \equiv L_1(0,T;\mathcal{H})$,
- $A_0 \equiv L_2(0,T;\mathcal{D}(\mathcal{A}^{1/2})) \cap L_\infty(0,T;\mathcal{H})$,
- $B_1 \equiv L_2(0,T;\mathcal{D}(\mathcal{A}^{1/2}))$,
- $B_0 \equiv L_2(0,T;\mathcal{D}(\mathcal{A}))$,
- $\mathcal{T}(f) \equiv z(\cdot)$, where z satisfies (3.135).

Indeed, inequalities (3.151), (3.153) and Interpolation Theorem 3.53 give the desired estimate with a constant $C_T \sim T^{1/2-\theta} \to 0$ when $T \to 0$ and $\theta < 1/2$. The proof of Proposition 3.55 is thus completed. □

To continue with the proof of Proposition 3.54 we proceed very much as before with the only difference that this time we must take advantage of the fact that the small size of the constant C_T is controlled. Indeed, applying Proposition 3.55 with

$$f \equiv \Pi'(w) - \Pi'(w-z) + \Pi'(0) + \gamma(w-z) + F^*(w, w_t), \ \theta = \epsilon,$$

and recalling Assumptions imposed on Π and F^* in Theorem 3.44 yields (see (3.146)) the following inequalities valid for solutions to equation (3.121):

$$|\mathcal{A}^\epsilon f| \leq C(\mathcal{B})\left[1 + |\mathcal{A}^{1/2+\epsilon} z|\right] \quad \text{and} \quad |f|_{L_\infty(0,T;\mathcal{H})} \leq C(\mathcal{B}).$$

Hence

$$(3.154) \qquad \int_0^T |\mathcal{A}^{1/2+\epsilon} z(t)|^2 \leq C_\mathcal{B}\left[1 + C_T \int_0^T |\mathcal{A}^{1/2+\epsilon} z|^2 dt\right],$$

where, we recall that $C_T \to 0$ as $T \to 0$. Taking T sufficiently small gives the estimate

$$(3.155) \qquad \int_0^T |\mathcal{A}^{1/2+\epsilon} z(t)|^2 \leq C_\mathcal{B}$$

Since from (3.148) we also have

$$(3.156) \qquad \int_0^T |\mathcal{A}^{1/2+\epsilon} z(t)|^2 + \sup_{t \in [0,T]} |\mathcal{A}^\epsilon z_t(t)|^2 \leq C_\mathcal{B} \int_0^T |\mathcal{A}^{1/2+\epsilon} z(t)|^2 dt + C_\mathcal{B},$$

combining with (3.155) yields

$$(3.157) \qquad \int_0^T |\mathcal{A}^{1/2+\epsilon} z(t)|^2 dt + \sup_{t \in [0,T]} |\mathcal{A}^\epsilon z_t(t)|^2 \leq C_\mathcal{B}$$

as desired for the proof of Proposition 3.54. $\square$

Finally, we shall prove the compactness statement in Lemma 3.50. For this we consider initial conditions (u_0, u_1) in the absorbing set. The corresponding sequence of solutions (z_n, z_{nt}) is bounded in $C([0,T];H)$. By Proposition 3.54 we also have that z_{nt} are uniformly bounded in $C(0,T^*;\mathcal{D}(\mathcal{A}^\epsilon))$ for some $T^* > 0$. Denote

$$\mathcal{F}(u_0, u_1) \equiv \Pi'(w) - \Pi'(w-z) + \Pi'(0) - \gamma(w-z) - F^*(w, w_t)$$

and consider $(u_0, u_1) \mapsto \mathcal{F}(u_0, u_1)$ as a mapping from $\mathcal{B}$ into $L_{1/2\epsilon}\left(0,T;[\mathcal{D}(\mathcal{A}^\epsilon)]'\right)$ and prove it is compact.

It follows from Assumption 1.1(F) that the set

$$\mathcal{F}(\mathcal{B}) = \{\mathcal{F}(u_0, u_1) : (u_0, u_1) \in \mathcal{B}\}$$

is bounded in $C([0,T];\mathcal{H})$. Using (3.116) and (3.117) it is also easy to see that for any $G \in \mathcal{F}(\mathcal{B})$ we have that

$$\left|\mathcal{A}^{-1}\left(G(t+h) - G(t)\right)\right| \leq C_\mathcal{B}\left(\left|\mathcal{A}^{1/2-\epsilon}\left(z(t+h) - z(t)\right)\right|\right.$$
$$\left. + \left|\mathcal{A}^{1/2-\epsilon}\left(w(t+h) - w(t)\right)\right| + \left|\mathcal{A}^{-\epsilon}\left(w_t(t+h) - w_t(t)\right)\right|\right).$$

By interpolation we have that

$$|\mathcal{A}^{1/2-\epsilon}\psi| \leq C \cdot |\psi|^{2\epsilon} \cdot |\mathcal{A}^{1/2}\psi|^{1-2\epsilon}, \quad \psi \in \mathcal{D}(\mathcal{A}^{1/2})$$

and

$$|\mathcal{A}^{-\epsilon}\psi| \leq C \cdot |\mathcal{A}^{-1/2}\varphi|^{2\epsilon} \cdot |\varphi|^{1-2\epsilon}, \quad \psi \in \mathcal{H}.$$

Therefore we can obtain that

$$\left|\mathcal{A}^{-1}\left(G(t+h) - G(t)\right)\right| \leq C_\mathcal{B}\left\{|h|^{2\epsilon} + \left[\int_0^h |\mathcal{A}^{-1/2} w_{tt}(t+\tau)| d\tau\right]^{2\epsilon}\right\}.$$

In the same way as it was shown in the proof of Theorem 3.26 (see (3.86)) we can conclude that
$$\int_0^T |\mathcal{A}^{-1/2} M v_{tt}(t)| dt \leq C_{\mathcal{B},T} \quad \text{for all} \quad (u_0, u_1) \in \mathcal{B}.$$
Therefore
$$\int_0^{T-h} |\mathcal{A}^{-1}(G(t+h) - G(t))|^{1/2\epsilon} dt \leq C_{\mathcal{B}} \cdot h.$$
Thus by Aubin's Lemma (see [96, Theorem 5]) the set of images $\mathcal{F}(\mathcal{B})$ is compact in the space $L_{1/2\epsilon}(0, T; \mathcal{D}(\mathcal{A}^\epsilon)')$.

Now, standard energy inequality applied to equation for z yields compactness of the set $(z(T), z_t(T))$ in H for any $0 < T \leq T^*$. Indeed, by considering appropriate weakly convergent subsequences $\{z_n\}$ and $\{z_{nt}\}$ and taking the differences of solutions $\tilde{z} \equiv z_n - z_m$ we obtain

$$|\mathcal{A}^{1/2}\tilde{z}(T)|^2 + |\tilde{z}_t(T)|^2 \leq 2 \int_0^T (\mathcal{F}(u_{0n}, u_{1n}) - \mathcal{F}(u_{0n}, u_{1n}), \tilde{z}_t) ds$$

$$\leq 2 \int_0^T |\mathcal{A}^{-\epsilon}(\mathcal{F}(u_{0n}, u_{1n}) - \mathcal{F}(u_{0n}, u_{1n})||\mathcal{A}^\epsilon \tilde{z}_t| ds$$

$$\leq C_{\mathcal{B}} \left(\int_0^T |\mathcal{A}^{-\epsilon}(\mathcal{F}(u_{0n}, u_{1n}) - \mathcal{F}(u_{0n}, u_{1n})|^{1/2\epsilon} ds \right)^{2\epsilon} \to 0.$$

Therefore the operator K is compact for every $0 < T \leq T^*$. For $T > T^*$ we can present this operator in the form
$$(3.158) \qquad K(T)x = U(T, T^*)[K(T^*)x], \quad x = (u_0, u_1).$$
Here $U(t, s)$, $t > s$, is the evolution operator for the differential equation from (3.121) which is defined by the formula
$$U(t, s)(z_0, z_1) = (z(t), z_t(t)), \quad t > s,$$
where $z(t)$ is the solution to the differential equation in (3.121) with initial data at the moment s: $z(s) = z_0$, $z_t(s) = z_1$. This operator depends on $w(t)$ and, hence, on $x = (u_0, u_1)$. However, by the standard energy argument one can prove that for every fixed t and s the operator $U(t, s)$ is a continuous mapping from H into itself uniformly with respect to $(u_0, u_1) \in \mathcal{B}$. Thus by (3.158) $K(T)$ is a compact operator for every $T > 0$. This completes the proof of Lemma 3.50. □

Now using Lemmas 3.49 and 3.50 and also Proposition 2.11 we obtain the statement of Theorem 3.44.

Proof of Theorem 3.47. As in the previous case, the proof is based on a decomposition of the flow into stable and compact part. We thus have $S_t x = (u(t), u_t(t))$ and $S_t = S(t) + K(t)$, where the "stable" part $S(t)x \equiv (v(t), v_t(t))$ and compact part $K(t)x \equiv (z(t), z_t(t))$ are defined as follows:

$$(3.159) \qquad \begin{aligned} v_{tt} + \mathcal{A}v + k(D(z_t + v_t) - D(z_t)) &= 0, \\ (v(0), v_t(0)) &= (u_0, u_1), \end{aligned}$$

$$(3.160) \qquad \begin{aligned} z_{tt} + \mathcal{A}z + kD(z_t) + \Pi'(u) - F^*(u, u_t) &= 0, \\ (z(0), z_t(0)) &= (0, 0). \end{aligned}$$

Since $\Pi'(u) \in C(0,\infty,\mathcal{H})$, equation (3.160) is just a forced dissipative system. On the other hand, solutions $(v(t), v_t(t))$ belong to the absorbing set. Since the same is true for $(u(t), u_t(t))$, we conclude the same bounds for all positive time solutions to the system $(z(t), z_t(t))$. This allows us to restrict the entire analysis to the absorbing set, as the entire dynamics of v and z system remains in the absorbing set. In order to prove the Theorem 3.47 it suffices to establish that the evolution operator $S(t)$ is uniformly stable and the operator $K(t)$ is compact for positive t. The proof of uniform stability consists of a strict subset of arguments used in the case of Theorem 3.44. Indeed, we have

LEMMA 3.56. *The evolution operator $S(t)$ is uniformly stable. This is to say, there exists a continuous function $k : [0,\infty) \times [0,\infty) \to R^+$, $k(t,r) \to 0, t \to \infty$, $r > 0$ such that $|S(t)x|_H \leq k(t,r), \forall x \in B_H(0,r)$.*

As for compactness of $K(t)$ we also have

LEMMA 3.57. *The operator $K(T)$ is compact on H for any $T > 0$.*

PROOF. To prove the Lemma it suffices to show that the map
$$(u_0, u_1) \to u \to -\Pi'(u) + F^*(u, u_t) \to (z(T), z_t(T))$$
is compact from H into itself. Since the map
$$(u_0, u_1) \to (u, u_t)$$
is continuous, $H \to C([0,T]; H) \subset L_r(0,T; H)$, and by the Assumption 3.46(iii):
$$(u_0, u_1) \to u \to -\Pi'(u) + F^*(u, u_t)$$
is compact from H into $L_2(0,T; V_0')$. Thus, it suffices to establish continuity of the map

(3.161) $$f \to (z(T), z_t(T))$$

from $L_2(0,T; V_0')$ into H, where z satisfies

(3.162) $$z_{tt} + \mathcal{A}z + kD(z_t) = f, \ z(0) = z_t(0) = 0$$

Energy relation applied to (3.162) gives

$$|\mathcal{A}^{1/2}z(T)|^2 + |z_t(T)|^2 + 2k \int_0^T (D(z_t), z_t)dt \leq C \int_0^T |f(t)|_{V_0'}|z_t(t)|_{V_0}dt$$

(3.163) $$\leq C_\epsilon \int_0^T |f(t)|_{V_0'}^2 + \epsilon \int_0^T |z_t(t)|_{V_0}^2 dt$$

From the coercivity Assumption 3.46(i) imposed on D and from (3.163) we infer

$$\int_0^T |z_t|_{V_0}^2 dt \leq C_\delta \int_0^T (D(z_t), z_t)dt + \delta$$

(3.164) $$\leq C_{k,\delta}^\epsilon \int_0^T |f(t)|_{V_0'}^2 dt + C_\delta \epsilon \int_0^T |z_t(t)|_{V_0}^2 dt + \delta.$$

Choosing ϵ suitably small we obtain

$$\int_0^T |z_t|_{V_0}^2 dt \leq C_{\delta,k} \int_0^T |f|_{V_0'}^2 dt + \delta$$

for each solution z. Denoting by $\tilde{f} = f_n - f_m$ and $\tilde{z} = z_n - z_m$, where z_n corresponds to a solution of (3.162) with a forcing term f_n, we obtain

$$|\mathcal{A}^{1/2}\tilde{z}(T)|^2 + |\tilde{z}_t(T)|^2 \leq 2\int_0^T |\tilde{f}|_{V_0'}|\tilde{z}_t|_{V_0} dt$$

and by (3.164)

$$|\mathcal{A}^{1/2}\tilde{z}(T)|^2 + |\tilde{z}_t(T)|^2 \leq C\int_0^T |\tilde{f}|_{V_0'}[|z_{nt}|_{V_0} + |z_{mt}(t)|_{V_0}]dt$$

$$\leq C\left(\int_0^T |\tilde{f}|_{V_0'}^2 dt\right)^{1/2} \left(1 + \int_0^T [|f_n|_{V_0'}^2 + |f_m|_{V_0'}^2]dt\right)^{1/2},$$

which implies the desired continuity of the map (3.161). □

To complete the proof of Theorem 3.47 it remains to apply Proposition 2.11 and Theorem 2.3.

3.4.3. The case of large damping. The two theorems stated the previous subsection are rather special, as they require certain structure of nonlinearities. A situation which is more demanding is when the "damping" is nonlinear and unstructured (Theorem 3.47 does not apply) and nonlinear term in the equation does not have a structure leading to the requirement (ii) in Assumption 3.42(F) or to the compactness hypothesis in Assumption 3.33(F)(i). In such cases, existence of global attractors can be proved under the assumption that the damping parameter k is sufficiently large. The corresponding result is stated below:

THEOREM 3.58. *Let Assumptions 1.1 and 3.21 with $\tilde{\eta} = 0$ in (3.61) hold. Moreover, we assume that $H_0(s) = m_0^{-1} \cdot s$, relation (3.60) holds without the term $\epsilon |\mathcal{A}^{1/2}w|^2$ and*

(i) $\mathcal{D}(\mathcal{A}^{1/2-\delta}) \subset V$ *is compact embedding for some $\delta > 0$,*
(ii) *there exists $\alpha > 0$ such that the system (H, S_t) generated by problem (1.1) in $H = \mathcal{D}(\mathcal{A}^{1/2}) \times V$ is dissipative for every $k \geq \alpha$ and the corresponding absorbing set $\mathcal{B}$ is forward invariant with the size independent of k for $k \geq \alpha$.*

Then there exists $k_0 > 0$ such that the dynamical system (H, S_t) generated by problem (1.1) possesses a compact global attractor A provided the damping parameter k satisfies the condition $k \geq k_0$.

The proof of Theorem 3.58 based on generalization of Lopez-Ceron criterion for asymptotic smoothness is in the style of the proof Theorem 3.26.

Proof of Theorem 3.58. The argument now depends on the size of the damping. The following inequality is critical for the proof.

LEMMA 3.59. *Let $w(t)$ and $v(t)$ be two generalized solutions to (1.1) corresponding to initial data in the forward invariant absorbing set $\mathcal{B}$:*

$$(w(t), w_t(t)) \equiv S_t y_0, \ (v(t), v_t(t)) \equiv S_t y_1, \ y_0, y_1 \in \mathcal{B}.$$

Let $z(t) = w(t) - v(t)$. Then under the assumptions of Theorem 3.58 there is $k_0 > 0$ such that for every $k \geq k_0$ there exists a positive constant $\omega_\mathcal{B} > 0$ and a constant

86   3. EXISTENCE OF COMPACT GLOBAL ATTRACTORS

$C(\mathcal{B})$ *such that*

$$E_0(z(t), z_t(t)) \leq Ce^{-\omega_{\mathcal{B}} t} E_0(z(0), z_t(0))$$
$$(3.165) \qquad + C(\mathcal{B}) \left\{ \sup_{0 \leq s \leq t} |\mathcal{A}^{1/2-\delta} z(s)|^\kappa + \int_0^t e^{-\omega_{\mathcal{B}}(t-s)} n_V^2(z_t(s)) ds \right\},$$

for some positive δ, where $E_0(z, z_t) = \frac{1}{2} \left(|z_t|_V^2 + |\mathcal{A}^{1/2} z|^2 \right)$.

PROOF. We shall use the same notation as in the proof of Theorem 3.26 and consider strong solutions first.

Step 1: reconstruction of kinetic energy. Recalling energy relation (3.76) and the hypotheses imposed on F we obtain:

$$|E_z(T) + k \int_t^T (D(z_t + v_t) - D(v_t), z_t) dt - E_z(t)|$$
$$(3.166) \qquad \leq C_{\mathcal{B}}\epsilon \int_t^T |\mathcal{A}^{1/2} z|^2 dt + (C_{\mathcal{B}} + C\epsilon^{-1}) \int_t^T |M^{1/2} z_t|^2 dt$$

for any $\epsilon > 0$ and $t \in [0, T]$. On the other hand, by applying coercivity property (3.59) with $H_0(s) = m_0^{-1} \cdot s$ and (3.166) with $t = 0$ we also obtain

$$m_0 k \int_0^T |M^{1/2} z_t|^2 dt \leq k \int_0^T (D(z_t + v_t) - D(v_t), z_t) dt$$
$$\leq E_z(0) + C_{\mathcal{B}}\epsilon \int_0^T |\mathcal{A}^{1/2} z|^2 dt + (C_{\mathcal{B}} + C\epsilon^{-1}) \int_0^T |M^{1/2} z_t|^2 dt$$

Let $k_1 > 2 C_{\mathcal{B}} m_0^{-1}$ be fixed, Taking $\epsilon > 0$ such that $C_{\mathcal{B}} + C\epsilon^{-1} = m_0 k_1 / 2$ yields:

$$(3.167) \quad k \int_0^T |M^{1/2} z_t|^2 dt \leq C E_z(0) + C_{\mathcal{B}} \int_0^T |\mathcal{A}^{1/2} z|^2 dt \quad \text{for every } k > k_1.$$

Relation (3.166) with $C_{\mathcal{B}} + C\epsilon^{-1} = m_0 k$ yields for any $t \in [0, T]$

$$|E_z(t) - E_z(T)| \leq 2k \int_0^T (D(z_t + v_t) - D(v_t), z_t) dt$$
$$(3.168) \qquad + \frac{C_{\mathcal{B}}}{k} \int_0^T |\mathcal{A}^{1/2} z|^2 dt, \quad k > k_1.$$

Step 2: potential energy. Since $\mathcal{D}(\mathcal{A}^{1/2-\delta}) \subset V$, we have that

$$(F(z+v, z_t+v_t) - F(v, v_t), z) \leq C_{\mathcal{B}} |\mathcal{A}^{1/2-\delta} z| \left[|\mathcal{A}^{1/2} z| + n_V(z_t) \right]$$
$$\leq \epsilon |\mathcal{A}^{1/2} z|^2 + C_{\mathcal{B},\epsilon} \left[|\mathcal{A}^{1/2-\delta} z|^2 + n_V^2(z_t) \right].$$

Therefore multiplying equation (3.75) by z and integrating by parts, yields:

$$\int_0^T |\mathcal{A}^{1/2} z|^2 dt \leq \int_0^T |M^{1/2} z_t|^2 dt + C[E_z(0) + E_z(T)]$$
$$+ C \cdot k \int_0^T |(D(z_t + v_t) - D(v_t), z)| dt$$
$$+ C_{\mathcal{B},\epsilon} \int_0^T [|\mathcal{A}^{1/2-\delta} z|^2 + n_V^2(z_t)] dt + \epsilon \int_0^T |\mathcal{A}^{1/2} z|^2 dt.$$

From here, after taking $\epsilon = 1/2$ we have that

$$\int_0^T |\mathcal{A}^{1/2}z|^2 dt \leq 2\int_0^T |M^{1/2}z_t|^2 dt + C[E_z(0) + E_z(T)]$$
$$+C\cdot k\int_0^T |(D(z_t+v_t) - D(v_t), z)|dt + C_{\mathcal{B}}\int_0^T [|\mathcal{A}^{1/2-\delta}z|^2 + n_V^2(z_t)]dt.$$

Therefore using (3.60) we obtain that

$$\int_0^T |\mathcal{A}^{1/2}z|^2 dt \leq 2\int_0^T |M^{1/2}z_t|^2 dt + CE_z(0) + CE_z(T)$$
$$+C_{\mathcal{B}}\cdot k\int_0^T (D(z_t+v_t) - D(v_t), z_t)dt$$
$$+C_{\mathcal{B},k}\int_0^T [1 + (D(w_t), w_t) + (D(v_t), v_t)]|\mathcal{A}^{1/2-\delta}z|^\kappa dt$$
$$+C_{\mathcal{B}}\int_0^T [|\mathcal{A}^{1/2-\delta}z|^2 + n_V^2(z_t)]dt.$$

As in the subcritical case, we obtain

$$\frac{1}{2}\int_0^T |\mathcal{A}^{1/2}z|^2 dt \leq 3\int_0^T |M^{1/2}z_t|^2 dt + CE_z(0) + CE_z(T)$$
$$+C_{\mathcal{B},T}\cdot k\int_0^T (D(z_t+v_t) - D(v_t), z_t)dt$$
$$+C_{\mathcal{B},k}\int_0^T [1 + (D(w_t), w_t) + (D(v_t), v_t)]|\mathcal{A}^{1/2-\delta}z|^\kappa dt$$
$$+C_{\mathcal{B}}\int_0^T |\mathcal{A}^{1/2-\delta}z|^2 + n_V^2(z_t)]dt.$$

Combining with (3.167) and (3.83) and taking without loss of generality $k - 2(1 + 3C_{\mathcal{B}}) > k/2$ we obtain that

$$k\int_0^T |M^{1/2}z_t|^2 dt + \int_0^T E_z(t)dt$$
$$\leq C(E_z(0) + E_z(T)) + C_{\mathcal{B},T}k\int_0^T (D(z_t+v_t) - D(v_t), z_t)dt$$
(3.169)
$$+C_{\mathcal{B},k,T}\left\{\sup_{t\in[0,T]}|\mathcal{A}^{1/2-\delta}z(t)|^\kappa + \int_0^T n_V^2(z_t)dt\right\}$$

for $k > k_2 \geq k_1$.

Step 4: completion of the argument. It follows from (3.168) that

$$E_z(0) \leq E_z(T) + 2k\int_0^T (D(z_t+v_t) - D(v_t), z_t)dt$$
$$+\frac{C_{\mathcal{B}}}{k}\int_0^T |\mathcal{A}^{1/2}z|^2 dt$$

and
$$TE_z(T) \leq \int_0^T E_z(t)dt + 2kT\int_0^T (D(z_t+v_t) - D(v_t), z_t)dt$$
$$+ \frac{C_{\mathcal{B},T}}{k}\int_0^T |\mathcal{A}^{1/2}z|^2 dt.$$

Therefore (3.169) implies that
$$k\int_0^T |M^{1/2}z_t|^2 dt + TE_z(T) + \int_0^T E_z(t)dt$$
$$\leq CE_z(T) + C_{\mathcal{B},T}k\int_0^T (D(z_t+v_t)-D(v_t),z_t)dt + \frac{C_{\mathcal{B},T}}{k}\int_0^T |\mathcal{A}^{1/2}z|^2 dt$$
$$+ C_{\mathcal{B},k,T}\cdot\left\{\sup_{t\in[0,T]}|\mathcal{A}^{1/2-\delta}z(t)|^\kappa + \int_0^T n_V^2(z_t)dt\right\}.$$

Taking $T > C+1$ yields
$$k\int_0^T |M^{1/2}z_t|^2 dt + E_z(T) + \int_0^T E_z(t)dt$$
$$\leq C_{\mathcal{B},T}k\int_0^T (D(z_t+v_t)-D(v_t),z_t)dt + \frac{C_{\mathcal{B},T}}{k}\int_0^T |\mathcal{A}^{1/2}z|^2 dt$$
$$(3.170) \qquad + C_{\mathcal{B},k,T}\cdot\left\{\sup_{t\in[0,T]}|\mathcal{A}^{1/2-\delta}z(t)|^\kappa + \int_0^T n_V^2(z_t)dt\right\}.$$

From (3.166) applied with $t=0$ we have
$$C_{\mathcal{B},T}k\int_0^T (D(z_t+v_t)-D(v_t),z_t)dt \leq C'_{\mathcal{B},T}(E_z(0) - E_z(T))$$
$$+ \frac{C'_{\mathcal{B},T}}{k}\int_0^T |\mathcal{A}^{1/2}z|^2 dt + k\int_0^T |M^{1/2}z_t|^2 dt$$

for $k > k_1$. Therefore from (3.170) we obtain
$$E_z(T) + \int_0^T E_z(t)dt \leq C_{\mathcal{B},T}(E_z(0) - E_z(T))$$
$$+ C_{\mathcal{B},k,T}\cdot\left\{\sup_{t\in[0,T]}|\mathcal{A}^{1/2-\delta}z(t)|^\kappa + \int_0^T n_V^2(z_t)dt\right\} + \frac{C_{\mathcal{B},T}}{k}\int_0^T |\mathcal{A}^{1/2}z|^2 dt.$$

Taking k large enough and recalling that $C_{\mathcal{B},T}$ does not depend on k, the inequality above implies
$$E_z(T) \leq C_{\mathcal{B},T}(E_z(0) - E_z(T)) + C_{\mathcal{B},k,T}\cdot\left\{\sup_{t\in[0,T]}|\mathcal{A}^{1/2-\delta}z(t)|^\kappa + \int_0^T n_V^2(z_t)dt\right\}$$

for some $T > 1$. The remaining part of the proof is now identical to that in the case of Theorem 3.26 (see Remark 3.30). Lemma 3.59 has been proved completely. □

The inequality in Lemma 3.59 establishes asymptotic smoothness of the flow (see Corollary 2.7). Indeed, it follows from (3.165) that
$$E_0(w(t)-v(t), w_t(t)-v_t(t)) \leq Ce^{-\omega_{\mathcal{B}}t}E_0(w(0)-v(0), w_t(0)-v_t(0)) + \varrho_{\mathcal{B}}^t(\{w\},\{w\})$$

for every couple of solutions $(w(t), w_t(t))$ and $(v(t), v_t(t))$ from the absorbing set, where
$$\varrho_{\mathcal{B}}^t(\{w\}, \{v\}) = C(\mathcal{B}) \left\{ \sup_{0 \leq s \leq t} |w(s) - v(s)|^{\kappa^*} + \int_0^t n_V^2(w_t(s) - v_t(s)) ds \right\}$$
with some $\kappa^* > 0$. By the same argument as in the proof of Theorem 3.26 the pseudometric $\varrho_{\mathcal{B}}^t$ is precompact on trajectories from the invariant absorbing set. Thus $\varrho_{\mathcal{B}}^t$ satisfies the hypotheses of Corollary 2.7 with $K_{\mathcal{B}}(t) = Ce^{-\omega_{\mathcal{B}} t}$. This combined with dissipativity of (H, S_t) and Theorem 2.3 implies the assertion of the Theorem 3.58. □

CHAPTER 4

Properties of global attractors for evolutions of the second order in time

The goal of this chapter is to study properties of global attractors for dynamics governed by equations of the form (1.1), i.e. by the problem:

(4.1) $$\begin{cases} Mu_{tt}(t) + \mathcal{A}u(t) + k \cdot D(u_t(t)) = F(u(t), u_t(t)), \\ u|_{t=0} = u_0 \in \mathcal{D}(\mathcal{A}^{1/2}), \ u_t|_{t=0} = u_1 \in V = \mathcal{D}(M^{1/2}). \end{cases}$$

We rely on the abstract results presented in Chapter 2. The main technical tool which we use here are stabilizability estimates (3.87) and (3.165) and some generalizations thereof. These estimates will allow us to verify conditions postulated by abstract results of Chapter 2.

4.1. Finite dimensionality of attractors

We start with finiteness of fractal dimension of global attractors in the context of problem (4.1).

THEOREM 4.1. *Assume the hypotheses of either Corollary 3.28 or Theorem 3.58 with $\kappa = 2$ in (3.60) and $H_0(s) = m_0^{-1}s$ in (3.59). The following assertions hold.*

- *If the nonlinear term $F(u, u_t) \equiv F(u)$ does not depend on the velocity u_t, then the global attractor A of the system (H, S_t) generated by (4.1) has a finite fractal dimension.*
- *In the general case the same statement remains true if we assume in addition that there exists $l \geq 1/2$ such that*

(4.2) $$|\mathcal{A}^{-l}(D(u+v) - D(v))| \leq C(r)|u|_V$$

for any $r > 0$ and $u, v \in V$ such that $|u|_V \leq r$ and $|v|_V \leq r$, where $C(r) > 0$ is a non-decreasing function of r.

REMARK 4.2. The theorem formulated above relies on either Corollary 3.28 or Theorem 3.58. As a consequence of this, the applicability of this Theorem is restricted to cases which are either "subcritical" ($\eta > 0$) or the damping parameter is sufficiently large. On the other hand, one can formulate more general results that do not require a-priori the above restrictions. In fact, as we shall see below, Theorem 4.1 follows from a such more general result.

The first part of Theorem 4.1 follows from Proposition 1.15, the inequality in Remark 3.30 or Lemma 3.59 (with $n_V \equiv 0$) and from the following more general assertion which we shall state and prove below.

4.1. FINITE DIMENSIONALITY OF ATTRACTORS

THEOREM 4.3. *Let the operator $\mathcal{A}^{-1}$ be compact. Assume that the system (H, S_t) generated by (4.1) in the space $H = \mathcal{D}(\mathcal{A}^{1/2}) \times V$ possesses the global attractor A and there exist non-negative scalar functions $a(t)$, $b(t)$ and $c(t)$ on $\mathbb{R}_+$ such that (i) $a(t)$ and $c(t)$ are locally bounded on $[0, \infty)$, (ii) $b(t) \in L_1(\mathbb{R}_+)$ possesses the property $\lim_{t \to \infty} b(t) = 0$ and (iii) for every $y_1, y_2 \in A$ and $t > 0$ the following relations*

$$|S_t y_1 - S_t y_2|_H^2 \leq a(t) \cdot |y_1 - y_2|_H^2 \tag{4.3}$$

and

$$|S_t y_1 - S_t y_2|_H^2 \leq b(t) \cdot |y_1 - y_2|_H^2 + c(t) \cdot \sup_{0 \leq s \leq t} |\mathcal{A}^\sigma(u^1(s) - u^2(s))|^2 \tag{4.4}$$

hold for some $\sigma \in [0, 1/2)$. Here we denote $S_t y_i = (u^i(t); u_t^i(t))$, $i = 1, 2$. Then the attractor A has a finite fractal dimension.

PROOF. We apply Theorem 2.15 in the space $H_T = H \times W_1(0, T)$ with an appropriate choice of T. Here

$$W_1(0, T) = \left\{ z(t) : |z|_{W_1(0,T)}^2 \equiv \int_0^T \left(|\mathcal{A}^{1/2} z(t)|^2 + |M^{1/2} z_t(t)|^2 \right) dt < \infty \right\}. \tag{4.5}$$

The norm in H_T is given by

$$\|U\|_{H_T}^2 = |\mathcal{A}^{1/2} u_0|^2 + |M^{1/2} u_1|^2 + |z|_{W_1(0,T)}^2, \quad U = (u_0, u_1, z). \tag{4.6}$$

Let $z(t) = u^1(t) - u^2(t)$. Integrating (4.4) from T to $2T$ with respect to t we obtain

$$\int_T^{2T} |S_t y_1 - S_t y_2|_H^2 dt \leq \widetilde{b}_T |y_1 - y_2|_H^2 + \widetilde{c}_T \sup_{0 \leq s \leq 2T} |\mathcal{A}^\sigma z(s)|^2, \tag{4.7}$$

where

$$\widetilde{b}_T = \int_T^{2T} b(t) dt, \text{ and } \widetilde{c}_T = \int_T^{2T} c(t) dt. \tag{4.8}$$

It also follows from (4.4) that

$$|S_T y_1 - S_T y_2|_H^2 \leq b(T) \cdot |y_1 - y_2|_H^2 + c(T) \cdot \sup_{0 \leq s \leq T} |\mathcal{A}^\sigma z(s)|^2.$$

Therefore (4.7) implies that

$$\begin{aligned}
|S_T y_1 - S_T y_2|_H^2 + \int_T^{2T} |S_t y_1 - S_t y_2|_H^2 dt \\
\leq b_T |y_1 - y_2|_H^2 + c_T \sup_{0 \leq s \leq 2T} |\mathcal{A}^\sigma z(s)|^2,
\end{aligned} \tag{4.9}$$

where

$$b_T = b(T) + \int_T^{2T} b(t) dt, \text{ and } c_T = c(T) + \int_T^{2T} c(t) dt. \tag{4.10}$$

Let A be the global attractor for (H, S_t). Consider in the space H_T the set

$$A_T := \{ U \equiv (u(0); u_t(0); u(t), t \in [0, T]) : (u(0); u_t(0)) \in A \},$$

where $u(t)$ is the solution to the initial problem with initial data $(u(0); u_t(0))$, and define operator $V : A_T \mapsto H_T$ by the formula

$$V : (u(0); u_t(0); u(t)) \mapsto (u(T); u_t(T); u(T + t)).$$

It is clear from (4.3) that V is Lipschitz on A_T. Since A is a strictly invariant set (with respect to S_t) and it consists of full trajectories, we also have that $VA_T = A_T$.

Since $|\mathcal{A}^\sigma u|^2 \leq \varepsilon |\mathcal{A}^{1/2} u|^2 + C_\varepsilon |u|^2$ for any $\varepsilon > 0$, it follows from (4.3) that
$$\sup_{0 \leq s \leq 2T} |\mathcal{A}^\sigma z(s)|^2 \leq \varepsilon |y_1 - y_2|_H^2 + C_\varepsilon(T) \sup_{0 \leq s \leq 2T} |z(s)|^2$$
for any $\varepsilon > 0$. Consequently, from (4.9) we obtain
$$\|VU_1 - VU_2\|_{H_T}^2 \leq \eta_T^\varepsilon \|U_1 - U_2\|_{H_T}^2 + K_T^\varepsilon \cdot (n_T^2(U_1 - U_2) + n_T^2(VU_1 - VU_2)),$$
for any $U_1, U_2 \in A_T$, where $K_T^\varepsilon > 0$ is a constant, $n_T(U) := \sup_{0 \leq s \leq T} |u(s)|$ and

(4.11) $$\eta_T^\varepsilon = b(T) + \int_T^{2T} b(t) dt + \varepsilon \cdot \left(c(T) + \int_T^{2T} c(t) dt \right).$$

Here $T > 1$ and $\varepsilon > 0$ are arbitrary. Since $W_1(0, T)$ is compactly embedded into $C(0, T; \mathcal{H})$, $n_T(U)$ is a compact seminorm on H_T and we can choose T and $\varepsilon > 0$ such that $\eta_T^\varepsilon < 1$. Therefore we can apply Theorem 2.15 which implies that A_T is a compact set in H_T of finite fractal dimension.

Let $\mathcal{P} : H_T \mapsto H$ be the operator defined by the formula
$$\mathcal{P} : (u_0; u_1; z(t)) \mapsto (u_0; u_1).$$
Since $A = \mathcal{P} A_T$ and $\mathcal{P}$ is Lipshitz continuous, we have that
$$\dim_{frac}^H A \leq \dim_{frac}^{H_T} A_T < \infty.$$
Here $\dim_{frac}^Y$ stands for fractal dimension of a set in the space Y. This concludes the proof of Theorem 4.3. $\square$

The proof of the second part of Theorem 4.1 employs, again, Proposition 1.15, inequalities either (3.87) or (3.165) and the following more general assertion which is a generalization of Theorem 4.3.

THEOREM 4.4. *Let (4.2) hold, the operator $\mathcal{A}^{-1}$ be compact, and Assumption 1.1 be in force. Assume that the system (H, S_t) generated by (4.1) possesses global attractor A and there exist a compact seminorm n_V on V and non-negative scalar functions $a(t)$, $b(t)$ and $c(t)$ on $\mathbb{R}_+$ such that (i) $a(t)$ and $c(t)$ are locally bounded on $[0, \infty)$, (ii) $b(t) \in L_1(\mathbb{R}_+)$ possesses the property $\lim_{t \to \infty} b(t) = 0$ and (iii) for every $y_1, y_2 \in A$ and $t > 0$ we have relation (4.3) and also*

(4.12) $$|S_t y_1 - S_t y_2|_H^2 \leq b(t) \cdot |y_1 - y_2|_H^2$$
$$+ c(t) \cdot \left\{ \sup_{0 \leq s \leq t} |\mathcal{A}^\sigma (u^1(s) - u^2(s))|^2 + \int_0^t n_V^2(u_t^1(s) - u_t^2(s)) ds \right\}$$

for some $\sigma \in [0, 1/2)$. Here we denote $S_t y_i = (u^i(t); u_t^i(t))$, $i = 1, 2$. Then the attractor A has a finite fractal dimension.

PROOF. We apply Theorem 2.15 in the space $X_T = H \times W_2(0, T)$ with an appropriate T. Here $W_2(0, T)$ is the space of functions $z(t)$ for which the norm
$$|z|_{W_2(0,T)} \equiv \left(\int_0^T \left(|\mathcal{A}^{1/2} z(t)|^2 + |M^{1/2} z_t(t)|^2 + |\mathcal{A}^{-l} M z_{tt}(t)|^2 \right) dt \right)^{1/2}$$
is finite. The norm in X_T is given by
$$\|U\|_{X_T}^2 = |\mathcal{A}^{1/2} u_0|^2 + |M^{1/2} u_1|^2 + |z|_{W_2(0,T)}^2, \quad U = (u_0, u_1, z).$$

4.1. FINITE DIMENSIONALITY OF ATTRACTORS

Let $z(t) = u^1(t) - u^2(t)$. As in the proof of Theorem 4.3 we can obtain that

$$|S_T y_1 - S_T y_2|_H^2 + \int_T^{2T} |S_t y_1 - S_t y_2|_H^2 dt$$

(4.13)
$$\leq b_T |y_1 - y_2|_H^2 + c_T \left\{ \sup_{0 \leq s \leq 2T} |\mathcal{A}^\sigma z(s)|^2 + \int_0^{2T} n_V^2(z_t(s)) dt \right\},$$

where b_T and c_T is given by (4.10). Using equation (4.1) and relation (4.2) we have that

(4.14) $\quad |\mathcal{A}^{-l} M z_{tt}(t)|^2 \leq C_A \left(|\mathcal{A}^{1/2} z(t)|^2 + |M^{1/2} z_t(t)|^2 \right) \leq c_A^2 |S_t y_1 - S_t y_2|_H^2$

for every $y_1, y_2 \in A$. Therefore, it follows from the corresponding analog (with n_V) of relation (4.7) that

$$\int_T^{2T} |\mathcal{A}^{-l} M z_{tt}(t)|^2 dt$$

(4.15)
$$\leq c_A^2 \widetilde{b}_T |y_1 - y_2|_H^2 + c_A^2 \widetilde{c}_T \left\{ \sup_{0 \leq s \leq 2T} |\mathcal{A}^\sigma z(s)|^2 + \int_0^{2T} n_V^2(z_t(s)) dt \right\},$$

where $\widetilde{b}_T$ and $\widetilde{c}_T$ is defined by (4.8).

As above consider in the space X_T the set

$$A_T := \{ U \equiv (u(0); u_t(0); u(t), t \in [0, T]) \, : \, (u(0); u_t(0)) \in A \},$$

where $u(t)$ is the solution to the initial problem with initial data $(u(0); u_t(0))$, and define operator $V : A_T \mapsto X_T$ by the formula

$$V : (u(0); u_t(0); u(t)) \mapsto (u(T); u_t(T); u(T+t)).$$

It is clear from (4.3) and (4.14) that V is Lipschitz on A_T. Since A is a strictly invariant set (with respect to S_t) and it consists of full trajectories, we also have that $V A_T = A_T$.

Similarly to the proof of Theorem 4.3, from (4.13) and (4.15) we obtain

$$\|VU_1 - VU_2\|_{X_T}^2 \leq \eta_T^\varepsilon \|U_1 - U_2\|_{H_T}^2 + K_T^\varepsilon \cdot (n_T^2(U_1 - U_2) + n_T^2(VU_1 - VU_2)),$$

for any $U_1, U_2 \in A_T$, where $K_T^\varepsilon > 0$ is a constant,

$$n_T(U) := \sup_{0 \leq s \leq T} |u(s)| + \left(\int_0^T n_V^2(u_t(s)) ds \right)^{1/2}$$

and

$$\eta_T^\varepsilon = b(T) + (1 + c_A^2) \int_T^{2T} b(t) dt + \varepsilon \cdot \left(c(T) + (1 + c_A^2) \int_T^{2T} c(t) dt \right).$$

We can choose $T > 1$ and $\varepsilon > 0$ such that $\eta_T^\varepsilon < 1$. Therefore to complete the proof we only need to prove that $n_T(U)$ is a compact seminorm on X_T.

Let V_1 be the completion of V with respect to the norm $|\cdot|_{V_1}$ given by

$$|v|_{V_1} = |\mathcal{A}^{-1/2} M v| + n_V(v)$$

and W be the completion of V with respect to the norm $|\cdot|_W = |\mathcal{A}^{-l}M\cdot|$. It is clear that $V \subset V_1 \subset W$ and the embedding $V \subset V_1$ is compact. Therefore by [**96**, Corollary 4] the embedding

$$X_T \subset C([0,T], \mathcal{H}) \cap L_2([0,T], V_1)$$

is compact. This implies that the seminorm $n_T(U)$ is compact on X_T.

Applying Theorem 2.15 we conclude the proof in the same way as it was done in Theorem 4.3. □

REMARK 4.5. The idea to use a space of "pieces" of trajectories in the proof of finite dimension goes back to the method of the so-called short trajectories suggested in the paper [**84**] and developed in [**85**]. For applications of this method we also refer to [**27, 89**] in the case of wave dynamics with interior dissipation, to [**22, 26**] for wave and plate systems with boundary damping, and to the paper [**34**] which involves the same idea in the study metric properties of trajectory attractors.

We conclude this section with the following proposition which provides more specific conditions implying the validity of hypotheses of Theorem 4.4 imposed on the system (H, S_t) generated by (4.1). The main trust of this result is that elements on the attractor may be more regular. This regularizing effect of the attractor may produce nonlinear terms in the equation that behave "subcritically" rather than "critically". This is a rather natural idea that has been exploited in the past in the context of hyperbolic-like dynamics (where there is no natural smoothing effect generated by dynamics) [**49, 4, 58, 77, 81**]. A variant of this idea, presented below, axiomatizes needed regularity properties, so that the previous arguments given in Lemma 3.59 can be repeated without the necessity of assuming large damping parameter. In fact, this particular setup has been used recently in [**31**] for proving finite dimensionality of attractors arising in von Karman equations. Theorem 4.16, formulated at the end of this section, provides an explicit statement, along with explicitly stated sufficient conditions insuring finite dimensionality of attractors in a critical case.

To begin the exposition, we shall introduce the following hypothesis imposed on

$$(4.16) \qquad G_{u^1, u^2}(t) \equiv F(u^1(t), u^1_t(t)) - F(u^2(t), u^2_t(t)),$$

where $y^1(t) = (u^1(t); u^1_t(t))$ and $y^2(t) = (u^2(t); u^2_t(t))$ are two trajectories of the system.

ASSUMPTION 4.6. Given any two trajectories lying on the attractor $y^1(t) = (u^1(t); u^1_t(t))$ and $y^2(t) = (u^2(t); u^2_t(t))$, let $z(t) = u^1(t) - u^2(t)$. Let $T_0 > 0$ be given. Assume that for any $\varepsilon > 0$ and for some $T \geq T_0$, there exists $b_\varepsilon(T) > 0$ such that

$$\sup_{t \in [0,T]} \left| \int_{s+t}^{s+T} (G_{u^1, u^2}(\tau), z_t(\tau)) d\tau \right| \leq \varepsilon \int_s^{s+T} E_z(\tau) d\tau$$

$$(4.17) \qquad + b_\varepsilon(T) \left\{ \sup_{0 \leq \tau \leq T} |\mathcal{A}^\sigma z(s+\tau)|^2 + \int_s^{s+T} n_V^2(z_t(s)) d\tau \right\}$$

for all $s \in \mathbb{R}$, where $\sigma < 1/2$, n_V is a compact seminorm on V and the energy function $E_z(t)$ corresponding to $(z; z_t)$ is defined as follows

$$(4.18) \qquad E_z(t) \equiv \frac{1}{2}\left[(Mz_t(t), z_t(t)) + (\mathcal{A}z(t), z(t))\right].$$

PROPOSITION 4.7. *Let Assumption 1.1 and Assumption 3.21 with $H_0(s) = m_0^{-1}s$ in (3.59) and $\kappa = 2$ in (3.60) be in force. Let $\mathcal{D}(\mathcal{A}^{1/2-\beta}) \subset V$ be compact for some $\beta > 0$. Assume that the system (H, S_t) generated by (4.1) possesses the global attractor A and for every pair $y^1(t) = (u^1(t); u_t^1(t))$ and $y^2(t) = (u^2(t); u_t^2(t))$ - trajectories from the attractor- Assumption 4.6 holds true. Then relation (4.12) holds with $b(t) = b_0 e^{-\omega t}$, $\omega > 0$. Thus, if in addition (4.2) holds, then the global attractor A has a finite fractal dimension.*

We note that the Assumption 4.6 expresses certain smoothing mechanism for the source term $F(u)$ with trajectories $y(t) = (u(t); u_t(t))$ on the attractor. Indeed, all the terms on the right side of inequality (4.17) are "subcritical". The mechanism for smoothing is related to the fact that trajectories on the attractor may converge as $t \to -\infty$ to the set of equilibria and this set is "smooth". Thus, it can be expected that the inequality in (4.17) holds at least for s approaching $-\infty$. The fact that this inequality should also hold for all $s \in \mathbb{R}$ results -roughly speaking- from the fact that "smooth" states near $-\infty$ propagate smooth trajectories in the forward direction. Thus, the additional hypothesis imposed in Proposition 4.7 is natural for systems with conservative loads.

REMARK 4.8. We can provide the following sufficient condition for (4.17) which sometimes may be easier to verify. Assume that the value $G_{u^1,u^2}(t)$ given by (4.16) admits the representation

$$(4.19) \qquad \left(G_{u^1,u^2}(t), u_t^1(t) - u_t^2(t)\right) = \frac{d}{dt}Q(t) + R(t),$$

where the scalar functions $Q(t) \equiv Q(u^1(t); u^2(t))$ and $R(t) \equiv R(y^1(t); y^2(t))$ enjoy the following properties: (i) there exist a number $0 \leq \sigma < 1/2$ and a locally bounded function $c(t)$ on $[0, \infty)$ such that

$$(4.20) \qquad |Q(t)| \leq c(t)|\mathcal{A}^\sigma(u^1(t) - u^2(t))|^2,$$

and (ii) for every $\varepsilon > 0$ there exists $b_\varepsilon(T) > 0$ such that
(4.21)
$$\int_0^T |R(t)|dt \leq \varepsilon \int_0^T E_z(t)dt + b_\varepsilon(T)\left\{\sup_{0 \leq s \leq T} |\mathcal{A}^\sigma z(s)|^2 + \int_0^T n_V^2(z_t(s)))ds\right\}$$

for every $T > 0$, where $z(t) = u^1(t) - u^2(t)$ and $E_z(t)$ is given by (4.18). Then it is clear that (4.19) implies (4.17).

We note that Theorem 4.25, given in the next section, provides concrete conditions imposed on the semilinear source F which guarantee representation (4.19) with properties (4.20) and (4.21) and applicability of Proposition 4.7, hence finite dimensionality of attractor.

Another sufficient condition for (4.17) on the attractor can be described as follows. Let A^c be the convex hull of the global attractor A. Assume that $(u_0; u_1) \mapsto F(u_0, u_1)$ is a C^1-mapping from H into $\mathcal{H}$ on the hull A^c and for every $\varepsilon > 0$ there

exists $C_\varepsilon > 0$ such that
(4.22)
$$\sup_{(u_0;u_1)\in A^c} |\langle F'(u_0,u_1); w_0, w_1\rangle|_{\mathcal{H}} \leq \varepsilon\left(|\mathcal{A}^{1/2}w_0| + |w_1|_V\right) + C_\varepsilon\left(|\mathcal{A}^\sigma w_0| + n_V(w_1)\right)$$

for every $(w_0; w_1) \in H = \mathcal{D}(\mathcal{A}^{1/2}) \times V$, where $\langle F'(u_0, u_1); w_0, w_1\rangle$ denotes the value of the first derivative F' at $(u_0; u_1)$ on the element $(w_0; w_1)$, $0 \leq \sigma < 1/2$ and n_V is a compact seminorm on V. Then relation (4.17) holds for any couple of trajectories from the attractor. Indeed, one can see that

$$G_{u^1, u^2}(t) = \int_0^1 \langle F'(\lambda u^1 + (1-\lambda)u^2, \lambda u_t^1 + (1-\lambda)u_t^2); z, z_t\rangle d\lambda$$

and hence by (4.22) we have that

$$|G_{u^1, u^2}(t)| \leq \varepsilon\left(|\mathcal{A}^{1/2}z| + |z_t|_V\right) + C_\varepsilon\left(|\mathcal{A}^\sigma z| + n_V(z_t)\right).$$

This implies (4.17). We use this idea in a bit different form in Theorem 4.27 below.

Proof of Proposition 4.7. The argument is basically the same as in Remark 3.30 or in the proof of Lemma 3.59 and relies on the estimate given in Lemma 3.23. However, due to the new assumption (4.17) we will not need the requirement that $\eta > 0$ (subcriticality) or a restriction on the size of the damping. For readers convenience we repeat the main steps.

The proof follows through a sequence of lemmas.

LEMMA 4.9. *Let Assumption 1.1 and Assumption 3.21 with $H_0(s) = m_0^{-1}s$ in (3.59) and $\kappa = 2$ in (3.60) be in force. Assume that w and v are strong solutions to (4.1) possessing properties*

(4.23) $$\max_{s\in[0,T]}\left\{|\mathcal{A}^{1/2}w(s)|^2 + |M^{1/2}w_t(s)|^2 + |\mathcal{A}^{1/2}v(s)|^2 + |M^{1/2}v_t(s)|^2\right\} \leq R^2$$

for some $R > 0$. Then for the difference $z(t) = w(t) - v(t)$ we have that

(4.24) $$TE_z(T) + \int_0^T E_z(t)dt$$
$$\leq C(R,T)\left[\int_0^T (D(t, z_t), z_t)dt + \max_{s\in[0,T]}|\mathcal{A}^{1/2-\delta}z(s)|^2\right] + c_0\Psi_T(w, v),$$

for $T \geq T_0$, where $c_0 > 0$ does not depend on R and T. Here above $E_z(t)$ is given by (4.18) and $D(t, z_t)$ and $\Psi_T(w, v)$ are the same as in Lemma 3.23, i.e.

$$D(t, z_t) = D(v_t(t) + z_t(t)) - D(v_t(t)),$$
$$\Psi_T(w,v) = \left|\int_0^T (G_{w,v}(\tau), z_t(\tau))d\tau\right| + \left|\int_0^T (G_{w,v}(t), z(t))dt\right|$$
(4.25)
$$+ \left|\int_0^T dt \int_t^T (G_{w,v}(\tau), z_t(\tau))d\tau\right|$$

with $G_{w,v}(t)$ given by (4.16).

PROOF. It follows from Lemma 3.25 and from (3.60) with $\kappa = 2$ that

$$\int_0^T (D(t, z_t), z)dt \leq C(R,T) \max_{s\in[0,T]} |\mathcal{A}^{1/2-\delta}z(s)|^2$$
$$+ C_\epsilon(R,T) \int_0^T (D(t,z_t), z_t)dt + \epsilon\left[E_z(T) + \int_0^T E_z(t)dt\right]$$

for any $\epsilon > 0$. Therefore using (3.78) with $H_0(s) = m_0^{-1}s$ and inequality (3.62) in Lemma 3.23 we obtain (4.24). $\square$

To complete the proof of Proposition 4.7 we need the following assertion.

LEMMA 4.10. *Assume that hypotheses of Lemma 4.9 are in force. In addition assume (4.17) with T_0 as in Lemma 4.9 and with trajectories $y^1(t) = (u^1(t); u_t^1(t))$ and $y^2(t) = (u^2(t); u_t^2(t))$ taken from the global attractor A. Then there exists $\gamma = \gamma(T) < 1$ such that*

$$(4.26) \quad E_z(s+T) \leq \gamma E_z(s) + C_T\left\{\sup_{0\leq\tau\leq T}|\mathcal{A}^{\sigma_*}z(s+\tau)|^2 + \int_s^{s+T} n_V^2(z_t(s))d\tau\right\}$$

for all $s \in \mathbb{R}$ with $\sigma_* < 1/2$.

PROOF. It follows from Lemma 4.9 that

$$TE_z(s+T) + \int_s^{s+T} E_z(\tau)d\tau$$
$$(4.27) \quad \leq C_T\left[\int_s^{s+T}(D(t, z_t), z_t)dt + \max_{s\in[0,T]}|\mathcal{A}^{1/2-\delta}z(s)|^2\right] + c_0\Psi_T^s(u^1, u^2),$$

where
(4.28)
$$\Psi_T^s(u^1, u^2) = (1+T)\sup_{t\in[0,T]}\left|\int_{s+t}^{s+T}(G_{u^1,u^2}(\tau), z_t(\tau))d\tau\right| + \left|\int_s^{s+T}(G_{u^1,u^2}(t), z(t))dt\right|.$$

By (4.17) we obtain that

$$\Psi_T^s(u^1, u^2) \leq \varepsilon\int_s^{s+T} E_z(\tau)d\tau + C_T^\varepsilon\left\{\sup_{0\leq\tau\leq T}|\mathcal{A}^\sigma z(s+\tau)|^2 + \int_s^{s+T} n_V^2(z_t(s))d\tau\right\}$$

for all s. From energy relation (3.66) we also have that

$$(4.29) \quad k\int_s^{s+T}(D(t,z_t), z_t)dt \leq E_z(s) - E_z(s+T) + \Psi_T^s(u^1, u^2).$$

Therefore we arrive at

$$E_z(s+T) \leq C_T(E_z(s) - E_z(s+T))$$
$$+ C_T\left\{\sup_{0\leq\tau\leq T}|\mathcal{A}^{\sigma_*}z(s+\tau)|^2 + \int_s^{s+T} n_V^2(z_t(s))d\tau\right\}$$

for all $s \in \mathbb{R}$, where $\sigma_* = \min\{\sigma, 1/2 - \delta\} < 1/2$. This implies (4.26). $\square$

To complete the proof of Proposition 4.7 we proceed as in the proof of Remark 3.30. Indeed, we note that (4.26) applied with $s = 0$ is a counterpart of inequality just below (3.88). The fact that (4.17) is assumed for all $s \in \mathbb{R}$ allows to reiterate the estimates on each subinterval $(mT, (m+1)T)$, $m = 1, 2\ldots$ leading to the conclusion in (3.87). This, in turn, implies (4.12). Theorem 4.4 yields the final result stated in Proposition 4.7. $\square$

Hypotheses in Assumption 4.6 may be complicated to verify in general. However in the case of conservative forces ($F^*(u,v) \equiv 0$) one can take an advantage of the finiteness of the dissipation integral

$$(4.30) \qquad \int_0^\infty (D(u_t(t)), u_t(t)) dt$$

evaluated along trajectories. As we shall see below, finiteness of the integral in (4.30) on trajectories may compensate for the lack of sufficient smoothness. This is in line with a general philosophy, not unusual in quasihyperbolic dynamics, that *damping → smoothness*. In order to exploit beneficial effects of the damping, we formulate the following hypothesis imposed on the nonlinear term F.

ASSUMPTION 4.11. Assume that
- $F(u,v) = F(u)$ is independent of velocities and does not contain non-conservative terms, i.e. $F(u) = -\Pi'(u)$.
- The functional $\Pi : \mathcal{D}(\mathcal{A}^{1/2}) \mapsto \mathbb{R}$ is a Frechet C^3-mapping.
- The second $\Pi^{(2)}(u)$ and the third $\Pi^{(3)}(u)$ Frechet derivatives of $\Pi(u)$ satisfy the conditions

$$(4.31) \qquad \left|\langle \Pi^{(2)}(u); v, v\rangle\right| \leq C_\rho |\mathcal{A}^\sigma v|^2, \quad v \in \mathcal{D}(\mathcal{A}^{1/2}),$$

for some $\sigma < 1/2$, and

$$(4.32) \qquad \left|\langle \Pi^{(3)}(u); v_1, v_2, v_3\rangle\right| \leq C_\rho |\mathcal{A}^{1/2} v_1| |\mathcal{A}^{1/2} v_2| |v_3|_V, \quad v_i \in \mathcal{D}(\mathcal{A}^{1/2}),$$

for all $u \in \mathcal{D}(\mathcal{A}^{1/2})$ such that $|\mathcal{A}^{1/2} u| \leq \rho$, where $\rho > 0$ is arbitrary and C_ρ is a positive constant. Here above $\langle \Pi^{(k)}(u); v_1, \ldots, v_k\rangle$ denotes the value of the derivative $\Pi^{(k)}(u)$ on elements $v_1, \ldots, v_k$,

We recall that the Frechet derivatives $\Pi^{(k)}(u)$ of the functional Π are *symmetric* k-linear continuous forms on $\mathcal{D}(\mathcal{A}^{1/2})$ (see, e.g., [10]). Moreover, if $\Pi \in C^3$, then $(F(u), v) \equiv -\langle \Pi'(u); v\rangle$ is C^2-functional for every fixed $v \in \mathcal{D}(\mathcal{A}^{1/2})$ and the following Taylor's expansion holds

$$(4.33) \quad (F(u+w)-F(u), v) = -\langle \Pi^{(2)}(u); w, v\rangle - \int_0^1 (1-\lambda)\langle \Pi^{(3)}(u+\lambda w); w, w, v\rangle d\lambda$$

for any $u, v \in \mathcal{D}(\mathcal{A}^{1/2})$ [10]. This structure of $F(u)$ is important for our subsequent considerations because it leads to a representation of the form (4.19). Indeed, assume that $u(t)$ and $z(t)$ belong to the class $C^1(a, b; \mathcal{D}(\mathcal{A}^{1/2}))$ for some interval $[a, b] \subseteq \mathbb{R}$. Then by the differentiation rule for composition of mappings [10] using the symmetry of the form $\Pi^{(2)}(u)$ we have that

$$\frac{d}{dt}\langle \Pi^{(2)}(u); z, z\rangle = \langle \Pi^{(3)}(u); u_t, z, z\rangle + 2\langle \Pi^{(2)}(u); z, z_t\rangle.$$

Therefore from (4.33) we obtain the representation

(4.34) $$(F(u(t)+z(t)) - F(u(t)), z_t(t)) = \frac{d}{dt}Q(t) + R(t), \quad t \in [a,b] \subseteq \mathbb{R},$$

with

(4.35) $$Q(t) = -\frac{1}{2}\langle \Pi^{(2)}(u(t)); z(t), z(t)\rangle$$

and

(4.36) $$R(t) = -\frac{1}{2}\langle \Pi^{(3)}(u); u_t, z, z\rangle - \int_0^1 (1-\lambda)\langle \Pi^{(3)}(u+\lambda z); z, z, z_t\rangle d\lambda.$$

REMARK 4.12. *Instead of Assumption 4.11 one could also postulate the decomposition as in (4.34) with $Q(t)$ and $R(t)$ given by (4.35) and (4.36) subject to (4.31) and (4.32).*

We also note that under Assumption 4.11 it is easy to see from (4.34) that for any $(u(t); u_t(t)), (z(t); z_t(t)) \in C(a,b; \mathcal{D}(\mathcal{A}^{1/2}) \times V)$ such that

(4.37) $$\sup_{t\in[a,b]} \left\{ |\mathcal{A}^{1/2}u(t)| + |\mathcal{A}^{1/2}z(t)| \right\} \leq R$$

we have the relation

(4.38) $$\left| \int_s^t (G_{u+z,u}(\tau), z_t(\tau))d\tau \right| \leq C_1(R) \sup_{s\leq \tau \leq t} |\mathcal{A}^\sigma z(\tau)|^2 + C_2(R) \int_s^t (|u_t(\tau)|_V + |z_t(\tau)|_V) |\mathcal{A}^{1/2}z(\tau)|^2 d\tau$$

for all $a < s \leq t < b$, where $\sigma < 1/2$ and $G_{u^1,u^2}(t)$ is given by (4.16), i.e. $G_{u^1,u^2}(t) = F(u^1(t)) - F(u^2(t))$ in the case considered. This property implies the following assertion, which constututes the main ingredient of the proof of stabilizability estimate. The latter is critical for both finite-dimensionality and regularity of attractors.

PROPOSITION 4.13. *Let Assumption 4.11 be in force. Assume that functions $(u^1(t); u_t^1(t))$ and $(u^2(t); u_t^2(t))$ from $C(\mathbb{R}_+; \mathcal{D}(\mathcal{A}^{1/2}) \times V)$ possess the property*

(4.39) $$\max_{s\in\mathbb{R}_+} \left\{ |\mathcal{A}^{1/2}u^1(s)|^2 + |M^{1/2}u_t^1(s)|^2 + |\mathcal{A}^{1/2}u^2(s)|^2 + |M^{1/2}u_t^2(s)|^2 \right\} \leq R^2$$

for some $R > 0$. Then for any $\varepsilon > 0$ and $T > 0$ there exist $a_\varepsilon(R,T)$ and $b_\varepsilon(R,T)$ such that for $z(t) = u^1(t) - u^2(t)$ we have the following relation

(4.40) $$\sup_{t\in[0,T]} \left| \int_{s+t}^{s+T} (G_{u^1,u^2}(\tau), z_t(\tau))d\tau \right| \leq \varepsilon \int_s^{s+T} E_z(\tau)d\tau + a_\varepsilon(R,T) \int_s^{s+T} d(\tau; u^1, u^2) E_z(\tau)d\tau + b_\varepsilon(R,T) \sup_{0\leq \tau \leq T} |\mathcal{A}^\sigma z(s+\tau)|^2$$

for all $s \geq 0$, where $\sigma < 1/2$. Here above $E_z(t)$ and $G_{u^1,u^2}(t)$ is given by (4.18) and (4.16), and

(4.41) $$d(t; u^1, u^2) = |u_t^1(t)|_V^2 + |u_t^2(t)|_V^2.$$

The main tool for proving finite dimensionality and regularity of attractors is stabilizability estimate formulated below.

PROPOSITION 4.14. *Let Assumption 1.1 and Assumption 3.21 with $H_0(s) = m_0^{-1}s$ in (3.59) and $\kappa = 2$ in (3.60) be in force. Assume that $u^1(t)$ and $u^2(t)$ are two generalized solutions to problem (4.1) possessing properties (4.39) and satisfying relation (4.40) with some non-negative function $d(\tau; u^1, u^2)$ such that*

$$(4.42) \qquad d_\infty = \int_0^\infty d(t; u^1, u^2)\, dt \leq C_R < \infty.$$

Then for $S_t y_i = (u^i(t), u_t^i(t))$ we have the following relation
(4.43)
$$\|S_t y_1 - S_t y_2\|_H^2 \leq C_1(R) e^{-\omega_R t} \|y_1 - y_2\|_H^2 + C_2(R) \max_{[0,t]} |\mathcal{A}^\sigma(u^1(\tau) - u^2(\tau))|^2, \quad t > 0,$$

where $C_1(R)$, $C_2(R)$ and ω_R are positive constants, $0 \leq \sigma < 1/2$.

PROOF. Without loss of generality we can assume that $u^1(t)$ and $u^2(t)$ are strong solutions. As above we denote $z(t) = u^1(t) - u^2(t)$.

As in the proof of Lemma 4.10 we have relation (4.27) with the constant C_T which may also depend on R. Now using (4.40) for $\Psi_T^s(u^1, u^2)$ given by (4.28) we obtain the following estimate

$$\Psi_T^s(u^1, u^2) \leq \varepsilon \int_s^{s+T} E_z(\tau) d\tau$$
(4.44)
$$+ \widetilde{a}_\varepsilon(R, T) \int_s^{s+T} d(\tau; u^1, u^2) E_z(\tau) d\tau + \widetilde{b}_\varepsilon(R, T) \sup_{0 \leq \tau \leq T} |\mathcal{A}^\sigma z(s+\tau)|^2$$

for every $\varepsilon > 0$. Consequently, using (4.27) and (4.29) we arrive at

$$E_z(s+T) \leq C_1(R, T)\left(E_z(s) - E_z(s+T)\right)$$
$$+ C_2(R, T) \int_s^{s+T} d(\tau; u^1, u^2) E_z(\tau) d\tau + C_3(R, T) \sup_{0 \leq \tau \leq T} |\mathcal{A}^\sigma z(s+\tau)|^2$$

for all $s > 0$ and $T \geq T_0$. This implies that

$$E_z((m+1)T) \leq \gamma E_z(mT) + c_{R,T} b_m, \quad m = 0, 1, 2, \ldots,$$

with $0 < \gamma \equiv \gamma_{T,R} < 1$, where

$$b_m \equiv \sup_{\tau \in [mT, (m+1)T]} |\mathcal{A}^\sigma z(\tau)|^2 + \int_{mT}^{(m+1)T} d(\tau; u^1, u^2) E_z(\tau) d\tau.$$

This yields

$$E_z(mT) \leq \gamma^m E_z(0) + c \sum_{l=1}^m \gamma^{m-l} b_{l-1}.$$

Since $\gamma < 1$, by the same argument as in Remark 3.30 along with the definition of b_l given above we obtain that there exists $\omega > 0$ such that

$$E_z(t) \leq C_1 e^{-\omega t} E_z(0)$$
$$+ C_2 \left\{ \sup_{\tau \in [0,t]} |\mathcal{A}^\sigma z(\tau)|^2 + \int_0^t e^{-\omega(t-\tau)} d(\tau; u^1, u^2) E_z(\tau) d\tau \right\}$$

for all $t \geq 0$. Therefore, applying Gronwall's lemma to a function $E_z(t) e^{\omega t}$ we find that

$$E_z(t) \leq \left[C_1 E_z(0) e^{-\omega t} + C_2 \sup_{\tau \in [0,t]} |\mathcal{A}^\sigma z(\tau)|^2 \right] \exp\left\{ C_2 \int_0^t d(\tau; u^1, u^2)\, d\tau \right\}.$$

Now using (4.42) we obtain estimate (4.43). This concludes the proof of Proposition 4.14. □

A direct consequence of Proposition 4.14 is the following assertion.

PROPOSITION 4.15. *Let Assumption 1.1 and Assumption 3.21 with $H_0(s) = m_0^{-1} s$ in (3.59) and $\kappa = 2$ in (3.60) be in force. Let $\mathcal{D}(\mathcal{A}^{1/2-\beta}) \subset V$ be compact for some $\beta > 0$. Assume that the system (H, S_t) generated by (4.1) possesses the global attractor A and for every pair $y^1(t) = (u^1(t); u_t^1(t))$ and $y^2(t) = (u^2(t); u_t^2(t))$ - trajectories from the attractor- relation (4.40) holds true with $d(t; u_1, u_2)$ possessing property (4.42). Then the global attractor A has a finite fractal dimension.*

PROOF. By Proposition 4.14 relation (4.4) holds for semi-flow S_t with $b(t) = b_0 e^{-\omega t}$, $\omega > 0$. Therefore the application of Theorem 4.3 along with Proposition 1.15 leads to the result desired. □

Now we are in position to conclude this section with the following theorem which recapitulates the ideas presented above and supplies the main result pertaining to finite-dimensionality of attractors in the critical and conservative case.

THEOREM 4.16. *Let Assumption 1.1 with $F^*(u,v) \equiv 0$, Assumption 3.21 with $H_0(s) = m_0^{-1} s$ in (3.59) and $\kappa = 2$ in (3.60) and Assumption 3.39 be in force. Assume that $\mathcal{D}(\mathcal{A}^{1/2-\beta})$ is compactly embedded in V for some $\beta > 0$, the nonlinear mapping $F(u) = -\Pi'(u)$ satisfies Assumption 4.11, and the dissipativity condition in (3.114) holds. Then the system (H, S_t) generated by (1.1) possesses a compact global attractor A of a finite fractal dimension.*

PROOF. The existence of a compact global attractor A follows from Theorem 3.40. By (3.59) we have that

$$(4.45) \quad d(t; u^1, u^2) \equiv |u_t^1(t)|_V^2 + |u_t^2(t)|_V^2 \leq m_0^{-1} \left[D(u_t^1(t), u_t^1(t)) + D(u_t^2(t), u_t^2(t)) \right].$$

Therefore it follow from (3.115) that relation (4.42) holds. Thus we can apply Proposition 4.15 to prove that A has a finite dimension. □

4.2. Regularity of elements from attractors

In this section we address a question of additional smoothness of elements in the attractor. With some additional hypotheses we will be able to prove that elements in the attractor are more regular than indicated by the topology of the phase space. While this kind of result is typical for parabolic-like flows, it is more unusual in hyperbolic dynamics with a nonlinear damping. This is due to an inherent to hyperbolic problems lack of smoothing that is particularly acute in the presence of strongly nonlinear dissipation.

For the second order in time evolution equations the problem of regularity of elements from a global attractor has been studied in abstract setting in [**57**] for the case of *scalar linear* damping and $M = I$.

We start with the following assertion which shows how stabilizability estimates can be used for the regularity of elements of the global attractor and which gives us the result in the case of the attractors constructed in Theorem 4.1 or in Theorem 4.16.

THEOREM 4.17. *Let the hypotheses of either Theorem 4.1 or Theorem 4.16 be in force. In addition, we assume that*

- if the seminorm n_V is not absent in (3.61), then (i) it possesses the property

(4.46) $$n_V^2(v) \leq \epsilon |v|_V^2 + C_\epsilon |Mv|^2_{[\mathcal{D}(\mathcal{A}^{1/2})]'}, \quad v \in V,$$

with arbitrary $\epsilon > 0$, and (ii) the operator D maps V into $[\mathcal{D}(\mathcal{A}^{1/2})]'$ and is bounded on bounded sets.

Then any full trajectory $\{(u(t); u_t(t)) : t \in \mathbb{R}\}$ which belongs to the global attractor enjoys the following regularity properties:

(4.47) $$\mathcal{A}u + kD(u_t) \in C_r(\mathbb{R}; V'), \quad u_t \in C_r(\mathbb{R}; \mathcal{D}(\mathcal{A}^{1/2})), \quad u_{tt} \in C_r(\mathbb{R}; V),$$

where C_r denotes the space right continuous functions. Moreover, there exists $R > 0$ such that

(4.48) $$|u_{tt}(t)|_V^2 + |\mathcal{A}^{1/2} u_t(t)|^2 + |\mathcal{A}u(t) + kD(u_t(t))|_{V'}^2 \leq R^2, \quad t \in \mathbb{R}.$$

PROOF. It follows from the inequality in Remark 3.30 or Lemma 3.59 (in the case of Theorem 4.1) and from Proposition 4.14 (in the case of Theorem 4.16) that for any two trajectory

$$\gamma = \{y(t) \equiv (u(t); u_t(t)) : t \in \mathbb{R}\} \quad \text{and} \quad \gamma^* = \{y^*(t) \equiv (u^*(t); u_t^*(t)) : t \in \mathbb{R}\}$$

from the global attractor we have that there exist positive constants ω, δ and C such that

(4.49) $$\begin{aligned} E_0(z(t), z_t(t)) &\leq Ce^{-\omega(t-s)} E_0(z(s), z_t(s)) \\ &\quad + C \left\{ \sup_{s \leq \tau \leq t} |\mathcal{A}^{1/2 - \delta} z(\tau)|^2 + \int_s^t e^{-\omega(t-\tau)} n_V^2(z_t(\tau)) d\tau \right\}, \end{aligned}$$

for all $s \leq t$, $s, t \in \mathbb{R}$, where $z(t) = u^*(t) - u(t)$ and $E_0(z, z_t) = \frac{1}{2}(|z_t|_V^2 + |\mathcal{A}^{1/2} z|^2)$. In the limit $s \to -\infty$ relation (4.49) gives us that

$$E_0(z(t), z_t(t)) \leq C \left\{ \sup_{-\infty \leq \tau \leq t} |\mathcal{A}^{1/2 - \delta} z(\tau)|^2 + \int_{-\infty}^t e^{-\omega(t-\tau)} n_V^2(z_t(\tau)) d\tau \right\},$$

for every $t \in \mathbb{R}$ and for every couple of trajectories γ and γ^*. Using the interpolation and property (4.46) we can conclude that

(4.50) $$\sup_{-\infty \leq \tau \leq t} E_0(z(\tau), z_t(\tau)) \leq C \sup_{-\infty \leq \tau \leq t} \left\{ |z(\tau)|_V^2 + |Mz_t(\tau)|^2_{[\mathcal{D}(\mathcal{A}^{1/2})]'} \right\},$$

or every $t \in \mathbb{R}$ and for every couple of trajectories γ and γ^* from the attractor.

Now we fix the trajectory γ and for $0 < |h| < 1$ we consider the shifted trajectory

$$\gamma_* \equiv \gamma_h = \{y(t+h) : t \in \mathbb{R}\}.$$

Applying for this pair of trajectories relation (4.50) and using the fact that all terms (4.50) are quadratic with respect to z we obtain that

(4.51) $$\sup_{-\infty \leq \tau \leq t} E_0(u^h(\tau), u_t^h(\tau)) \leq C \sup_{-\infty \leq \tau \leq t} \left\{ |u^h(\tau)|_V^2 + |Mu_t^h(\tau)|^2_{[\mathcal{D}(\mathcal{A}^{1/2})]'} \right\},$$

where $u^h(t) = h^{-1} \cdot [u(t+h) - u(t)]$. On the attractor we obviously have that

$$|u^h(t)|_V \leq \frac{1}{h} \cdot \int_0^h |u_t(\tau + t)|_V d\tau \leq C, \quad t \in \mathbb{R},$$

and
$$|Mu_t^h(t)|_{[\mathcal{D}(\mathcal{A}^{1/2})]'} \le \frac{1}{h} \cdot \int_0^h |Mu_{tt}(\tau+t)|_{[\mathcal{D}(\mathcal{A}^{1/2})]'} d\tau \le C, \quad t \in \mathbb{R},$$
with uniformity in h. Therefore (4.51) implies that
$$|u_t^h(t)|_V^2 + |\mathcal{A}^{1/2} u^h(t)|^2 \le C, \quad t \in \mathbb{R}.$$
Passing with the limit on h yields then
(4.52) $$|u_{tt}(t)|_V^2 + |\mathcal{A}^{1/2} u_t(t)|^2 \le C, \quad t \in \mathbb{R}.$$
Return to the original equation we obtain that
(4.53) $$|\mathcal{A}u(t) + kD(u_t(t))|_{V'} \le C|u_{tt}|_V + |F(u(t), u_t(t))|_{V'} \le C, \quad t \in \mathbb{R}.$$
Thus, we have that
$$(\mathcal{A}u + kD(u_t); u_t; u_{tt}) \in L_\infty(\mathbb{R}; V' \times \mathcal{D}(\mathcal{A}^{1/2}) \times V).$$
In particular $\mathcal{A}u(t) + kD(u_t(t)) \in V'$ for almost all $t \in \mathbb{R}$. Therefore by Theorem 1.5 $u(t)$ is a strong solution to the original equation and hence relations (4.47) hold. The estimate in (4.48) follows from (4.52) and (4.53). $\square$

COROLLARY 4.18. *In addition to the hypotheses of Theorem 4.17 we assume that the damping operator D maps $\mathcal{D}(\mathcal{A}^{1/2})$ into V' and is bounded on bounded sets. Then the global attractor A is a closed bounded set in the space $W \times \mathcal{D}(\mathcal{A}^{1/2})$, where $W = \{u \in \mathcal{D}(\mathcal{A}^{1/2}) : \mathcal{A}u \in V'\}$.*

PROOF. Under the additional assumption imposed on the damping operator D we have that $|D(u_t(t))|_{V'} \le C$ for all $t \in \mathbb{R}$ on the attractor. Therefore the conclusion follows from (4.48). $\square$

In the reminder of this section we restrict ourselves to the case when nonlinear forces does not depend on velocities, i.e. we consider the following problem
$$\begin{cases} Mu_{tt}(t) + \mathcal{A}u(t) + k \cdot D(u_t(t)) = F(u(t)), \\ u|_{t=0} = u_0 \in \mathcal{D}(\mathcal{A}^{1/2}), \ u_t|_{t=0} = u_1 \in V = \mathcal{D}(M^{1/2}). \end{cases}$$
The result provided below extends *linear* result of [**57**] to the case of *nonlinear dissipation* and more general vectorial structure of the model.

THEOREM 4.19. *Assume that Assumption 1.1 holds and the system (H, S_t) generated by (1.1) possesses a global attractor A. Additionally assume that*

- *The damping operator D maps $\mathcal{D}(\mathcal{A}^{1/2})$ into V' continuously and is bounded on bounded sets.*
- *D is Frechet differentiable as a mapping from $\mathcal{D}(\mathcal{A}^{1/2})$ into $[\mathcal{D}(\mathcal{A}^{1/2})]'$, the derivative $D'(u)$ generates a symmetric bilinear form on $\mathcal{D}(\mathcal{A}^{1/2})$, and there exist positive constants k_1 and k_2 such that*

(4.54) $$\begin{aligned} k_1(Mw, w) &\le (\langle D'(\lambda u + (1-\lambda)v), w\rangle, w) \\ &\le k_2 \left[1 + (D(u), u) + (D(v), v)\right] |\mathcal{A}^{1/2} w|^2 \end{aligned}$$

for any $0 \le \lambda \le 1$ and $u, v, w \in \mathcal{D}(\mathcal{A}^{1/2})$ such that $|u|_V, |v|_V \le 1 + R_0$, where R_0 is a dissipativity radius of the system.

- *Concerning the nonlinear mapping $F(u,v)$ we assume that $F(u,v) = F(u)$ is a C^1 mapping from $\mathcal{D}(\mathcal{A}^{1/2})$ into V' and*

(4.55) $$|\langle F'(u), w\rangle|_{V'} \leq C_\rho \cdot |\mathcal{A}^{1/2-\delta}w|$$

for every $\rho > 0$, where $|\mathcal{A}^{1/2}u| \leq \rho$, $w \in \mathcal{D}(\mathcal{A}^{1/2})$ and $\delta \in (0, 1/2]$.

Then the global attractor is bounded set in $W \times \mathcal{D}(\mathcal{A}^{1/2}))$, where $W = \{u \in \mathcal{D}(\mathcal{A}^{1/2}) : \mathcal{A}u \in V'\}$ and any full trajectory $\{(u(t); u_t(t)) : t \in \mathbb{R}\}$ which belongs to the global attractor possesses the following regularity properties

(4.56) $$u \in C(\mathbb{R}; W), \quad u_t \in C(\mathbb{R}; \mathcal{D}(\mathcal{A}^{1/2})), \quad u_{tt} \in C(\mathbb{R}; V).$$

Moreover, there exists $R > 0$ such that

(4.57) $$|u_{tt}(t)|_V^2 + |\mathcal{A}^{1/2}u_t(t)|^2 + |\mathcal{A}u(t)|_{V'}^2 \leq R^2, \ t \in \mathbb{R}.$$

REMARK 4.20. A natural way to establish an additional regularity of attractors is to differentiate (with respect to time) the original dynamics and to consider regularity of the resulting solutions originating with initial conditions taken a priori in a larger dual (distributional) spaces. However, this method works well for *linear damping* only (see [**57**]), as only in that case principal part of evolution obtained is *time independent*. Thus, standard semigroup theory provides a meaning to trajectories originating with initial conditions taken in the duals to the domains of the generator. Instead, in a more general case of nonlinear damping, the evolution becomes time dependent and the trajectories are not generally defined for initial conditions that are taken in the dual spaces. Thus, the difficulty is intrinsic to poorly characterized regularity of domains of the generators corresponding to time dependent evolutions. These domains are no longer dense in a phase space, as it is the case with the semigroups (time independent case). In order to alleviate this difficulty, we shall consider approximation scheme for which we derive certain uniform estimates. This method is more technical and elaborate than in the time independent case, but as we shall see, it leads to the sought-after estimates. Passing with the limit on the final quantities will yield the desired inequality.

PROOF. In order to proceed with the program outlined above we define

(4.58) $$u^h(t) = \frac{u(t+h) - u(t)}{h}, \quad 0 < h < 1,$$

where $\{(u(t); u_t(t)) : t \in \mathbb{R}\}$ is a trajectory from the attractor. It is easy to verify that u^h satisfies the equation

(4.59) $$Mu_{tt}^h + D^h(t)u_t^h + \mathcal{A}u^h = F^h(t),$$

where

$$D^h(t)w = k\int_0^1 \langle D'(u_t(t) + \xi(u_t(t+h) - u_t(t))), w\rangle d\xi$$

and

(4.60) $$F^h(t) = \int_0^1 \langle F'(u(t) + \xi(u(t+h) - u(t))), u^h(t)\rangle d\xi.$$

We consider the following (time-dependent) evolution

(4.61) $$Mv_{tt} + D^h(t)v_t + \mathcal{A}v = 0.$$

The above equation, when considered for a fixed trajectory $u(t)$, is linear in the variable v, however non-autonomous. Standard evolution theory argument applies

in order to assert an existence of evolution operator $U^h(t,s) : H \to H$ such that for $t > s$
$$(v(t), v_t(t)) = U^h(t,s)(v(s), v_t(s)) \in H.$$
Our first step is to ensure that this evolution is exponentially stable with the parameters that are uniform (with respect to h).

LEMMA 4.21. *There exist positive constants C_R and ω_R such that for all $s < t$ the following estimate*
$$\|U^h(t,s)\|_{\mathcal{L}(H)} \leq C_R e^{-\omega_R(t-s)}$$
holds uniformly in $0 < h < 1$.

PROOF. We begin by defining the energy function associated with the system (4.61). Let
$$E(t) \equiv |M^{1/2}v_t(t)|^2 + |\mathcal{A}^{1/2}v(t)|^2 = |U^h(t,s)(v(s), v_t(s))|_H^2.$$
The standard energy calculations give:
$$(4.62) \qquad E(t) + 2\int_s^t (D^h(\tau)v_t(\tau), v_t(\tau))d\tau = E(s).$$
From (4.54) and the definition of D^h we have
$$(4.63) \qquad (D^h(t)v_t, v_t) \geq k_1 \cdot k \cdot |M^{1/2}v_t|^2.$$
Multiplying equation (4.61) by v and integrating from s to $s+T$ give
$$-\int_s^{T+s} |M^{1/2}v_t|^2 dt + \int_s^{T+s} |\mathcal{A}^{1/2}v|^2 dt + \int_s^{T+s} (D^h(t)v_t, v) dt$$
$$\leq C[E(s) + E(T+s)],$$
and by (4.63), (4.62) we have that
$$\int_s^{T+s} E(t)dt + \sup_{s \leq t \leq T+s} E(t)$$
$$(4.64) \qquad \leq C_1 \int_s^{T+s} (D^h(t)v_t, v_t) dt + C_2 E(T+s) + \int_s^{T+s} |(D^h(t)v_t, v)| dt.$$
Recalling the assumption imposed on D' along with the fact that
$$\int_s^{s+T} \left[(D(u_t(\tau+h)), u_t(\tau+h)) + (D(u_t(\tau)), u_t(\tau))\right] dt \leq C_{R,T},$$
we obtain
$$\int_s^{T+s} |(D^h(t)v_t, v)| dt \leq C_\epsilon \int_s^{T+s} (D^h(t)v_t, v_t) dt + \epsilon \int_s^{T+s} (D^h(t)v, v) dt$$
$$\leq C_\epsilon \int_s^{T+s} (D^h(t)v_t, v_t) dt + C_R \epsilon \sup_{s \leq t \leq T+s} |\mathcal{A}^{1/2}v(t)|^2.$$
Combining the above inequality with (4.64) and selecting ϵ suitably small give
$$(4.65) \qquad \int_s^{T+s} E(t)dt \leq CE(T+s) + C_R \int_s^{T+s} (D^h(t)v_t, v_t) dt.$$

Since by (4.62) $E(t)$ is not increasing, the above inequality implies:
$$TE(T+s) \leq \int_s^{T+s} E(t)dt \leq CE(T+s) + C_R \int_s^{T+s} (D^h(t)v_t, v_t)dt.$$

After taking T large enough and using once more (4.62) yields
$$E(T+s) \leq C_R \int_s^{T+s} (D^h(t)v_t, v_t)dt \leq C_R(E(s) - E(T+s)).$$

From here we find that
$$E(T+s) \leq \frac{C_R}{C_R+1} E(s) \equiv \gamma_R E(s), \quad \text{where} \quad \gamma_R < 1, \ s \in \mathbb{R},$$

and γ_R does not depend on s. Reiterating this argument on subintervals $(mT, (m+1)T)$ and recalling evolution property of the operator $U^h(t,s)$ yield the desired conclusion. □

To continue with the proof of Theorem 4.19 we write

(4.66) $\quad y^h(t) = U^h(t,s)y^h(s) + \int_s^t U^h(t,\tau)(0; M^{-1}F^h(\tau))d\tau, \ t \geq s,$

with $y^h(t) = (u^h(t); u_t^h(t))$. Since $y^h(t)$ belongs to the attractor, for each fixed h the function $|y^h(t)|_H$ is bounded for all $t \in \mathbb{R}$ with the bound possibly depending on h. Sending $s \to -\infty$ in (4.66) and applying Lemma 4.21 we obtain with a fixed h the following representation:

(4.67) $\quad y^h(t) = \int_{-\infty}^t U^h(t,\tau)(0; M^{-1}F^h(\tau))d\tau, \quad t \geq s.$

From condition (4.55) we also obtain
$$|F^h(t)|_{V'} \leq C_R |\mathcal{A}^{1/2-\delta} u^h(t)| \leq \epsilon |\mathcal{A}^{1/2} u^h(t)| + C_{R,\epsilon}|u_h(t)|$$
$$\leq \epsilon |\mathcal{A}^{1/2} u^h(t)| + C_{R,\epsilon},$$

where in the last step we account for the fact that

(4.68) $\quad |u^h(t)| \leq \frac{1}{h} \cdot \int_0^h |u_t(\tau+t)|d\tau \leq C_R, \quad t \in \mathbb{R},$

with uniformity in h. Hence combining with (4.67) and Lemma 4.21

(4.69) $\quad |y^h(t)|_H \leq C_R \int_{-\infty}^t e^{-\omega_R(t-\tau)} |F^h(\tau)|_{V'} d\tau \leq \epsilon \sup_{s \leq t} |y^h(s)|_H + C_{\epsilon,R}.$

Taking ϵ suitably small gives

(4.70) $\quad |y^h(t)|_H \leq C_R \quad \text{for all} \quad t \in \mathbb{R}.$

Passing with the limit on h yields then

(4.71) $\quad |u_{tt}(t)|_V^2 + |\mathcal{A}^{1/2} u_t(t)|^2 \leq C_R, \quad t \in \mathbb{R}.$

It is clear that $v(t) = u_t(t)$ solves the linear problem

(4.72) $\quad Mv_{tt} + \langle D'(u_t(t)), v_t \rangle + \mathcal{A}v = \langle F'(u(t)), v \rangle$

and hence by the standard energy method, Frechet differentiability of F in (4.55) (where we can take $\delta = 0$) and (4.71) we have that

$$\int_0^T (\langle D'(u_t), v_t \rangle, v_t) dt \leq C.$$

Thus $(\langle D'(u_t(t)), v_t(t) \rangle, v_t(t))$ and $(\langle F'(u(t)), v(t) \rangle v_t(t))$ are in $L_1^{loc}(\mathbb{R})$. This implies by standard energy considerations that $E(t)$ is continuous in time. Since (4.71) implies weak continuity in time of (v, v_t) with values in H, we obtain that

$$u_t \in C(\mathbb{R}; \mathcal{D}(\mathcal{A}^{1/2})), \quad u_{tt} \in C(\mathbb{R}; V).$$

In order to obtain the improved regularity for $u(t)$ we return to the original equation, which gives the relation

(4.73) $\qquad |\mathcal{A}u(t)|_{V'} \leq C|u_{tt}|_{V'} + |F(u(t))|_{V'} + k|D(u_t)|_{V'},$

and account for the improved regularity in (4.71) along with the assumptions imposed. $\qquad\square$

The following observation is important in the study of smoothness properties of an attractor in Theorem 4.25 below.

REMARK 4.22. In Theorem 4.19 the we need the continuity and boundedness of the map $D : \mathcal{D}(\mathcal{A}^{1/2}) \to V'$ only to conclude, via "elliptic" regularity argument, that $u \in C_{bnd}(\mathbb{R}; W)$ on the attractor (from (4.71) via (4.73) we have that $|\mathcal{A}u(t)| < C$). This condition imposes certain restrictions on the "growth" of the damping operator. Without this requirement imposed on D we can only conclude that $t \mapsto \mathcal{A}u(t) + kD(u_t(t))$ is bounded and right continuous in V' (see the argument given in the proof of relation (4.47) in Theorem 4.17). In particular, there exits $R > 0$ such that for any trajectory $(u(t); u_t(t))$ from the attractor we have the relation

$$\mathcal{A}u(t) \in W_D(R) \equiv \left\{ w - kD(v) \,:\, w \in V', \, v \in \mathcal{D}(\mathcal{A}^{1/2}), \, |w|_{V'} + |\mathcal{A}^{1/2}v| \leq R \right\}$$

for all $t \in \mathbb{R}$. This implies that, if, for instance, D maps $\mathcal{D}(\mathcal{A}^{1/2})$ into $[\mathcal{D}(\mathcal{A}^{1/2-\delta})]'$, $\delta \geq 0$, and is bounded on bounded sets, then one can still conclude some additional regularity of elements from the global attractor. Namely, in the case when $M = I$ from (4.71) we can conclude that $|\mathcal{A}^{1/2+\delta}u(t)| < C$ on the attractor. In this case the argument given in Theorem 4.19 allow us to state that the attractor A is a bounded set in $\mathcal{D}(\mathcal{A}^{1/2+\delta}) \times \mathcal{D}(\mathcal{A}^{1/2})$.

Assuming more regularity on the damping D one can reiterate the argument of Theorem 4.19 in order to obtain even higher regularity of trajectories lying in the attractor. In fact, this was done in the case of wave equation in [**81**]. However, the method necessitates rather strong limitations imposed on the nonlinear damping. For this reason we shall not pursue this line of argument, but rather limit ourselves to a linear damping where "infinite smoothness" of attractors can be easily deduced as a corollary of Theorem 4.19.

COROLLARY 4.23. *Assume that Assumption 1.1 holds and the system* (H, S_t) *generated by (1.1) possesses a global attractor A. Additionally assume that*

- the damping operator D is a linear positive self-adjoint operator in $\mathcal{H}$ which is bounded from $\mathcal{D}(\mathcal{A}^{1/2})$ into V' and possesses the property

(4.74) $$k_1(Mw, w) \leq (Dw, w) \leq k_2|\mathcal{A}^{1/2}w|^2$$

for any $w \in \mathcal{D}(\mathcal{A}^{1/2})$, where k_1 and k_2 are positive constants.

- Concerning the nonlinear mapping $F(u)$ we assume that $F(u)$ belongs to C^m as a mapping from $\mathcal{D}(\mathcal{A}^{1/2})$ into V' and

(4.75) $$|\langle F^{(k)}(u); w_1, \ldots, w_k\rangle|_{V'} \leq C_\rho \cdot \prod_{j=1}^{k} |\mathcal{A}^{1/2-\delta}w_j|,$$

for every $\rho > 0$, where $k = 0, 1, \ldots, m$, $|\mathcal{A}^{1/2}u| \leq \rho$, $w_j \in \mathcal{D}(\mathcal{A}^{1/2})$ and $\delta \in (0, 1/2]$. Here $F^{(k)}(u)$ is the k-order Frechet derivative of $F(u)$ and $\langle F^{(k)}(u); w_1, \ldots, w_k\rangle$ is the value of $F^{(k)}(u)$ on elements $w_1, \ldots, w_k$.

Then any full trajectory $\{(u(t); u_t(t)) : t \in \mathbb{R}\}$ which belongs to the global attractor possesses the following regularity properties:

(4.76)
$$u^{(k)}(t) \in C(\mathbb{R}; W), \ k = 0, 1, 2, \ldots, m-1,$$
$$u^{(m)}(t) \in C(\mathbb{R}; \mathcal{D}(\mathcal{A}^{1/2})), \quad u^{(m+1)}(t) \in C(\mathbb{R}; V),$$

where $W = \{u \in \mathcal{D}(\mathcal{A}^{1/2}) : \mathcal{A}u \in V'\}$. Here and below $u^{(k)}(t) = \partial_t^k u(t)$. Moreover

(4.77) $$|u^{(k+1)}(t)|_V^2 + |\mathcal{A}^{1/2}u^{(k)}(t)|^2 + |\mathcal{A}u^{(k-1)}(t)|_{V'}^2 \leq R^2, \ t \in \mathbb{R},$$

where $k = 0, 1, \ldots, m$.

REMARK 4.24. The result stated above, in the special case of abstract wave equation with linear scalar damping ($M = D = I$), has been proved in [**57**]. Thus Corollary 4.23 extends this result to more general vectorial structures.

PROOF. By Theorem 4.19 the statement of this corollary is true for $m = 1$.

Now we apply induction argument. Assume that (4.76) and (4.77) holds true for $m = n$. Then $v(t) = u^{(n)}(t)$ satisfies the equation

$$Mv_{tt} + Dv_t + \mathcal{A}v = \Phi(t) \equiv \Phi(v(t); u'(t), \ldots, u^{(n-1)}(t)),$$

where

$$\Phi(v(t); u'(t), \ldots, u^{(n-1)}(t)) = \frac{d^n}{dt^n} F(u(t)).$$

As above the function $v^h(t) = h^{-1}(v(t+h) - v(t))$, $h > 0$, satisfies the equation

$$Mv_{tt}^h + Dv_t^h + \mathcal{A}v^h = \Phi^h(t) \equiv \frac{1}{h}\left(\Phi(t+h) - \Phi(t)\right).$$

Using (4.75) and the induction hypothesis one can show that

$$|\Phi^h(t)|_{V'} \leq \varepsilon |\mathcal{A}^{1/2}v^h|^2 + C_{R,\varepsilon}.$$

Thus the same argument as in Theorem 4.19 allows us to prove (4.76) and (4.77) for $m = n + 1$. □

Under some conditions it is also possible to transfer time regularity of trajectories lying in the attractor into spatial smoothness of its elements. For details we refer to [**57**], where the case $M = D = I$ is discussed.

The most restrictive assumption in Theorem 4.19 is inequality (4.55), which requires that the nonlinear term F is - in some sense - *subcritical*. However, in the

4.2. REGULARITY OF ELEMENTS FROM ATTRACTORS

special case of gradient systems subcriticality restriction can be removed. We have seen this already in Proposition 4.14 and Theorem 4.16. Subcriticality allows (due to compactness of Sobolev's emebeddings) to introduce a small parameter into the estimates controlling principal part of the energy function. On the other hand, the velocity damping may also play a role of "small parameter" - at least for large values of time. Thus, the key idea is that the "damping" implies hidden smoothness and compensated compactness. In fact, this is precisely the key idea behind stabilizability estimate in Proposition 4.14 where the function d_∞ is controlled by the damping in the system according to (4.45). This philosophy leads, in a simple manner, to a powerful criterion for regularity (and also finite dimensionality) of attractors in the "critical" and conservative case. Indeed, theorem stated below, which builds on the results of Theorem 4.16 and Theorem 4.17, recapitulates with an explicit statement conditions sufficient for both regularity and finite-dimensionality.

THEOREM 4.25. *Let Assumption 1.1 with $F^*(u,v) \equiv 0$ and Assumption 3.21 with $H_0(s) = m_0^{-1} s$ in (3.59) and $\kappa = 2$ in (3.60) be in force. Assume $\mathcal{D}(\mathcal{A}^{1/2-\beta}) \subset V$ is compact for some $\beta > 0$ and the nonlinear mapping $F(u) = -\Pi'(u)$ satisfies Assumption 4.11. If the system (H, S_t) generated by (1.1) possesses a compact global attractor A, then the following assertions hold.*

- *The attractor A is a closed bounded set in $\mathcal{D}(\mathcal{A}^{1/2}) \times \mathcal{D}(\mathcal{A}^{1/2})$, more precisely, we have that*

$$(4.78) \quad A \subset \left\{ (u;v) \in \mathcal{D}(\mathcal{A}^{1/2}) \times \mathcal{D}(\mathcal{A}^{1/2}) \;\middle|\; \begin{array}{l} w \equiv \mathcal{A}u + kD(v) \in V', \\ |w|_{V'} + |\mathcal{A}^{1/2}v| + |\mathcal{A}^{1/2}u| \leq R \end{array} \right\}$$

  for some $R > 0$.
- *Any full trajectory $\gamma = \{(u(t); u_t(t)) : t \in \mathbb{R}\}$ lying in the global attractor A satisfies the regularity properties in (4.47) and (4.48).*
- *The attractor A has a finite fractal dimension.*

PROOF. We note that the assumptions imposed by the present theorem imply validity of hypotheses assumed in Proposition 4.14. Indeed, Assumption 4.11 implies the validity of integral relation (4.40) and $F^* \equiv 0$ implies that (4.42) holds true (see (4.45)). Thus by Proposition 4.14 the stabilizability inequality (4.43) holds true. On the other hand, this stabilizability inequality implies finite dimensionality of the attractor via Theorem 4.3. The same inequality proves also regularity of the attractor by the argument used in Theorem 4.17. The quantitative description of the regularity follows from obvious calculations involving differential equation. □

The following corollary gives additional smoothness of the attractor under the same smoothness hypotheses concerning the damping operator as in Theorem 4.19.

COROLLARY 4.26. *In addition to hypotheses of Theorem 4.25 assume that the damping operator D maps $\mathcal{D}(\mathcal{A}^{1/2})$ into V' continuously and is bounded on bounded sets. Then*

- *The attractor A is a closed bounded set in the space $H_1 = W \times \mathcal{D}(\mathcal{A}^{1/2})$, where $W = \{u \in \mathcal{D}(\mathcal{A}^{1/2}) : \mathcal{A}u \in V'\}$.*
- *Assume in addition that (i) $D : \mathcal{D}(\mathcal{A}^{1/2}) \mapsto V'$ is continuous and (ii) D is Frechet differentiable as a mapping from $\mathcal{D}(\mathcal{A}^{1/2})$ into $[\mathcal{D}(\mathcal{A}^{1/2})]'$ and the derivative $D'(u)$ generates a symmetric bilinear form on $\mathcal{D}(\mathcal{A}^{1/2})$ such*

that (4.54) holds. Then any full trajectory $\gamma = \{(u(t); u_t(t)) : t \in \mathbb{R}\}$ lying in the global attractor A satisfies the following regularity properties:

(4.79) $$u(t) \in C(\mathbb{R}; W), \quad u_t(t) \in C(\mathbb{R}; \mathcal{D}(\mathcal{A}^{1/2})), \quad u_{tt}(t) \in C(\mathbb{R}; V),$$

where $W = \{u \in \mathcal{D}(\mathcal{A}^{1/2}) : \mathcal{A}u \in V'\}$. Moreover, there exists $R > 0$ such that

(4.80) $$|u_{tt}(t)|_V^2 + |\mathcal{A}^{1/2} u_t(t)|^2 + |\mathcal{A}u(t)|_{V'}^2 \leq R^2, \ t \in \mathbb{R}.$$

PROOF. The first statement follows from relation (4.78) and from the fact that under additional boundedness imposed on the damping operator D we have that $|D(u_t(t))|_{V'} \leq C$ for all $t \in \mathbb{R}$ on the attractor.

To prove the second statement we need to consider $v = u_t$ as a solution to the non-autonomous problem in (4.72) and apply the same argument as in Theorem 4.19. □

Although the Theorem 4.25 is all that is needed for our purposes, for sake of completeness of presentation we shall present another approach to the study of smoothness and finite dimensionality of attractors, which is rooted in more classical arguments of dynamical system theory originally introduced in [**2**] (see also [**4**]). In fact, a variant of this method was used recently by [**67**] in the study of attractors arising in von Karman evolutions. What we shall present below is an "abstract version" of that approach. The main idea is to take advantage of the structure of the attractor and, in particular, of additional smoothness displayed by the stationary solutions. In that case, the nonlinear term acting on stationary solutions behaves subcritically. Thus, the same method as used in the proof of Theorem 4.19 gives the result which covers nonlinear terms F that are critical. In short, one can relax assumption (4.55) in the case when the system (H, S_t) generated by (1.1) is *gradient* and possesses a *compact* global attractor. The main assumption in this case is relation (4.55), but imposed on the convex hull of the set of equilibria. Precise formulation is given below.

THEOREM 4.27. *Assume that Assumption 1.1 holds with $F^*(u,v) \equiv 0$, $\mathcal{A}^{-1}$ is compact, and the system (H, S_t) generated by (1.1) possesses a compact global attractor A. Additionally, assume that*

- *The damping operator D is a continuous mapping from $\mathcal{D}(\mathcal{A}^{1/2})$ into V' and is bounded on bounded sets.*
- *D is Frechet differentiable as a mapping from $\mathcal{D}(\mathcal{A}^{1/2})$ into $[\mathcal{D}(\mathcal{A}^{1/2})]'$ and the derivative $D'(u)$ generates a symmetric bilinear form on $\mathcal{D}(\mathcal{A}^{1/2})$ such that (4.54) holds.*
- *Concerning the nonlinear mapping $F(u) = -\Pi'(u)$ we assume that*
  (i) *F is a C^1 mapping from $\mathcal{D}(\mathcal{A}^{1/2})$ into V' and*

(4.81) $$|\langle F'(u_1) - F'(u_2), w \rangle|_{V'} \leq C_\rho \cdot |\mathcal{A}^{1/2}(u_1 - u_2)| |\mathcal{A}^{1/2} w|$$

  for every $\rho > 0$, where $|\mathcal{A}^{1/2} u_i| \leq \rho$, $w \in \mathcal{D}(\mathcal{A}^{1/2})$;
  (ii) *we have*

(4.82) $$\sup \{|\langle F'(e), w \rangle|_{V'} : e \in co(\mathcal{N}_*)\} \leq C \cdot |\mathcal{A}^{1/2-\delta} w|,$$

  where $co(\mathcal{N}_*)$ is the convex hull of the set $\mathcal{N}_*$ of stationary solutions, $\mathcal{N}_* = \{v \in \mathcal{D}(\mathcal{A}^{1/2}) : \mathcal{A}v = F(v)\}$;

(iii) *for every fixed $u, u^* \in W = \{u \in \mathcal{D}(\mathcal{A}^{1/2}) : \mathcal{A}u \in V'\}$ there exits a constant $C = C_{u,u^*}$ such that*

(4.83) $$|\langle F'(\lambda u + (1-\lambda)u^*), w\rangle|_{V'} \leq C \cdot |\mathcal{A}^{1/2-\delta}w|$$

for all $w \in \mathcal{D}(\mathcal{A}^{1/2})$ and $0 \leq \lambda \leq 1$.

Then, any full trajectory $\gamma = \{(u(t); u_t(t)) : t \in \mathbb{R}\}$ lying in the global attractor A enjoys the regularity properties in (4.79) and (4.80).

PROOF. Follows the line of argument given in the proof of Theorem 4.19. However we need to overcome some additional difficulties related to the fact that the regularity assumptions imposed on F are weaker -allowing critical exponent in (4.81). This is done by exploiting gradient structure of the system along with the assumed regularity of the set of equilibria -see [2] and most recently [67], where the same idea was applied to some von Karman equations.

Step 1: proof of relation (4.79). This property is implied by the following lemma which establishes the additional smoothness of attractor's trajectories along the negative time scale $(-\infty, T]$. More specifically we have:

LEMMA 4.28. *For any full trajectory $\gamma = \{(u(t); u_t(t)) : t \in \mathbb{R}\}$ which belongs to the global attractor there exists time T_γ such that*

(4.84) $$\begin{cases} u(t) \in C((-\infty, T_\gamma]; W), \quad u_t(t) \in C((-\infty, T_\gamma]; \mathcal{D}(\mathcal{A}^{1/2})), \\ u_{tt}(t) \in C((-\infty, T_\gamma]; V). \end{cases}$$

Moreover, there exists $R_* > 0$ independent of the trajectory γ such that

(4.85) $$|u_{tt}(t)|_V^2 + |\mathcal{A}^{1/2}u_t(t)|^2 + |\mathcal{A}u(t)|_{V'}^2 \leq R_*^2 \quad \text{for all} \quad t \in (-\infty, T_\gamma].$$

PROOF. We use the same idea as in the proof of Theorem 4.19. Let u^h be given by (4.58) and $y^h(t) = (u^h(t); u_t^h(t))$. This function solves (4.67) with $F^h(t)$ given by (4.60).

Now we estimate $F^h(t)$. Since $F^* = 0$, by Theorem 2.28 we have that $A = \mathcal{M}^u(\mathcal{N})$, where $\mathcal{N}$ is the set of equilibria, $\mathcal{N} = \{(v; 0) : v \in \mathcal{N}_*\}$. Thus

(4.86) $$y(t) \equiv (u(t); u_t(t)) \to \mathcal{N} \quad \text{as} \quad t \to -\infty.$$

Therefore, for any $\epsilon > 0$ there exists $T_\epsilon = T_\epsilon(\gamma)$ such that

$$\text{dist}_H(y(t), \mathcal{N}) \leq \epsilon/2 \quad \text{for all} \quad t \in (-\infty, T_\epsilon].$$

This implies that for any $t \in (-\infty, T_\epsilon]$ there exists $\psi^t \in \mathcal{N}_*$ such that

(4.87) $$|\mathcal{A}^{1/2}(u(t) - \psi^t)| \leq \epsilon \quad \text{for all} \quad t \in (-\infty, T_\epsilon].$$

We denote

$$F_\psi^h(t) = \int_0^1 \langle F'(\psi^t + \xi(\psi^{t+h} - \psi^t)), u^h(t)\rangle d\xi.$$

Since $\psi^t + \xi(\psi^{t+h} - \psi^t) \in co(N^*)$, By (4.82) we have that

$$|F_\psi^h(t)|_{V'} \leq C|\mathcal{A}^{1/2-\delta}u^h(t)|, \quad t \in (-\infty, T_\epsilon].$$

Therefore using (4.68) by interpolation we have that

(4.88) $$|F_\psi^h(t)|_{V'} \leq \eta|\mathcal{A}^{1/2}u^h(t)| + C_\eta, \quad t \in (-\infty, T_\epsilon],$$

for every $\eta > 0$. On the other hand, by (4.81) and (4.87) we obtain that
$$|F^h(t) - F^h_\psi(t)|_{V'} \leq C\epsilon |\mathcal{A}^{1/2} u^h(t)|, \quad t \in (-\infty, T_\epsilon].$$
Consequently, using (4.88) we obtain the following estimate with the constants independent on h:
$$|F^h(t)|_{V'} \leq (C\epsilon + \eta)|\mathcal{A}^{1/2} u^h(t)| + C_\eta, \quad t \in (-\infty, T_\epsilon].$$
Taking ϵ and η suitably small and using Lemma 4.21 and (4.67) we can prove that
$$(4.89) \qquad |u^h_t(t)|^2_V + |\mathcal{A}^{1/2} u^h(t)|^2 \leq C \quad \text{for all} \quad t \in (-\infty, T_\gamma], \ 0 < h \leq h_0,$$
where C does not depend on h and on the trajectory γ, and conclude the proof in the same way as in Theorem 4.19. □

In order to extend the smoothness property from "negative" to "positive" times (as claimed in (4.79)) we appeal to standard energy argument. Consider u^h as a solution to problem (4.59) with iniitial data $(u^h(T_\gamma); u^h_t(T_\gamma))$ at at the moment T_γ and consider forward dynamics. For this we apply the energy inequality
$$E_{u^h}(t) \leq E_{u^h}(T_\gamma) + \int_{T_\gamma}^t (F^h(t), u^h_t) d\tau, \quad \text{for all} \quad t \geq T_\gamma,$$
where $E_{u^h}(t) = \frac{1}{2}\left(|M^{1/2} u^h_t|^2 + |\mathcal{A}^{1/2} u^h|^2\right)$. Since $|F^h(t)|_{V'} \leq C_R |\mathcal{A}^{1/2} u^h(t)|$ for all $t \in \mathbb{R}$ and for any trajectory from the attractor, Gronwall's lemma and (4.89) yield
$$E_{u^h}(t) \leq C_1 \exp\{C_R(t - T_\gamma)\} \quad \text{for all} \quad t \geq T_\gamma,$$
where the constants C_1 and C_R does not depend on h. Therefore the same argument as above implies (4.79).

Step 2: proof of relation (4.80). Relation (4.79) implies that the global attractor A is a subset in $H_1 = W \times \mathcal{D}(\mathcal{A}^{1/2})$. However it does not guarantee boundedness of A in H_1. For this we need an additional argument that exploits the compactness of the attractor. This step follows the argument inspired by [**67**]. We start with the following lemma.

LEMMA 4.29. *For any $\epsilon > 0$ there exists C_ϵ such that*
$$(4.90) \quad \sup\{|\langle F'(\lambda u + (1-\lambda)u^*), z\rangle|_{V'} : (u;v), (u^*;v^*) \in A\} \leq \epsilon |\mathcal{A}^{1/2} z| + C_\epsilon |z|$$
for all $z \in \mathcal{D}(\mathcal{A}^{1/2})$ and $0 \leq \lambda \leq 1$.

PROOF. Since A is a compact set in H, for any $\epsilon > 0$ there exists a finite set $\{(\varphi_i; \varphi'_i) : i = 1, 2, \ldots, N\} \subset A \subset H_1$ such that for any $(u; v), (u^*; v^*) \in A$ we can find number i and j with the property
$$|\mathcal{A}^{1/2}(u - \varphi_i)| + |\mathcal{A}^{1/2}(u^* - \varphi_j)| \leq \epsilon$$
In this case, by (4.81) and (4.83) we obtain
$$\begin{aligned}|\langle F'(\lambda u + (1-\lambda)u^*), z\rangle|_{V'} &\leq |\langle F'(\lambda u + (1-\lambda)u^*) - F'(\lambda \varphi_i + (1-\lambda)\varphi_j), z\rangle|_{V'} \\ &\quad + |\langle F'(\lambda \varphi_i + (1-\lambda)\varphi_j), z\rangle|_{V'} \\ &\leq \epsilon \cdot C_R |\mathcal{A}^{1/2} z| + C_{ij} |\mathcal{A}^{1/2 - \delta} z|\end{aligned}$$
This implies that
$$|\langle F'(\lambda u + (1-\lambda)u^*), z\rangle|_{V'} \leq \epsilon \cdot C_R |\mathcal{A}^{1/2} z| + \max_{ij} C_{ij} \cdot |\mathcal{A}^{1/2 - \delta} z|$$

for all $z \in \mathcal{D}(\mathcal{A}^{1/2})$ and $(u;v),(u^*;v^*) \in A$. We note that the constants C_{ij} depend on ϵ. However, using subcriticality $\delta > 0$ we can write
$$\max_{ij} C_{ij} \cdot |\mathcal{A}^{1/2-\delta} z| \leq \eta |\mathcal{A}^{1/2} z| + C_{\epsilon,\eta} |z|, \quad \forall \eta > 0,$$
which inequality leads to (4.90). □

Lemma 4.29 and relation (4.68) implies that
(4.91) $\qquad |F^h(t)|_{V'} \leq \epsilon |\mathcal{A}^{1/2} u^h(t)| + C_\epsilon \quad \text{for all} \quad t \in \mathbb{R},$
for any positive ϵ, where the constant $C_\epsilon > 0$ does not depend on the trajectory $\gamma = \{(u(t); u_t(t)) : t \in \mathbb{R}\}$ from the global attractor. Therefore we can return to relation (4.67) and complete the proof of (4.80) in the same way as in Theorem 4.19. □

4.3. Rate of stabilization to equilibria

In this section we consider properties of global attractors in the case when non-conservative forces are absent ($F^* \equiv 0$ in representation (1.4)), i.e. we deal with the problem
$$\begin{cases} Mu_{tt}(t) + \mathcal{A}u(t) + k \cdot D(u_t(t)) = -\Pi'(u(t)), \\ u|_{t=0} = u_0 \in \mathcal{D}(\mathcal{A}^{1/2}), \ u_t|_{t=0} = u_1 \in V = \mathcal{D}(M^{1/2}). \end{cases}$$
We shall show that in this case the attractor has a regular structure and any trajectory converges to the set $\mathcal{N}$ of equilibria. If $\mathcal{N}$ is finite and hyperbolic in some (weak) sense to be described below, we estimate the rate of this convergence. We note that the results on the rate of stabilization to equilibria which are available in literature rely usually on the study of the linearization near an equilibrium point and the corresponding local unstable manifold (see. e.g., [4]). This approach requires a rather strong hyperbolicity condition which may fail in the case of nonlinear damping. In contrast, our method is completely analytic and covers the case of a damping operator degenerated near zero. We note however that the fact that the energy is non-increasing is critical to our proofs.

We start with the following simple structural property.

PROPOSITION 4.30. *Let Assumption 1.1 hold with $F^* \equiv 0$ in representation (1.4). Assume also that property (3.59) is satisfied for $u = 0$, i.e. there exist a strictly increasing, concave function $H_0 \in C(\mathbb{R}^+)$ with the property $H_0(0) = 0$, such that*
(4.92) $\qquad H_0((D(v),v)) \geq (Mv,v) \quad \text{for any} \quad v \in \mathcal{D}(\mathcal{A}^{1/2}).$
If the dynamical system (H, S_t) generated by (1.1) in $H = \mathcal{D}(\mathcal{A}^{1/2}) \times V$ possesses a compact global attractor A, then
$$A = \mathcal{M}^u(\mathcal{N}),$$
where $\mathcal{M}^u(\mathcal{N})$ is the unstable manifold emanating form the set $\mathcal{N}$ of equilibria for (1.1), $\mathcal{N} = \{(u;0) : \mathcal{A}u = F(u)\}$. Moreover, the statements of Theorem 2.31 and Corollary 2.32 are in force. In particular, every trajectory $S_t y$ converges to the set $\mathcal{N}$ as $t \to \infty$.

PROOF. The energy $\mathcal{E}(u, u_t)$ given by (1.18) satisfies the relation
$$\mathcal{E}(u(t), u_t(t)) + k \int_0^t (Du_t(\tau), u_t(\tau)) d\tau = \mathcal{E}(u_0, u_1)$$

on strong solutions. This implies that

$$\mathcal{E}(u(t), u_t(t)) + k \int_0^t H_0^{-1}\left(|M^{1/2} u_t(\tau)|^2\right) d\tau \leq \mathcal{E}(u_0, u_1). \tag{4.93}$$

Thus after limit transition we obtain that (4.93) remains true for any generalized solution. This property implies that the energy $\mathcal{E}(u_0, u_1)$ is a strict Lyapunov function for (H, S_t) (see Definition 2.26). Therefore we can apply Theorem 2.28, Theorem 2.31 and Corollary 2.32 to conclude the proof. □

The main result of this section read as follows:

THEOREM 4.31. *Assume that Assumption 1.1 holds with $F^* \equiv 0$ in the representation (1.4), $\mathcal{A}^{-1}$ is compact, and the dynamical system (H, S_t) generated by (1.1) possesses a compact global attractor A. Assume additionally that*

- *Relation (4.92) holds and we have the inequality*

$$|(D(v), w)| \leq C_1(r) \cdot (D(v), v) + C_2(r) \cdot |\mathcal{A}^{1/2} w|^2 \tag{4.94}$$

for any $v, w \in \mathcal{D}(\mathcal{A}^{1/2})$ such that $|\mathcal{A}^{1/2} w| + |M^{1/2} v| \leq r$ with arbitrary $r > 0$, where $C_1(r)$ and $C_2(r)$ are non-decreasing functions of r.

- $F(u)$ *is Frechet differentiable and its derivative $F'(u)$ possesses the properties: there exists $\delta > 0$ such that*

$$|\langle F'(u), w \rangle|_{[\mathcal{D}(\mathcal{A}^{1/2})]'} \leq C_R |\mathcal{A}^{1/2 - \delta} w|, \quad w \in \mathcal{D}(\mathcal{A}^{1/2}), \tag{4.95}$$

and

$$|\langle F'(u) - F'(v), w \rangle|_{[\mathcal{D}(\mathcal{A}^{1/2})]'} \leq C_R |\mathcal{A}^{1/2 - \delta}(u - v)| \cdot |\mathcal{A}^{1/2} w| \tag{4.96}$$

for any $w \in \mathcal{D}(\mathcal{A}^{1/2})$ and $u, v \in \mathcal{D}(\mathcal{A}^{1/2})$ such that $|\mathcal{A}^{1/2} u| \leq R$ and $|\mathcal{A}^{1/2} v| \leq R$.

- *Any generalized solution $u(t)$ satisfies the energy inequality*

$$\mathcal{E}(u(t), u_t(t)) + k \int_s^t (D u_t(\tau), u_t(\tau)) d\tau \leq \mathcal{E}(u(s), u_t(s)) \tag{4.97}$$

for any $0 \leq s \leq t < \infty$, where the energy $\mathcal{E}(u, u_t)$ given by (1.18).

- *The set $\mathcal{N}$ of equilibrium points is finite and all equilibria are hyperbolic in the sense that the equation $\mathcal{A} u = \langle F'(w), u \rangle$ has only trivial solution for each $(w; 0) \in \mathcal{N}$.*

Then for any $y \in H$ there exists an equilibrium $y_0 \in \mathcal{N}$ such that

$$|S_t y - y_0|_H^2 \leq C \cdot \sigma\left([t T^{-1}]\right), \quad t > 0, \tag{4.98}$$

where C and $T = T_y$ are positive constants, $[a]$ denotes the integer part of a and $\sigma(t)$ satisfies the following ODE:

$$\frac{d\sigma}{dt} + Q(\sigma) = 0, \quad t > 0, \quad \sigma(0) = C(y, y_0). \tag{4.99}$$

Here $C(y, y_0)$ is a constant depending on y and y_0, $Q(s) = s - (I + G_0)^{-1}(s)$ with $G_0(s) = c_1 (I + H_0)^{-1} (c_2 s)$, where c_1 and c_2 are positive numbers.

In particular, if $H_0(s) = a_0 s$, then for any $y \in H$ there exist $\gamma > 0$, $C > 0$ and an equilibrium $y_0 \in \mathcal{N}$ such that

$$|S_t y - y_0|_H \leq C e^{-\gamma t}, \quad t > 0. \tag{4.100}$$

Proof of this theorem will be given below.

REMARK 4.32. Relation (4.94) holds if we assume that (3.60) in Assumption 3.21 is satisfied for $u = 0$ with $\kappa = 2$. We also note that the substitution $\epsilon \cdot w$ in (4.94) instead of w gives the relation

$$(4.101) \quad |(D(v), w)| \leq C_1(r)\epsilon^{-1} \cdot (D(v), v) + \epsilon \cdot C_2(r) \cdot |\mathcal{A}^{1/2} w|^2, \quad 0 < \epsilon < 1,$$

for any $v, w \in \mathcal{D}(\mathcal{A}^{1/2})$ such that $|\mathcal{A}^{1/2} w| + |M^{1/2} v| \leq r$ with arbitrary $r > 0$.

REMARK 4.33. The standard hyperbolicity conditions (see, e.g., [4]) requires that the spectrum of the linearization of the semiflow S_t around an equilibrium does not intersect the unit circumference in the complex plane. In Theorem 4.31 we do not assume the existence of this linearization. However, if the linearization exists, our hyperbolicity condition is equivalent to the requirement that $\lambda = 1$ does not belong to the spectrum. This allow us to apply Theorem 4.31 in the case of wave equation (1.8) with the damping function $g(s)$ possessing the property $g'(0) = 0$ (see Chapter 5). In this case the linearization of S_t, if it exists, is an unitary operator and, hence, its' spectrum is on the unit circumference.

REMARK 4.34. As one can see from the proof of Theorem 4.31 given below we need the existence of a global attractor (together with finiteness of the set $\mathcal{N}$) only for separating trajectories converging to different equilibria. Therefore we can state Theorem 4.31 in the conditional form: if some trajectory converges to an isolated equilibrium, then it converges with the rate prescribed in (4.98). Moreover, in the case of a single equilibrium ($\mathcal{N}$ is a single point set) there is no need to assume a priori convergence of a trajectory to some attractor. In this case Theorem 4.31 provides us with a rate of global stabilization to an equilibrium which, a posteriori, is a global attractor.

REMARK 4.35. It is clear that the function $Q(s)$ in (4.99) is continuous and possesses the properties (i) $Q(s) > 0$ for $s > 0$ and (ii) $Q(0) = 0$. This implies that any nonzero solution $\sigma(t)$ to problem (4.99) is decreasing and tends to zero when $t \to +\infty$. Moreover the function $\sigma(t)$ can be found as a solution to the functional equation

$$(4.102) \quad \int_{\sigma(t)}^{\sigma(0)} \frac{ds}{Q(s)} = t.$$

Thus the rate of stabilization in (4.98) is determined by the behaviour of the function $Q(s)$ (and hence $H_0(s)$) around zero. For an example, in the case when $H_0(s) \sim c_0 s^\alpha$ as $s \to +0$ with $0 < \alpha < 1$, one can see that $Q(s) \sim c_1 s^{1/\alpha}$ as $s \to +0$. This makes it possible to derive from (4.102) that

$$\sigma(t) \sim c t^{-\alpha/(1-\alpha)} \quad \text{as} \quad t \to +\infty.$$

PROOF OF THEOREM 4.31. By Proposition 4.30 for any $y \in H$ there exists an equilibrium $y_0 \in \mathcal{N}$ such that

$$(4.103) \quad |S_t y - y_0|_H \to 0, \ t \to \infty.$$

Thus we need only prove that $S_t y$ tends to y_0 with the stated rate.

If $S_t y = (u(t); u_t(t))$ and $y_0 = (w; 0)$, then the function $v(t) = u(t) - w$ satisfies the following equation

$$(4.104) \quad M v_{tt} + k D(v_t) + \mathcal{A} v = F(w + v(t)) - F(w), \ t > 0.$$

We first establish some auxiliary inequality for the function v.

LEMMA 4.36. *Under the hypotheses of Theorem 4.31 any (generalized) solution $v(t)$ to problem (4.104) possessing the property*

(4.105) $$\sup_{t\geq 0}|\mathcal{A}^{1/2}v(t)| < R \quad \text{for some} \quad R > 0$$

satisfies the inequality

(4.106) $$T\widetilde{E}(T) + \int_0^T E_0(t)dt \leq C_1(I+\mathcal{H}_0)\left(k^{-1}\left[\widetilde{E}(0) - \widetilde{E}(T)\right]\right) + C_2 \sup_{t\in[0,T]}|v(t)|^2$$

for all $T > 1$ large enough, where the constants C_1 and C_2 may depend on R and T and

$$\widetilde{E}(t) \equiv \widetilde{E}(v(t), v_t(t)) = E_0(v(t), v_t(t)) + \Phi(v(t))$$

with

$$E_0(t) \equiv E_0(v(t), v_t(t)) = \frac{1}{2}\left(|M^{1/2}v_t(t)|^2 + |\mathcal{A}^{1/2}v(t)|^2\right)$$

and

$$\Phi(v) = \Pi(u) - \Pi(w) + (F(w), v) \equiv -\int_0^1 (F(w+zv) - F(w), v)dz.$$

PROOF. It is sufficient to prove (4.106) for a strong solution with the property (4.105). The desired inequality remains true for generalized (weak) solutions because they can be approximate by a sequence of strong solutions.

Multiplying equation (4.104) in $\mathcal{H}$ by v_t we obtain:

(4.107) $$\frac{1}{2} \cdot \frac{d}{dt}\left(|M^{1/2}v_t(t)|^2 + |\mathcal{A}^{1/2}v(t)|^2\right) + k(D(v_t), v_t) = (F(u) - F(w), v_t).$$

It is not difficult to see that

$$(F(u) - F(w), v_t) = (F(u), u_t) - (F(w), v_t) = -\frac{d}{dt}\Phi(v(t)).$$

Consequently from (4.107) we obtain the equality:

(4.108) $$\frac{d}{dt}\widetilde{E}(v(t), v_t(t)) + k(D(v_t), v_t) = 0.$$

This implies that

(4.109) $$\widetilde{E}(v(t), v_t(t)) + k\int_0^t (Dv_t(\tau), v_t(\tau))d\tau = \widetilde{E}(v_0, v_1).$$

In particular we have that $\widetilde{E}(t) \equiv \widetilde{E}(v(t), v_t(t))$ is non-increasing. Moreover, it follows from (4.103) that $\widetilde{E}(t) \geq 0$ for all $t > 0$. It is also clear from (4.95) that

(4.110) $$|\widetilde{E}(v(t), v_t(t)) - E_0(v(t), v_t(t))| = |\Phi(v(t))| \leq \varepsilon|\mathcal{A}^{1/2}v(t)|^2 + C_{R,\varepsilon}|v(t)|^2$$

for any $\varepsilon > 0$.

Multiplying equation (4.104) by v we find

$$\frac{d}{dt}(Mv_t, v) = |M^{1/2}v_t|^2 - |\mathcal{A}^{1/2}v|^2 + (F(v+w) - F(w), v) - k(D(v_t), v).$$

Therefore using (4.95) and also (4.92) and (4.101) one can prove that

$$\int_0^T E_0(t)dt \leq C_1(E_0(0) + E_0(T)) + C_2 \int_0^T (H_0 + I)((Dv_t(t), v_t(t)))\, dt$$
$$+ C_3 \int_0^T |v(t)|^2 dt,$$

where C_1 does not depend on T. We also obviously have that
$$T\widetilde{E}(T) \leq \int_0^T \widetilde{E}(t)dt \leq C_R \int_0^T E_0(t)dt.$$
Thus, using Jensen's inequality we obtain that
$$T\widetilde{E}(T) + \int_0^T E_0(t)dt \leq C_1(E_0(0) + E_0(T))$$
$$+ C_2(I + \mathcal{H}_0)\left(\int_0^T (Dv_t(t), v_t(t))dt\right) + C_3 \int_0^T |v(t)|^2 dt,$$
where $\mathcal{H}_0(s) = TH_0(s/T)$. It follows from (4.110) that
(4.111) $$E_0(t) \leq 2\widetilde{E}(t) + C_R \sup_{t \in [0,T]} |v(t)|^2, \; t \in [0,T].$$
Therefore using the relation (see (4.109))
$$\widetilde{E}(0) = \widetilde{E}(T) + k\int_0^T (Dv_t(t), v_t(t))dt$$
we find that
$$T\widetilde{E}(T) + \int_0^T E_0(t)dt \leq C_1 \widetilde{E}(T)$$
$$+ C_2(I + \mathcal{H}_0)\left(\int_0^T (Dv_t(t), v_t(t))dt\right) + C_3 \sup_{t \in [0,T]} |v(t)|^2.$$
If we choose $T > 2C_1$ we obtain that
(4.112)
$$T\widetilde{E}(T) + \int_0^T E_0(t)dt \leq C_1(I + \mathcal{H}_0)\left(\int_0^T (Dv_t(t), v_t(t))dt\right) + C_2 \sup_{t \in [0,T]} |v(t)|^2.$$
Therefore by (4.109) we obtain (4.106) for every strong solution v to problem (4.104). □

LEMMA 4.37. *Assume that $T > 1$ is such that (4.106) holds. Let $v(t)$ be a solution to (4.104) such that*
(4.113) $$\int_{T-1}^T E_0(v(t), v_t(t))dt \leq \delta \; \text{and} \; \sup_{t \in \mathbb{R}_+} E_0(v(t), v_t(t)) \leq \varrho.$$
Then there exists $\delta_0 > 0$ such that
(4.114) $$\max_{[0,T]} |v(t)|^2 \leq C(I + \mathcal{H}_0)\left(k^{-1}\left[\widetilde{E}(0) - \widetilde{E}(T)\right]\right)$$
for every $0 < \delta \leq \delta_0$ with some constant C depending on δ, ϱ and $T > 1$. Here $\mathcal{H}_0(s) = TH_0(s/T)$ and $\widetilde{E}(t)$ is the same as in Lemma 4.36.

PROOF. Assume that (4.114) is not true. Then for some $\delta > 0$ small enough there exists a sequence of solutions $\{v^n(t)\}$ satisfying (4.113) and such that
(4.115) $$\lim_{n \to \infty} \frac{\max\{|v^n(t)|^2 : t \in [0,T]\}}{(I + \mathcal{H}_0)\left(k^{-1}\left[\widetilde{E}(0) - \widetilde{E}(T)\right]\right)} = \infty.$$

It follows from (4.97) that
$$\int_0^T (D(v_t^n), v_t^n) dt \le k^{-1} \left[\widetilde{E}(0) - \widetilde{E}(T) \right].$$

Therefore we also have that

(4.116) $$\lim_{n \to \infty} \frac{\max\{|v^n(t)|^2 : t \in [0, T]\}}{(I + \mathcal{H}_0) \left(\int_0^T (D(v_t^n), v_t^n) dt \right)} = \infty.$$

By (4.113) we have that $\max\{|v^n(t)|^2 : t \in [0, T]\} < C$ for all $n = 1, 2, \ldots$ and hence

(4.117) $$\lim_{n \to \infty} \left\{ \int_0^T (D(v_t^n), v_t^n) dt + \mathcal{H}_0 \left(\int_0^T (D(v_t^n), v_t^n) dt \right) \right\} = 0.$$

By (4.92) this implies that

(4.118) $$\int_0^T (M v_t^n, v_t^n) dt \le \mathcal{H}_0 \left(\int_0^T (D(v_t^n), v_t^n) dt \right) \to 0.$$

Therefore we can assume that there exist $v^* \in \mathcal{D}(\mathcal{A}^{1/2})$ such that

(4.119) $$(v^n; v_t^n) \to (v^*; 0) \quad \text{*-weakly in } L_\infty(0, T; \mathcal{D}(\mathcal{A}^{1/2}) \times V).$$

Now we prove that $u^* = w + v^* \in \mathcal{D}(\mathcal{A}^{1/2})$ solves the problem $\mathcal{A}u = F(u)$. Indeed, from (4.118) it follows that $v_t^n \to 0$ in $L_2(0, T; V)$. On the other hand, from (4.94) with $w = \varepsilon \cdot \varphi$ we can conclude that

$$\int_0^T |(D(v_t^n), \varphi)| dt \le \frac{C_1}{\varepsilon} \int_0^T (D(v_t^n), v_t^n) dt + C_T \cdot \varepsilon \cdot \sup_{t \in [0,T]} |\mathcal{A}^{1/2} \varphi(t)|^2$$

for any $\varphi(t) \in L_\infty(0, T; \mathcal{D}(\mathcal{A}^{1/2}))$. Thus from (4.118) we infer that

$$\limsup_{n \to \infty} \int_0^T |(D(v_t^n), \varphi)| dt \le C_T \cdot \varepsilon \cdot \sup_{t \in [0,T]} |\mathcal{A}^{1/2} \varphi(t)|^2$$

for any $\varphi(t) \in L_\infty(0, T; \mathcal{D}(\mathcal{A}^{1/2}))$ and $\varepsilon > 0$. This implies that

$$D(v_t^n) \to 0 \quad \text{weakly in } L_1(0, T; [\mathcal{D}(\mathcal{A}^{1/2})]'),$$

which allows us to make the limit transition in (4.104) and to conclude that $\mathcal{A}u^* = F(u^*)$. Our next step is to show that $v^* = 0$. From (4.113) we have that $|\mathcal{A}^{1/2}(u^* - w)|^2 \le 2\delta$. If we choose $\delta_0 > 0$ such that $|\mathcal{A}^{1/2}(w_1 - w_2)|^2 > 2\delta$ for every couple w_1 and w_2 of stationary solutions (we can do it because the set $\mathcal{N}$ is finite), then we can conclude that $u^* = w$ provided $\delta \le \delta_0$. Thus we have $v^* = 0$ in (4.119).

Now we normalize sequence v^n by defining

$$\hat{v}^n \equiv \frac{v^n}{c_n}, \quad c_n = \max\{|v^n(t)| : t \in [0, T]\},$$

where we account only for a suitable subsequence of nonzero terms in c_n (it is clear that $c_n \to 0$ as $n \to \infty$). It follows from (4.116) and (4.118) that

$$\int_0^T |M^{1/2} \hat{v}_t^n(t)|^2 dt \to 0 \text{ as } n \to \infty.$$

Relations (4.106), (4.111) and (4.115) imply that
$$\sup_{t\in[0,T]} \left\{ |M^{1/2}\hat{v}^n_t(t)|^2 + |\mathcal{A}^{1/2}\hat{v}^n(t)|^2 \right\} \leq C, \ n = 1, 2, \ldots.$$
Thus we can suppose that

(4.120) $\quad (\hat{v}^n, \hat{v}^n_t) \to (\hat{v}^*; 0)$ *-weakly in $L_\infty(0, T; \mathcal{D}(\mathcal{A}^{1/2}) \times V)$.

The function $\hat{v}^n$ satisfies the equation

(4.121) $\quad M\hat{v}^n_{tt} + \dfrac{k}{c_n} D(v^n_t) + \mathcal{A}\hat{v}^n = \dfrac{1}{c_n}[F(w + v^n) - F(w)],$

As above, from (4.94) we conclude that
$$\frac{1}{c_n}\int_0^T |(D(v^n_t), \varphi)| dt \leq \frac{C_1}{c_n\varepsilon_n}\int_0^T (D(v^n_t), v^n_t) dt + C_2(T)\frac{\varepsilon_n}{c_n}\sup_{t\in[0,T]} |\mathcal{A}^{1/2}\varphi(t)|^2$$
for any $\varphi(t) \in L_\infty(0, T; \mathcal{D}(\mathcal{A}^{1/2}))$. If we choose $\varepsilon_n \equiv c_n \cdot \varepsilon$, then from (4.116) we infer that
$$\limsup_{n\to\infty} \frac{1}{c_n}\int_0^T |(D(v^n_t), \varphi)| dt \leq C\varepsilon \sup_{t\in[0,T]} |\mathcal{A}^{1/2}\varphi(t)|^2$$
for any $\varphi(t) \in L_\infty(0, T; \mathcal{D}(\mathcal{A}^{1/2}))$ and $\varepsilon > 0$. Thus

(4.122) $\quad \dfrac{1}{c_n}(D(v^n_t)) \to 0$ weakly in $L_1(0, T; [\mathcal{D}(\mathcal{A}^{1/2})]')$.

It also follows from (4.95) and (4.96) that
$$\frac{1}{c_n}[F(w + v^n) - F(w)] \to \langle F'(w), \hat{v}^* \rangle \text{ weakly in } L_2(0, T; [\mathcal{D}(\mathcal{A}^{1/2})]').$$
Therefore, using (4.122) after taking the limits in (4.121) we conclude that $\hat{v}^*$ satisfies $\mathcal{A}\hat{v}^* = \langle F'(w), \hat{v}^* \rangle$ and, by hyperbolicity of w we conclude that $\hat{v}^* = 0$. Thus (4.120) implies that
$$\max_{t\in[0,T]} |\hat{v}^n(t)| \to 0 \text{ as } n \to \infty,$$
which is impossible. $\square$

COMPLETION OF THE PROOF OF THEOREM 4.31. By (4.103) we choose $T_0 > 0$ such that (4.113) holds with $\delta \leq \delta_0$ and $T > T_0$. Therefore Lemma 4.37 and relation (4.106) in Lemma 4.36 imply that

(4.123) $\quad \widetilde{E}(T) \leq C(I + \mathcal{H}_0)\left(k^{-1}\left[\widetilde{E}(0) - \widetilde{E}(T)\right]\right).$

This relation implies that
$$\widetilde{E}((m+1)T) + G_0\left(\widetilde{E}((m+1)T)\right) \leq \widetilde{E}(mT), \ m = 0, 1, 2 \ldots,$$
where $G_0(s) = k(I + \mathcal{H}_0)^{-1}(s/C)$. Therefore Lemma 3.3 [80] implies that

(4.124) $\quad \widetilde{E}(mT) \leq \sigma(m), \ m = 0, 1, 2 \ldots,$

where $\sigma(t)$ solves (4.99) with $\sigma(0) = \widetilde{E}(0)$. Using (4.111) and (4.114) we have
$$E_0(mT) \leq 2\widetilde{E}(mT) + C \sup_{t\in[mT,(m+1)T]} |v(t)|^2 \leq C\widetilde{E}(mT).$$

Therefore (4.124) implies that
$$E_0(mT) \le C\sigma(m), \ m = 0, 1, 2 \ldots$$
Now by Proposition 1.15 we obtain (4.98).

In the case $H_0(s) = a_0 s$ we obviously have that $Q(s) = \alpha s$ for some positive α. Therefore solving (4.99) we obtain (4.100). This completes the proof of Theorem 4.31.

REMARK 4.38. One of crucial hypotheses of Theorem 4.31 is energy inequality (4.97) for generalized solutions. We use it in the proof of Lemma 4.37 to derive (4.116) from (4.115). In general, the validity of (4.97) for all generalized solutions may require some additional assumptions concerning the damping operator D (see, e.g., Proposition 1.12 and Remark 1.13). However in the applications considered in Chap.5 and Chap.7 the requirement in (4.97) can be derived under the same hypotheses which we need for the existence of a compact global attractor (see, e.g., the proof of Theorem 5.10).

4.4. Determining functionals

Unfortunately a detailed study of the structure of the attractor is possible for a rather special class of problems. Because of this it becomes important to search for minimal (or close to minimal) sets of natural parameters of the problem that uniquely determine long-time behaviour of the system. This problem was first discussed by Foias and Prodi [54] and by Ladyzhenskaya [70] for the 2D Navier-Stokes equations. They proved that the long-time behaviour of the solutions is completely determined by dynamics of the first N Fourier modes, if N is sufficiently large. Later similar results were obtained for other parameters and equations (see, for example [37, 52, 55, 56, 72, 93] and the references quoted therein). The concepts of determining nodes [52, 55, 93] and determining local volume averages [56, 64, 65] were introduced. The question on the relation between the problem on existence of a finite number of determining parameters and some problems of interpolation theory has been discussed and a general concept of determining functionals was introduced (see [35, 36]). For further details we refer to the survey [18] and to the references quoted therein (see also [19, Chap.5]).

To characterize asymptotic behaviour of trajectories by means of determining functionals we need the notion of completeness defect.

Let $\mathcal{L} = \{l_j : j = 1, \ldots, N\}$ be a set of functionals on $\mathcal{D}(\mathcal{A}^{1/2})$. The completeness defect of this set with respect to the pair $(\mathcal{D}(\mathcal{A}^{1/2}), \mathcal{H})$ is defined by the formula:

$$\epsilon_\mathcal{L}(\mathcal{D}(\mathcal{A}^{1/2}), \mathcal{H}) = \sup\left\{|u| \ : \ u \in \mathcal{D}(\mathcal{A}^{1/2}), l_j(u) = 0, j = 1, \ldots, N, |\mathcal{A}^{1/2} u| \le 1\right\}.$$

We refer to [19, Chap.5] and [18] for details concerning completeness defect.

The following result is based on the relation of the type (3.165).

THEOREM 4.39. *Assume that the system (H, S_t) generated by (1.1) possesses a positively invariant absorbing set the $\mathcal{B}$ and there exists non-negative functions $a(t)$, $b(t)$ and $c(t)$ on $\mathbb{R}_+$ such that (i) $a(t)$ and $c(t)$ are locally bounded on $[0, \infty)$, (ii) $b(t) \in L_1(\mathbb{R}_+)$ possesses the property $\lim_{t \to \infty} b(t) = 0$ and (iii) for every $y_1, y_2 \in \mathcal{B}$ and $t > 0$ the following relations*

(4.125) $$|S_t y_1 - S_t y_2|_H^2 \le a(t) \cdot |y_1 - y_2|_H^2$$

and
$$|S_t y_1 - S_t y_2|_H^2 \leq b(t) \cdot |y_1 - y_2|_H^2 + c(t) \cdot \sup_{0 \leq s \leq t} |\mathcal{A}^\sigma(u^1(s) - u^2(s))|^2 \qquad (4.126)$$

hold for some $\sigma \in [0, 1/2)$. Here and below we denote $S_t y_i = (u^i(t); u_t^i(t))$, $i = 1, 2$. Let $\mathcal{L} = \{l_j : j = 1, ..., N\}$ be a set of functionals on $\mathcal{D}(\mathcal{A}^{1/2})$ with the completeness defect $\epsilon_\mathcal{L} \equiv \epsilon_\mathcal{L}(\mathcal{D}(\mathcal{A}^{1/2}), \mathcal{H})$. If there exists $\tau > 0$ such that
$$\eta_\tau \equiv b(\tau) + \epsilon_\mathcal{L}^{2-4\sigma} \cdot c(\tau) \cdot \sup_{s \in [0,\tau]} a(s) < 1, \qquad (4.127)$$

then the relation
$$\Delta_\mathcal{L}(t) \equiv \sup_{s \in [t, t+\tau]} \max_j |l_j(u^1(s) - u^2(s))| = 0, \ t \to \infty, \qquad (4.128)$$

implies that $\lim_{t \to \infty} |S_t y_1 - S_t y_2|_H = 0$.

PROOF. Assume that $S_t y_i = (u^i(t), u_t^i(t)) \in \mathcal{B}$ for $t \geq t_0$, $i = 1, 2$. From (4.126) we have that
$$|S_{t+\tau} y_1 - S_{t+\tau} y_2|_H^2 \leq b(\tau) \cdot |S_t y_1 - S_t y_2|_H^2 + c(\tau) \cdot \sup_{t \leq s \leq t+\tau} |\mathcal{A}^\sigma(u^1(s) - u^2(s))|^2$$

for any $t \geq t_0$. One can prove (see [**18**] or [**19**, Chap.5]) that
$$|\mathcal{A}^\sigma v|^2 \leq (1+\delta)\epsilon_\mathcal{L}^{2-4\sigma} |\mathcal{A}^{1/2} v|^2 + C_{\mathcal{L},\delta} \max_{j=1,...,N} |l_j(v)|^2$$

for each $\delta > 0$. Therefore using (4.125) we find that
$$\sup_{t \leq s \leq t+\tau} |\mathcal{A}^\sigma(u^1(s) - u^2(s))|^2 \leq \left[(1+\delta) \epsilon_\mathcal{L}^{2-4\sigma} \sup_{s \in [0,\tau]} a(s) \right] |S_t y_1 - S_t y_2|_H^2 + C_{\mathcal{L},\delta} \Delta_\mathcal{L}^2(t).$$

Consequently
$$|S_{t+\tau} y_1 - S_{t+\tau} y_2|_H^2 \leq \eta |S_t y_1 - S_t y_2|_H^2 + C_{\mathcal{L},\delta} c(\tau) \Delta_\mathcal{L}^2(t),$$

where
$$\eta = b(\tau) + (1+\delta)\epsilon_\mathcal{L}^{2-4\sigma} \cdot c(\tau) \cdot \sup_{s \in [0,\tau]} a(s).$$

Under condition (4.127) we can choose $\delta > 0$ such that $\eta < 1$ and find that
$$|S_{t_0+n\tau} y_1 - S_{t_0+n\tau} y_2|_H^2 \leq \eta^n \cdot |S_{t_0} y_1 - S_{t_0} y_2|_H^2 + C \sum_{m=0}^{n-1} \eta^{n-m-1} \Delta_\mathcal{L}^2(t_0 + m\tau).$$

It is easy to see now that
$$\lim_{n \to \infty} |S_{t_0+n\tau} y_1 - S_{t_0+n\tau} y_2|_H^2 = 0$$

under conditions (4.127) and (4.128). Using (4.125) we then obtain the conclusion. □

Theorem 4.39 implies the following assertion.

COROLLARY 4.40. *Let the nonlinear force $F(u, u_t) \equiv F(u)$ be independent of u_t and Assumption 3.21 be in force with $\kappa = 2$ in (3.60) and $H_0(s) = m_0^{-1} s$ in (3.59). Assume*

- *either the hypotheses of Corollary 3.28;*
- *or the hypotheses of Theorem 3.58;*
- *or else the hypotheses of Theorem 4.16.*

Let (H, S_t) be the system generated by (1.1) and $\mathcal{L} = \{l_j : j = 1, ..., N\}$ be a set of functionals on $\mathcal{D}(\mathcal{A}^{1/2})$ with the completeness defect $\epsilon_\mathcal{L} \equiv \epsilon_\mathcal{L}(\mathcal{D}(\mathcal{A}^{1/2}), \mathcal{H})$. Then there exists $\epsilon_0 > 0$ such that under the condition $\epsilon_\mathcal{L} < \epsilon_0$ the relation
$$\lim_{t \to \infty} l_j(u^1(t) - u^1(t)) = 0, \quad j = 1, 2, \ldots, N,$$
implies that $\lim_{t \to \infty} |S_t y_1 - S_t y_2|_H = 0$. Here $S_t y_i = (u^i(t), u_t^i(t))$, $i = 1, 2$.

PROOF. In the first two cases in follows immediately from the inequality (3.87) with $n_V \equiv 0$ in Remark 3.30 or from Lemma 3.59. In the third case we use Proposition 4.14. □

As we can see from Theorem 4.39 and Corollary 4.40, the smallness of the completeness defect is the main condition on a set of functionals to be determining. We refer to [18] and [19, Chap.5] for examples of sets of functionals with small completeness defect.

We also note that the problem on the existence of determining functionals for second order in time equations with nonlinear damping was considered earlier in [23, 25]. This papers have used another approach.

4.5. Exponential fractal attractors (inertial sets)

In this subsection we present a result on the existence of fractal exponential attractors.

The following concept has been introduced in [39].

DEFINITION 4.41. A compact set $A_{\exp} \subset X$ is said to be *inertial* (or a *fractal exponential attractor*) for the dynamical system (X, S_t) iff A is a positively invariant set of finite fractal dimension and for every bounded set $D \subset X$ there exist positive constants t_D, C_D and γ_D such that
$$(4.129) \quad d_X\{S_t D \,|\, A_{\exp}\} \equiv \sup_{x \in D} \text{dist}_X(S_t x, A_{\exp}) \leq C_D \cdot e^{-\gamma_D(t - t_D)}, \quad t \geq t_D.$$

Using stabilizability estimates and abstract results presented in Chap. 2 we can construct inertial sets for a class of systems subjected to *non-conservative* loads.

THEOREM 4.42. Let the nonlinear force $F(u, u_t) \equiv F(u)$ be independent of u_t. Assume the hypotheses of either Corollary 3.28 or Theorem 3.58 with $\kappa = 2$ in (3.60) and $H_0(s) = m_0^{-1} s$ in (3.59). We also assume that there exist $l \geq 0$ and $0 < \alpha < 1$ such that
$$(4.130) \qquad |\mathcal{A}^{-l} D(v)| \leq C(r) \left[1 + (D(v), v)\right]^\alpha, \quad v \in \mathcal{D}(\mathcal{A}^{1/2}), \ |v|_V \leq r.$$
Then the system (H, S_t) generated by (1.1) possesses a fractal exponential attractor whose dimension is finite in the space $\widetilde{H} = V \times W$, where W is a completion of V with respect to the norm $|\cdot|_W = |\mathcal{A}^{-l} M \cdot |$.

This theorem follows from the inequality in Remark 3.30 or Lemma 3.59 (with $n_V \equiv 0$) and from the following more general assertion.

THEOREM 4.43. Let $\mathcal{A}^{-1}$ be compact. Assume that the dynamical system (H, S_t) generated by equation (1.1) has a positively invariant absorbing set the $\mathcal{B}$ and there exists non-negative functions $a(t)$, $b(t)$ and $c(t)$ on $\mathbb{R}_+$ such that (i) $a(t)$ is locally bounded on $[0, \infty)$, (ii) $b(t) \in L_1(\mathbb{R}_+)$ possesses the properties $\lim_{t \to \infty} b(t) = 0$ and, and (iii) for every $y_1, y_2 \in \mathcal{B}$ and $t > 0$ relations (4.125) and (4.126) hold for some $\sigma \in [0, 1/2)$. We also assume that there exists a space $\widetilde{H} \supseteq H$ such that $t \mapsto S_t y$ is Hölder continuous in $\widetilde{H}$ for every $y \in \mathcal{B}$, i.e. there exist $0 < \gamma \leq 1$ and $C_{\mathcal{B},T} > 0$ such that

(4.131) $\qquad |S_{t_1} y - S_{t_2} y|_{\widetilde{H}} \leq C_{\mathcal{B},T} |t_1 - t_2|^{\gamma}, \quad t_1, t_2 \in [0, T], \ y \in \mathcal{B}.$

Then the dynamical system (H, S_t) possesses a fractal exponential attractor whose dimension is finite in the space $\widetilde{H}$.

PROOF. We apply the same idea as in the proof of Theorem 4.3.

In the space $H_T = H \times W_1(0, T)$ equipped with the norm (4.6), where $W_1(0, T)$ is given by (4.5), we consider the set

$$\mathcal{B}_T := \{ U \equiv (u(0); u_t(0); u(t), t \in [0, T]) \ : \ (u(0); u_t(0)) \in \mathcal{B} \},$$

where $u(t)$ is the solution to problem (1.1) with initial data $(u(0); u_t(0))$, and define operator $V : \mathcal{B}_T \mapsto H_T$ by the formula

$$V \ : \ (u(0); u_t(0); u(t)) \mapsto (u(T); u_t(T); u(T + t)).$$

It is clear that $\mathcal{B}_T$ is a closed bounded set in H_T which is forward invariant with respect to V.

It follows from (4.125) that

$$\|VU_1 - VU_2\|^2_{H_T} \leq \left(a(T) + \int_T^{2T} a(t) dt \right) \|U_1 - U_2\|^2_{H_T}, \quad U_1, U_2 \in \mathcal{B}_T.$$

As in the proof of Theorem 4.3 we can obtain that

$$\|VU_1 - VU_2\|^2_{H_T} \leq \eta^\varepsilon_T \|U_1 - U_2\|^2_{H_T} + K^\varepsilon_T \cdot (n_T^2(U_1 - U_2) + n_T^2(VU_1 - VU_2))$$

for any $U_1, U_2 \in \mathcal{B}_T$, where $K^\varepsilon_T > 0$ is a constant, $n_T(U) := \sup_{0 \leq s \leq T} |u(s)|$ and η^ε_T is given by (4.11). We can choose T and ε such that $\eta^\varepsilon_T < 1$. Therefore by Corollary 2.23 the mapping V possesses a fractal exponential attractor, i.e. there exists a compact set $\mathcal{A}_T \subset \mathcal{B}_T$ and a number $0 < q < 1$ such that $\dim_f^{H_T} \mathcal{A}_T < \infty$, $V \mathcal{A}_T \subset \mathcal{A}_T$ and

(4.132) $\qquad \sup \{ \operatorname{dist}_{H_T}(V^k U, \mathcal{A}_T) \ : \ U \in \mathcal{B}_T \} \leq q^k, \quad k = 1, 2, \ldots.$

In particular, this relation implies that

(4.133) $\qquad \sup \{ \operatorname{dist}_H(S_{kT} y, \mathcal{A}) \ : \ u \in \mathcal{B} \} \leq q^k, \quad k = 1, 2, \ldots,$

where $\mathcal{A}$ is the projection of $\mathcal{A}_T$ of the first two component, i.e.

$$\mathcal{A} = \{ (u(0); u_t(0)) \in \mathcal{B} \ : \ (u(0); u_t(0); u(t), t \in [0, T]) \in \mathcal{A}_T \}.$$

Here $u(t)$ is the solution to problem (1.1) with initial data $(u(0); u_t(0))$. It is clear that $\mathcal{A}$ is a compact forward invariant set with respect to S_T, i.e. $S_T \mathcal{A} \subset \mathcal{A}$. Moreover $\dim_f^H \mathcal{A} \leq \dim_f^{H_T} \mathcal{A}_T < \infty$. Let

$$A_{\exp} = \cup \{ S_t \mathcal{A} \ : \ t \in [0, T] \}.$$

Then $\mathcal{A}$ is a compact forward invariant set with respect to S_t, i.e. $S_t A_{\exp} \subset A_{\exp}$. Using (4.131) one can see that and $\dim_f^{\widetilde{H}} A_{\exp} \leq c \left[1 + \dim_f^H \mathcal{A} \right] < \infty$. We also have from (4.133) that

$$\sup \{ \operatorname{dist}_H(S_t y, A_{\exp}) \ : \ u \in \mathcal{B} \} \leq C e^{-\gamma t}, \quad t \geq 0,$$

for some $\gamma > 0$. Thus $A_{\exp}$ is a fractal exponential attractor. $\square$

PROOF OF THEOREM 4.42. It follows from (4.130) that
$$\int_0^T |\mathcal{A}^{-l}Mu_{tt}(\tau)|^{1/\alpha}d\tau \leq C_\mathcal{B}(T)$$
for any strong solution $u(t)$ from the absorbing set $\mathcal{B}$. This implies (4.131) with $\gamma = 1 - \alpha$ in the space $\widetilde{H} = V \times W$. □

Another application of Theorem 4.43 is the following assertion.

THEOREM 4.44. *Let the hypotheses of Theorem 4.16 be in force. If relation (4.130) holds, then the system (H, S_t) generated by (1.1) possesses a fractal exponential attractor whose dimension is finite in the space $\widetilde{H}$ defined in the statement of Theorem 4.42.*

PROOF. We apply Proposition 4.14 to reduce the case to Theorem 4.43. □

We note that the existence and properties of a fractal exponential attractor for wave equations with *linear damping* was studied earlier in [**41, 44, 45**]. Thus, again, our results in Theorem 4.42 and Theorem 4.44 provide extensions of the respective results to problems with *nonlinear* dissipation.

CHAPTER 5

Semilinear wave equation with a nonlinear dissipation

5.1. The model

Let $\Omega \subset \mathbb{R}^n$, $n = 2, 3$, be a bounded, connected domain with a sufficiently smooth boundary Γ. The exterior normal on Γ is denoted by ν. We consider the following wave equation

(5.1) $\qquad w_{tt} - \Delta w + kg(w_t) + f(w) = g^*(w_t) \quad \text{in} \quad Q = [0, \infty) \times \Omega$

subject to boundary condition either of Dirichlet type

(5.2) $\qquad w = 0 \quad \text{on} \quad \Sigma \equiv [0, \infty) \times \Gamma,$

or else of Robin type

(5.3) $\qquad \partial_\nu w + w = 0 \quad \text{on} \quad \Sigma.$

The initial conditions are given by $w(0) = w_0$ and $w_t(0) = w_1$. We shall consider the dynamics in a standard finite energy space $H = \mathcal{V} \times L_2(\Omega)$, where $\mathcal{V} = H_0^1(\Omega)$ in the case of the Dirichlet boundary conditions and $\mathcal{V} = H^1(\Omega)$ in the Robin case. We assume that k is a positive parameter and impose the following standing assumptions on the nonlinear damping $g(s)$, the anti-damping $g^*(s)$ and the function $f(s)$.

ASSUMPTION 5.1.
- $g \in C^1(\mathbb{R})$ is a monotone increasing function such that $g(0) = 0$, and there exist two positive constants m_1 and m_2 such that

(5.4) $\qquad m_1 \leq g'(s) \leq m_2 |s|^{p-1} \quad \text{for all} \quad |s| \geq 1,$

where $1 \leq p \leq 5$ when $n = 3$ and $1 \leq p < \infty$ when $n = 2$.
- $g^* : L_2(\Omega) \mapsto L_2(\Omega)$ possesses the properties

$$\|g^*(v_1) - g^*(v_2)\|_{L_2(\Omega)} \leq C \|K(v_1 - v_2)\|_{L_2(\Omega)}, \quad v_1, v_2 \in L_2(\Omega),$$

where $K : L_2(\Omega) \mapsto L_2(\Omega)$ is a compact operator, and

$$\int_\Omega g^*(w_t) w_t dx \leq \frac{1}{2} k \int_\Omega g(w_t) w_t dx + C_0$$

for some constant C_0.
- Function $f \in C^2(\mathbb{R})$ is of the following polynomial growth condition: there exists a positive constant $M > 0$ such that

(5.5) $\qquad |f''(s)| \leq M|s|^{q-1}, \; |s| \geq 1,$

where $q \leq 2$ when $n = 3$ and $q < \infty$ when $n = 2$. Moreover, the following dissipativity condition holds:

(5.6) $$\liminf_{|s| \to \infty} \frac{f(s)}{s} \equiv \mu > -\lambda_1,$$

where $\lambda_1 > 0$ is the first eigenvalue of the operator $-\Delta$ equipped with appropriate (Dirichlet or Robin) boundary conditions.

REMARK 5.2.
- We note that assumptions imposed on f and g allow for critical exponents of semilinear function f and strong nonlinearity of the damping which, in addition, may not be quantified at the origin. In fact, it is known [62] that for $n = 3$ the maximal power for the growth of nonlinearity which still allows the boundedness (in time) of the energy for a simple damped wave equation with $L_\infty(0,T; L_2(\Omega))$ forcing term is $p = 5$. Thus, the problem under consideration includes the "double critical exponent" - $(q,p) = (2,5)$. Treatment of this case is more delicate and requires special arguments.
- The presence of the "anti-damping" term g^* reflects possibility of including non-conservative forces depending on the velocity. These forces may tend to destabilize the dynamics. In fact, with an anti-damping present, the energy may not be decreasing (note that the last term in the energy relation (5.7) below has uncontrolled sign). An example of non-conservative term g^* can be given by taking $[g^*(v)](x) \equiv A(x) \int_\Omega v(y) B(y) dy$, where $A, B \in L_2(\Omega)$. The assumption imposed on g^* is satisfied provided that A and B are of suitable size or the damping $g(w_t)$ is minimally superlinear at infinity, i.e. $sg(s) \geq s^l$ with $l > 2$ for $|s|$ large enough.

By Theorem 1.5 under Assumption 5.1 problem (5.1) with the boundary condition either (5.2) or (5.3) has unique generalized solution for any initial data (w_0, w_1) from H (see the considerations in Example 1.2). Thus problem (5.1) generates a dynamical system (H, S_t) in the space $H = \mathcal{V} \times L_2(\Omega)$, where either $\mathcal{V} = H_0^1(\Omega)$ or $\mathcal{V} = H^1(\Omega)$ depending on the boundary conditions. The corresponding evolution operator S_t is given by the formula $S_t(w_0; w_1) = (w(t); w_t(t))$, where $w(t)$ is solves (5.1) with the initial data $(w_0; w_1)$. We also note that the energy function associate with (5.1) and (5.2) takes the form

$$\mathcal{E}(t) \equiv \mathcal{E}_D(w(t); w_t(t)) = \int_\Omega \left[\frac{1}{2} |\nabla w(t,x)|^2 + \frac{1}{2} |w_t(t,x)|^2 + \hat{f}(w(t,x)) \right] dx,$$

where $\hat{f}$ denotes antiderivative of f, and by Theorem 1.5 strong solutions to (5.1) satisfy the energy relation:

(5.7)
$$\mathcal{E}(t) + \int_0^t \int_\Omega g(w_t(\tau,x)) w_t(\tau,x) dx d\tau = \mathcal{E}(0) + \int_0^t \int_\Omega g^*(w_t)(\tau,x) w_t(\tau,x) dx d\tau.$$

A similar relation holds for the case of Robin type boundary condition (5.3) with the energy functional

$$\mathcal{E}(t) \equiv \mathcal{E}_R(w(t); w_t(t)) = \mathcal{E}_D(w(t); w_t(t)) + \frac{1}{2} \int_\Gamma |w(t,x)|^2 ds.$$

5.2. Main results

We begin with a preliminary result which asserts an existence of a global attractor associated with the semiflow S_t generated by (5.1). We recall that a closed bounded set A in H is a global attractor for a dynamical system (H, S_t) if A is strictly invariant and uniformly attracts every bounded set from H (see Definition 2.2). We first treat the case when nonlinearity is subcritical or else the damping is large.

THEOREM 5.3 (**Compact attractors**). *Assume that Assumption 5.1 holds. Then equation (5.1) generates a continuous semiflow S_t in the space $H = \mathcal{V} \times L_2(\Omega)$ by the formula $S_t(w_0; w_1) = (w(t); w_t(t))$, where $w(t)$ is a generalized solution to (5.1) with the initial data $(w_0; w_1)$. Moreover, in the case $n = 2$ and $n = 3, q < 2$ the dynamical system (H, S_t) possesses global compact attractor A. When $n = 3$, $q = 2$ and $g^* \equiv 0$ the same conclusion holds true under the additional assumptions that (i) $g'(s) \geq m_1$ for all $s \in \mathbb{R}$; (ii) the damping parameter k is sufficiently large; and (iii) $p > 5$.*

The theorem stated above is an application of Corollary 3.28 - in subcritical case, and Theorem 3.58 - in the critical case when the damping parameter k is sufficiently large. In the critical case, $n = 3$ and $q = 2$, and *without assuming large values for the damping k*, one can still establish existence of global attractor under some additional assumptions imposed on the the function g and g^*. The additional requirements depend whether damping parameter p is "critical" ($p = 5$) or "subcritical" ($p < 5$). In short, the following three scenarios will be considered below: (i) the nonlinearity of the damping is restricted to $p < 5$, (ii) when $p = 5$ additional bound $g'(s) \leq [1 + sg(s)]^{\frac{2}{3}}$ is required, (iii) the anti-damping term is absent, i.e. $g^* \equiv 0$. The corresponding results are formulated below:

THEOREM 5.4 (**Compact attractors, critical case revisited**). *Let $n = 3$ and $q = 2$. Assume that Assumption 5.1 is in force. In addition, we assume that one of the following conditions is satisfied:*

- *$p < 5$ and $g^* : L_2(\Omega) \to H^\epsilon(\Omega)$ is bounded for some $\epsilon > 0$;*
- *$p = 5$ and $g^* : L_2(\Omega) \to H^\epsilon(\Omega)$ is bounded for some $\epsilon > 0$, and*

(5.8) $$g'(s) \leq [1 + sg(s)]^{2/3} \quad \text{for all} \quad s \in \mathbb{R};$$

- *$g^* \equiv 0$.*

Then the dynamical system (H, S_t) represented by (5.1) with either Dirichlet boundary conditions (5.2) or Robin (5.3) boundary conditions possesses a compact global attractor.

Theorem 5.4 is an application of abstract Theorems 3.34, 3.40 and 3.47. In the fully critical case - $p = 5$ and $q = 2$ - it gives a stronger assertion in comparison with Theorem 5.3. Indeed, there is no requirement that the damping parameter should be large. We also refer to Remark 5.6 below for some comments concerning relation (5.8).

An application of Theorems 4.17, 4.27 and 4.25 makes it possible to obtain the following result on smoothness of elements from the global attractor.

THEOREM 5.5 (**Regularity of attractor**). *Let Assumption 5.1 be in force, $g^* \equiv 0$ and $g'(s) \geq m > 0$ for all $s \in \mathbb{R}$. We distinguish cases $n = 2$ and $n = 3$.*

- **[i]** Let $n = 2$ and assume that

(5.9) $\qquad g'(s) \leq C(1 + sg(s))^\beta \quad \text{for all} \quad s \in \mathbb{R} \text{ and for some } \beta < 1,$

- **[ii]** Let $n = 3$, $1 \leq p \leq 3$, $0 \leq q \leq 2$, and relation (5.8) holds.

Then the global attractor A given by Theorem 5.3 or by Theorem 5.4 is a closed bounded set in the space $H_1 = \mathcal{W} \times \mathcal{V}$, where $\mathcal{W} = (H^2 \cap H_0^1)(\Omega)$, $\mathcal{V} = H_0^1(\Omega)$ in the Dirichlet case and $\mathcal{W} = \{u \in H^2(\Omega) : \partial_\nu u + u = 0 \text{ on } \Gamma\}$, $\mathcal{V} = H^1(\Omega)$ in the Robin case.

- **[iii]** If $n = 3$, $3 < p \leq 5$, $0 \leq q \leq 2$, and (5.8) holds, then A is a bounded set in $W_{6/p}^2(\Omega) \times \mathcal{V}$, where $W_{6/p}^2(\Omega)$ is the $L_{6/p}$-based second order Sobolev space. In partucular, since $W_{6/p}^2(\Omega) \subset H^{1+\delta}(\Omega)$ for $\delta = (5-p)/2$, we have that A is bounded in $H^{1+\delta}(\Omega) \times \mathcal{V}$ with $\delta = (5-p)/2$.

REMARK 5.6. We note that the property (5.9) holds true, if we assume that either $p < 3$ or else $p \geq 3$ and $g_i(s)s \geq m|s|^l - m_1$ for all $s \in \mathbb{R}$ and for some $m_1 \in \mathbb{R}$ and $l > p - 1$. As for (5.8), we can guarantee this requirement, if either $p \leq 7/3$ or else $p > 7/3$ and there exists $l \in \left[\frac{3}{2}(p-1), p+1\right]$ such that $g(s)s \geq m|s|^l - m_1$, for all $s \in \mathbb{R}$ and for some $m_1 \in \mathbb{R}$.

REMARK 5.7. In reference to points (i) and (ii) in Theorem 5.5, one can obtain higher regularity of the attractor by assuming higher order differentiability of the damping g and of the source f. This can be accomplished by reiterating the argument used for the proof of $H^2 \times H^1$ regularity [**81**] (see also the idea presented in Corollary 4.23 in Chapter 4).

We turn next to finite-dimensionality of attractors and their properties. The corresponding results are formulated below.

THEOREM 5.8 (**Finite dimensionality**). *Let Assumption 5.1 be in force and $g'(s) \geq m > 0$ for $s \in \mathbb{R}$. The cases $g^* = 0$ and $g^* \neq 0$ cases are distinguished.*

- **[i]** $g^* \equiv 0$: *for $n = 2$ (resp. $n = 3$) relation (5.9) (resp. (5.8)) is assumed;*
- **[ii]** $g^* \not\equiv 0$: *in addition to part* **[i]** *above we assume that $p \leq 2$ and, in the case $n = 3$, also $q < 2$.*

Then, the global attractor A has a finite fractal dimension.

REMARK 5.9. Under Assumption 5.1 and with $g^* \equiv 0$ the dynamical system (H, S_t) generated by (5.1) possesses a strict Lyapunov function (see Definition 2.26):

$$\Phi(y) = \mathcal{E}(u_0, u_1) = \frac{1}{2} \int_\Omega \left(|u_1(x)|^2 + |\nabla u_0(x)|^2 \right) dx + \int_\Omega \hat{f}(u_0(x)) dx,$$

where $y = (u_0, u_1) \in H$ and $\hat{f}(u) = \int_0^u f(\xi) d\xi$ is an antiderivative for f. The formula above pertains to Dirichlet boundary conditions. In the Robin case one has an obvious modification,

$$\Phi(y) = \mathcal{E}(u_0, u_1) = \frac{1}{2} \int_\Omega \left(|u_1(x)|^2 + |\nabla u_0(x)|^2 \right) dx + \frac{1}{2} \int_\Gamma |u_0|^2 ds + \int_\Omega \hat{f}(u_0(x)) dx.$$

By Theorem 2.28 this implies that the global attractor A has the form

$$A = \mathcal{M}^u(\mathcal{N}),$$

where $\mathcal{M}^u(\mathcal{N})$ is the unstable manifold emanating from the set $\mathcal{N}$ (see the definition in Sect. 2.4) and $\mathcal{N}$ is the set of equilibria for the semiflow S_t (every point $y \in \mathcal{N}$

has the form $y = (u, 0)$, where u solves the elliptic equation $-\Delta u + f(u) = 0$ with the appropriate boundary conditions). Moreover, the statements of Theorem 2.31 and Corollary 2.32 are in force. In particular, every trajectory $\gamma = \{S_t y\}$ of (H, S_t) stabilizes to the set $\mathcal{N}$ of stationary points and, if $\mathcal{N}$ is finite, for each y there exists an equilibrium $w \in \mathcal{N}$ such that $S_t y \to w$ as $t \to \infty$.

Our next result deals with the rate of convergence to equilibrium. For this we introduce the following notation. Let $H_0(s)$ be a continuous, concave increasing function, zero at the origin and such that: $s^2 \leq H_0(sg(s))$, $|s| \leq 1$. Such function can be always constructed due to monotonicity assumption imposed on g [80]. The role of H_0 is to capture the behavior of g at the origin, which is critical in describing the rates of convergence of solutions to an equilibrium.

THEOREM 5.10 (**Decay rates to equilibrium**). *We assume Assumption 5.1 and also (i) $g^* = 0$, (ii) for $n = 2$ (resp. $n = 3$) relation (5.9) (resp. (5.8)) holds, and (iii) the equilibria are isolated and hyperbolic in the sense that the problem*
$$-\Delta u + f'(w)u = 0$$
with the corresponding boundary conditions has only trivial solution for every stationary solution w to (5.1). Then, for every $y \in H$ there exist an equilibrium $y_0 = (w; 0) \in H$ such that
$$|S_t y - y_0|_H \leq C\sigma\left([t/T]\right),$$
where C and $T = T_y$ are positive constants, $[a]$ is the integer part of a and $\sigma(t) \to 0$, $t \to \infty$, satisfies the following nonlinear ODE:

(5.10) $$\sigma_t + Q(\sigma) = 0, \; t > 0, \; \sigma(0) = \sigma_0.$$

Here $Q(s) = s - c_1^{-1}[I + c_2[I + H_0]^{-1}]^{-1}(c_1 s)$, where c_1 and c_2 are positive constants, and σ_0 depends on y.

REMARK 5.11. If $g'(s) \geq m$, $s \in \mathbb{R}$, then $H_0(s) = m^{-1}s$ and $Q(s) = \alpha s$, where $\alpha > 0$. Therefore the above decay rates are exponential. Otherwise, the rates are described by the above nonlinear ODE (5.10) which depends on the characteristics of g at the origin [80] (see also Remark 4.35).

REMARK 5.12. Under the conditions of Theorem 5.8 with $g^* \equiv 0$ we can also apply Theorem 4.44 and conclude the existence of fractal exponential attractor, i.e. a forward invariant finite dimensional exponentially attracting compact set (see Definition 4.41).

The following theorem shows that problem (5.1) possesses a finite number of determining functionals.

THEOREM 5.13 (**Determining functionals**). *Assume that the hypotheses of Theorem 5.8 hold and $g^* \equiv 0$. Let $\mathcal{L} = \{l_j\}$ be a set of linear functionals on $H^1(\Omega)$ with the completeness defect $\epsilon_\mathcal{L} \equiv \epsilon_\mathcal{L}(H^1(\Omega), L_2(\Omega))$ with respect to the pair $(H^1(\Omega), L_2(\Omega))$ (for the definition see [18], or [19], or Sect.4.4 in Chapter 4). Then there exists $\epsilon_0 > 0$ such that under the condition $\epsilon_\mathcal{L} < \epsilon_0$ the property*
$$l_j(u^1(t) - u^2(t)) \to 0, \; t \to \infty, \quad l_j \in \mathcal{L},$$
for two generalized solutions $u^1(t)$ and $u^2(t)$ to problem (5.1) with the corresponding boundary conditions implies that
$$\lim_{t \to \infty} \left\{ \|u_t^1(t) - u_t^2(t)\|_{L_2(\Omega)}^2 + \|u^1(t) - u^2(t)\|_{H^1(\Omega)}^2 \right\} = 0.$$

This theorem follows from Theorem 4.39. As examples of set $\mathcal{L}$ of functionals with small completeness defect $\epsilon_{\mathcal{L}}(H^1(\Omega), L_2(\Omega))$ we can consider local volume averages and modes (for details and further discussion we refer to [18, 19] and also to [36, 65] and the references therein).

5.2.1. Discussion of the results. We conclude this section with few historical remarks. The problem of existence of attractors for semilinear wave equations has attracted considerable attention in the literature [1, 4, 58, 59, 60, 61, 90, 92] and references therein. An excellent review of results pertaining to wave equations with *linear damping* is given in [5]. As said before, our main emphasis is on nonlinear dissipation. The nonlinearity of the damping in hyperbolic dynamics leads to significantly more difficult class of problems and, consequently, the literature is rather scarce. First pioneering papers include [48, 49, 50, 60, 61, 90] where an existence of global attractors under various hypotheses imposed on nonlinear damping and forcing terms is established. In [90] global attractors are established in the critical case subject to restriction $m_1 < g'(s) \leq [1 + |s|^{2/3}]$ with m_1 sufficiently large. In [51] the restriction imposed on large damping has been removed, but linear growth condition on g is imposed. In [50] global attractors with $n = 3$, $p < 5$ and supercritical nonlinear terms f were studied. The conditions assumed in [50] include strong coercivity condition imposed on g which is correlated to the growth conditions imposed on f. Thus, this is a very special class of problems. We also mention that the note [48] claims the result on the existence of the global attractor in the case of large damping, critical exponents and decreasing energy (i.e. $g^* = 0$). The result presented in [48] is close to a special case of ours Theorem 5.3 (critical case with large damping and with $g^* = 0$). However the detailed proof of this result, to our best knowledge, was not published.

Theorem 5.3, which is derived from abstract theory presented in Chapter 3, allows for the maximal values of polynomial powers p associated with dissipation $g(s)$ and also for critical exponents of nonlinear term f. This is the most demanding case with the so called "double critical exponents" (i.e. criticality in the source f ($q = 2$) and the damping g ($p = 5$). In the three-dimensional case $p \leq 5$ is a very natural restriction. Indeed, this is in view of the fact that the system is dissipative for $p \leq 5$ [62]. We also note that Theorem 5.3 does not require any compatibility between the growth of f and growth of g. In the double critical case ($n = 3$, $q = 2$, $p \leq 5$), Theorem 5.4 asserts an existence of global attractors without assuming *large damping*. In addition, it applies to non-dissipative waves ($g^* \neq 0$). This result may be contrasted with [51] where a *linear growth on g is assumed* and also [90] where both large damping and $p \leq 5/3$ are required. We also mention the recent paper [99] where subcritical ($p < 5$) case was considered. Thus, to our best knowledge, the result stated in Theorem 5.4 is the first one pertaining to *critical* case with a *damping not necessarily linearly bounded* and without necessity of assuming large values of the damping parameter.

Questions related to finite dimensionality, regularity and structure of attractors are technically more demanding than just existence of attractors. As recognized in the literature, finite-dimensionality of attractors in hyperbolic dyanamics is a very delicate issue. The flow is not C^1 on the phase space and the damping operator $g(u_t)$ is not Frechet differentiable. Thus, the standard methods do not apply. First result in this direction is in [49], where the problem was settled in the one-dimensional case. More recently in [81] the regularity of attractors and their finite dimensionality is shown for the two-dimensional case. Instead, in [89] wave equation is studied in *subcritical case only* in dimensions two and three. Finite dimensionality of attractors is shown in [89] in the subcritical case and under suitable conditions imposed on nonlinear functions g that control the growth of function from above and below. More specifically it is assumed in [89]:

$$m_1(1+|s|)^{p-1} \leq g'(s) \leq m(1+|s|)^{p-1}, \tag{5.11}$$

where p satisfies $0 \leq p < 3$ for $n = 2$ and $0 \leq p < 2\frac{1}{3}$ for $n = 3$.

Our contribution with respect to properties of attractors such as regularity and finite-dimensionality is presented by Theorem 5.5 and Theorem 5.8. The results presented there (i) deal with double critical cases and non-monotone energy, (ii) extend the range of nonlinearity, measured by parameters q and p, for which one can still claim finite dimensionality and additional regularity of the attractor. Indeed, by comparing results of Theorem 5.8 with those cited above we see that for $n = 2$ Theorem 5.8 allows to consider full range of $0 < p < \infty$ (see (5.9) and Remark 5.6). Moreover for $p < 3$ there is no need for superlinear coercivity condition. Indeed, (5.9) is satisfied for $p \geq 3$ under the coercivity condition in Remark 5.6 which is significantly milder than in (5.11). Similarly, by Remark 5.6 for $n = 3$ Theorem 5.8 allows for $p \leq 5$ with no superlinear coercivity for $p \leq 2\frac{1}{3}$. In conclusion, Theorem 5.8 and Theorem 5.5 provide statements on finite dimensionality and regularity of attractors under much weaker hypotheses imposed on *nonlinear terms than previously assumed in the literature*. In particular, "double critical" exponents (in the source and the damping) and non-monotone structures are allowed.

Theorem 5.10 describes the rates of convergence to equilibria. This result is, to best our knowledge, new in the literature in the case of *nonlinear damping*. For *linear damping*, i.e. when $g(s) = g_0 \cdot s$ is a linear function, the result given by Theorem 5.10 was obtained earlier in [4] by another method, which does not apply to nonlinear damping. The approach in [4] relies on geometrical consideration in the neighborhood of a stationary point. We also note that the existence of fractal exponential attractors was proved in [41] for the case of *linear damping* only. Thus, Remark 5.12 provides the first (to our best knowledge) extension of this property to the case of *nonlinear damping*.

Finally, concerning Theorem 5.13 on determining functionals we note that this result was proved in [18] (see also [19, Chap.4]) in the case of linear damping. For nonlinear damping the problem was considered in [23] and [25] by another method.

REMARK 5.14. We want to emphasize that all results stated above concerning wave equation (5.1) are *direct* corollaries of the abstract theory developed in the previous chapters. Involving more deeply the structure of the problem and relying on methods developed in this monograph one can improve (or suggest other versions of) the result concerning wave dynamics. As examples we point out the paper [9] where some results on attractors are established for the case of Neumann boundary conditions as a byproduct in the study of a coupled wave–plate structure, the paper [32] devoted to a wave dynamics with nonlinear interior/boundary dissipation, and also the paper [33] which deals with localized interior damping and a source term of critical exponent.

5.3. Proofs

5.3.1. Abstract setting for the wave equation. The system generated by (5.1) is a special case of evolutionary system (1.1). To see this we set

- $\mathcal{H} = L_2(\Omega)$, $\mathcal{A} = -\Delta - \mu_0$ with $\mathcal{D}(\mathcal{A}) \equiv H_0^1(\Omega) \cap H^2(\Omega)$ in the Dirichlet case, and $\mathcal{D}(\mathcal{A}) \equiv \{u \in H^2(\Omega); \frac{\partial}{\partial \nu}u + u = 0 \text{ on } \Gamma\}$ in the Robin case. The parameter μ_0 is chosen such that $-\mu < \mu_0 < \lambda_1$, where μ and λ_1 are the same as in (5.6).
- $V = \mathcal{H}$, $M = I$.
- $D(u) = g(u)$ and $F(u,v) = -f(u) - \mu_0 u - g^*(v)$.

The analysis given in Example 1.2 shows that the damping operator $D(v)$ and the operator $F(u)$ satisfy all the hypotheses in Assumption 1.1 with

$$\Pi_0(u) \equiv \int_\Omega \hat{f}(u)dx + M_f, \quad \Pi_1(u) = -M_f, \quad F^*(u,v) \equiv g^*(v),$$

where $\hat{f}$ denotes the antiderivative of $f(u) + \mu_0 u$ and M_f can be selected due to dissipativity condition (5.6) in Assumption 5.1. Thus (see Theorem 1.5) under the assumptions imposed the equation (5.1) generates dynamical system S_t on $H \equiv H_0^1(\Omega) \times L_2(\Omega)$ - in the the Dirichlet case - and $H \equiv H^1(\Omega) \times L_2(\Omega)$ - in the Robin case. We also have topological identifications $\mathcal{D}(\mathcal{A}^{1/2}) \sim H_0^1(\Omega)$ in Dirichlet case and $\mathcal{D}(\mathcal{A}^{1/2}) \sim H^1(\Omega)$ in Robin case.

5.3.2. Proof of Theorem 5.3. Theorem 5.3 is an application of abstract results stated in Corollary 3.28 and Theorem 3.58. To see this, we need to verify the dissipativity of the system and also the assumptions of Theorem 3.58 -for the case $n = 3, q = 2$ and of Corollary 3.28 -for the remaining cases.

Step 1: Ultimate dissipativity of the system, including the control of the size of absorbing set, follows from abstract Theorem 3.4, in the conservative case $g^* = 0$ and from Corollary 3.17 in the non-conservative case with anti-damping present. To see this, we begin with the first case $g^* = 0$. For this we need to check conditions (D) and (F) in Assumption 3.1. It is convenient to introduce the notation: $\Omega_1 = \{x \in \Omega : |v(x)| \geq 1\}$ and $\Omega_2 = \{x \in \Omega : |v(x)| < 1\}$. Since, in our case, $M = I$, Assumption 5.1 implies

$$\|v\|^2_{L_2(\Omega)} = \int_{\Omega_2} v^2 dx + \int_{\Omega_1} v^2 dx \leq meas(\Omega) + m_1^{-1} \int_\Omega g(v)v dx,$$

which is precisely condition (3.2). As for (3.3), it suffices to consider more delicate case $n = 3$ (the arguments for $n = 2$ are simpler). We begin with an obvious inequality

$$\int_\Omega ug(v)dx \leq \int_{\Omega_1} |u||g(v)|dx + \int_{\Omega_2} |u||g(v)|dx.$$

It follows from Assumption 5.1 and the embedding $H^1(\Omega) \subset L^6(\Omega)$ that

$$\int_{\Omega_1} |u||g(v)|dx \leq \|u\|_{L_6(\Omega)} \cdot \|g(v)\|_{L_{6/5}(\Omega_1)}$$

$$\leq \|u\|_{L_6(\Omega)} \cdot \left[\int_{\Omega_1} |g(v)| \cdot |g(v)|^{1/5} dx\right]^{5/6}$$

$$\leq C\|u\|_{L_6(\Omega)} \cdot \left[\int_{\Omega_1} vg(v)dx\right]^{5/6}$$

$$\leq \delta\|u\|^2_{H^1(\Omega)} + C_\delta \|u\|^{4/5}_{H^1(\Omega)} \cdot \int_\Omega vg(v)dx$$

for any $\delta > 0$. It is easy to see from (5.4) that $|s| \leq \bar{\delta} + C_{\bar{\delta}}^* [sg(s)]^{1/2}$ for every $s \in \mathbb{R}$ with arbitrary $\bar{\delta} > 0$. Therefore, since $g(0) = 0$, we also have that

$$\int_{\Omega_2} |u||g(v)|dx \leq C\int_{\Omega_2} |u||v|dx \leq \bar{\delta}\|u\|_{L_2(\Omega)} + C_{\bar{\delta}}\|u\|_{L_2(\Omega)} \cdot \left(\int_\Omega vg(v)dx\right)^{1/2}$$

$$\leq \bar{\delta}\left(1 + \|u\|^2_{L_2(\Omega)}\right) + \widetilde{C}_{\bar{\delta}} \cdot \int_\Omega vg(v)dx$$

for any $\bar{\delta} > 0$. Hence

(5.12) $$\left|\int_\Omega ug(v)dx\right| \leq \delta\left(1 + \|u\|^2_{H^1(\Omega)}\right) + C_\delta\left(1 + \|u\|^{4/5}_{H^1(\Omega)}\right) \cdot \int_\Omega vg(v)dx$$

for any $\delta > 0$. This inequality implies (3.3). Similar (and also simpler) argument applies in the case $n = 2$. As for (3.4), it follows from (5.6) that

(5.13) $$(u, F(u)) = -\int_\Omega \left(f(u(x))u(x) + \mu_0 u(x)^2\right) dx \leq C(\mu, \mu_0, \Omega).$$

Therefore (3.4) holds with $\eta = 0$. Consequently Theorem 3.4 applies and yields ultimate dissipativity for the conservative case.

For the non-conservative case ($g^* \neq 0$) we appeal to Corollary 3.17. To apply this Corollary it suffices to verify (3.51). Indeed, note that the second relevant condition condition (3.52) follows from the growth assumption imposed on g^* along with Lipschitz continuity. For (3.51) the following simple argument applies (we take here the most demanding case $n = 3$):

$$(g^*(v), u) \leq C\|u\|_{L_2(\Omega)} \cdot \left(1 + \|v\|_{L_2(\Omega)}\right) \leq \delta\|u\|^2_{H^1(\Omega)} + C_\delta\left(1 + \|v\|^2_{L_2(\Omega)}\right)$$

$$\leq \delta\|u\|^2_{H^1(\Omega)} + \widetilde{C}_\delta\left(1 + \int_\Omega g(v)v dx\right).$$

Therefore using (5.12) we obtain inequality (3.51) with $\gamma = 2/5$ Thus, ultimate dissipativity holds in both cases $g^* = 0$ and $g^* \neq 0$. Moreover, in the first case one has a control of disspativity radius independently of the damping parameter k.

Thus, to complete the proof of Theorem 5.3 it suffices to establish asymptotic smoothness. This is done below.

Step 2: Asymptotic smoothness - this follows from abstract results in Corollary 3.28 (subcritical case), Theorem 3.58 (critical case) and amounts to verification of inequalities in (3.59) and (3.60) in Assumption 3.21. Verification of the remaining hypothesis (3.61) (with $n_V(v) = \|Kv\|_{L_2(\Omega)}$) is straightforward.

Regarding (3.59) in the case $g'(s) \geq m_1$ for *all* $s \in \mathbb{R}$ we obviously have (3.59) with $H_0(s) = m_1^{-1}s$. In the general case we can take $H_0(s) = h(s) + m_1^{-1}s$, where function $h(s)$ a continuous, concave increasing function on $\mathbb{R}_+$, $h(0) = 0$. The role of the function $h(s)$ is to account for the behaviour of $g(s)$ for "small" values of s. The above follows from the construction in Proposition 4.3 in [20] (see also [80]). Indeed, it is shown there that under the assumptions that g is strictly increasing, $g'(s) \geq m_1, |s| \geq 1$ and $g(0) = 0$ there exists a monotone, concave function, $h : \mathbb{R}_+ \to \mathbb{R}_+$, $h(0) = 0$, and such that with

$$(5.14) \qquad s^2 \leq \frac{1}{m_1}(sg_a(s)) + h(sg_a(s)), \text{ for all } s \in \mathbb{R} \text{ and for every } a \in \mathbb{R},$$

where $g_a(s) \equiv g(s+a) - g(a)$. The above gives rise to the function $H_0(s)$ as stated above.

As to the inequality (3.60), we carry the following string of calculations. In what follows we shall use the notation $|\cdot|_s \equiv \|\cdot\|_{H^s(\Omega)}$.

Let $r = 1 + p^{-1}$ and $\bar{r} = 1 + p$ are Hölder conjugate exponents and δ is a suitably small constant. Then we have

$$\left|\int_\Omega (g(u+v) - g(u))w\,dx\right| \leq \|w\|_{L^{\bar{r}}(\Omega)} \left|\int_\Omega |(g(u+v) - g(u)|^r dx\right|^{1/r}$$

$$\leq C\|w\|_{L^{\bar{r}}(\Omega)} \left|\int_\Omega \left(|g(u+v)||u+v|^{p(r-1)} + |g(u)||u|^{p(r-1)}\right) dx + 1\right|^{1/r}$$

$$\leq C\|w\|_{L^{\bar{r}}(\Omega)} \left|\int_\Omega (g(u+v)(u+v) + g(u)u)\,dx + 1\right|^{1/r}$$

$$(5.15) \qquad \leq C|w|_{1-2\delta}\left(1 + \int_\Omega [g(u+v)(u+v) + g(u)u]\,dx\right),$$

where we have used restrictions (5.4) on the growth of g in Assumption 5.1 along with Sobolev's embedding theorems: for $n = 2$, $\bar{r} < \infty$ and for $n = 3$, $\bar{r} = p+1 < 6$. Thus, inequality (3.60) holds with $\kappa = 1$ (and without the term $\epsilon|\mathcal{A}^{1/2}w|^2$ in the right hand side), as desired. This completes the proof of Theorem 5.3. □

5.3.3. Proof of Theorem 5.4 -critical case. The conclusion of this theorem follows from Theorem 3.34 (when $p \leq 5$ and $g^* \not\equiv 0$) and from Theorem 3.40 (when $g^* \equiv 0$).

Step 1: case when $p \leq 5$ and $g^* \not\equiv 0$. We begin by recalling that ultimate dissipativity of the system has been already shown in Step 1 of the previous section. Assumption 3.21(D) (without the term $\epsilon|\mathcal{A}^{1/2}w|^2$ in the case $p < 5$) was established in the course of the proof of Theorem 5.3. When $p = 5$ and (5.8) holds, validity of (3.60) holds on the strength of calculations in (5.22) and (5.23) (see (5.24)). Moreover, relation (3.92) follows from (5.14). Therefore we need only to verify Assumption 3.33(F).

We recall that

$$\Pi(u) \equiv \int_\Omega \hat{f}(u)dx \quad \text{with} \quad \hat{f}(s) = \int_0^s (f(\xi) + \mu_0\xi)\,d\xi.$$

Since

$$|\hat{f}(s_1) - \hat{f}(s_2)| \leq C\left(1 + |s_1|^3 + |s_2|^3\right) \cdot |s_1 - s_2|,$$

we have that
$$|\Pi(u_1) - \Pi(u_2)| \leq C \left(1 + \|u_1\|_{L_4(\Omega)}^3 + \|u_2\|_{L_4(\Omega)}^3\right) \cdot \|u_1 - u_2\|_{L_4(\Omega)}.$$

Therefore by the embedding $H^s(\Omega) \subset L_p(\Omega)$ for $s = 3(1/2 - 1/p)$ we obtain (F)(i) in Assumption 3.33.

To prove (F)(ii) we note that the embeddings $L_1(\Omega) \subset H^{-2}(\Omega)$ and $H^1(\Omega) \subset L_4(\Omega)$ imply that
$$|\mathcal{A}^{-1}(F(u+z) - F(u))| \leq C \int_\Omega (|z| + |f(u+z) - f(u)|)\, dx$$
$$\leq C \int_\Omega |z|\left(1 + |u|^2 + |z|^2\right) dx \leq C\|z\|_{L_2(\Omega)} \left(1 + \|u\|_{H^1(\Omega)}^2 + \|z\|_{H^1(\Omega)}^2\right).$$

This yields (F)(ii) in Assumption 3.33.

Since $F^*(v) \equiv g^*(v)$, property (F)(iii) follows from the regularity assumption imposed on g^*.

Thus Theorem 5.4 proved in the case $p \leq 5$ and $g^* \not\equiv 0$.

Step 2: case when $g^* \equiv 0$. In this case we apply Theorem 3.40. Since g is increasing, it follows from (5.4) that
$$|s|^2 \leq \delta + C_\delta s \cdot g(s) \quad \text{for every } \delta > 0, \ s \in \mathbb{R}.$$

This implies relation (3.112) in Assumption 3.39. As for (3.113), this relation is the same as (3.3) in Assumption 3.1 and follows from (5.12). Thus the hypotheses concerning the damping function g listed in Assumption 3.39(D) holds true.

The hypotheses on the potential energy $\Pi(u)$ in Assumption 3.39(F)(i,ii) were checked at the first step. The requirement in Assumption 3.39(F)(iii) is obvious. At last the dissipativity relation in (3.114) follows from (5.13). Thus, the conclusion of Theorem 3.40 applies, yielding the final result in Theorem 5.4.

REMARK 5.15. In the case $p = 5$ and $g^* \equiv 0$ we can also we can also apply Theorem 3.47 provided g is superlinear, i.e. there exist $m > 0$ and $l > 2$ such that
(5.16) $$g(s)s \geq m|s|^l \quad \text{for all} \quad |s| \geq 1.$$

In this case we need to check Assumption 3.46. Since in the case when (5.16) holds, we have that
$$|s|^l \leq \delta + C_\delta s \cdot g(s) \quad \text{for every } \delta > 0, \ s \in \mathbb{R}$$
with $l > 2$. Therefore we can take $V_0 \equiv L_l(\Omega) \subset L_2(\Omega)$. The compactness of the mapping $u \mapsto \Pi'(u) = -f(u) - \mu_0 u$ considered as an operator from $L_r(0, T; H^1(\Omega)) \cap H^1(0, T; L_2(\Omega))$ into $L_2(0, T; L_{\frac{l}{l-1}}(\Omega))$ for $r \geq 2$ large enough follows now from the standard Sobolev embedding and Aubin's lemma. Indeed, let $l_* = l(l-1)^{-1} < 2$. Then
$$\|f(u_1) - f(u_2)\|_{L_{l_*}(\Omega)}^{l_*} \leq C \int_\Omega \left(1 + |u_1|^2 + |u_2|^2\right)^{l_*} |u_1 - u_2|^{l_*} dx$$
$$\leq C \left(1 + \|u_1\|_{L_{3l_*}(\Omega)}^{2l_*} + \|u_2\|_{L_{3l_*}(\Omega)}^{2l_*}\right) \|u_1 - u_2\|_{L_{3l_*}(\Omega)}^{l_*}.$$

Since $H^{1-\delta}(\Omega) \subset L_{3l_*}(\Omega)$ for $\delta = l_*^{-1} - 1/2 > 0$, we obtain that
(5.17)
$$\|f(u_1) - f(u_2)\|_{L_{l_*}(\Omega)} \leq C \left(1 + \|u_1\|_{H^{1-\delta}(\Omega)}^2 + \|u_2\|_{H^{1-\delta}(\Omega)}^2\right) \|u_1 - u_2\|_{H^{1-\delta}(\Omega)}$$

with $\delta = l_*^{-1} - 1/2 > 0$. By [**96**, Corollary 8] we have the compact embedding
$$\mathcal{L}_r \equiv L_r(0, T; H^1(\Omega)) \cap H^1(0, T; L_2(\Omega)) \subset C([0, T]; H^{1-\delta}(\Omega))$$
for $r > 2(1-\delta)\delta^{-1}$. Therefore (5.17) implies that $u \mapsto -f(u) - \mu_0 u$ is a compact operator from $\mathcal{L}_r$ into $L_2(0, T; L_{\frac{l}{l-1}}(\Omega))$ for some r large enough.

Finally, to establish Assumption 3.46(ii) we will show that the requirement given in Assumption 3.42(D)(iii) holds (with $l = 1$) for any $1 \leq p \leq 5$. For this, the following inequality is critical:

$$\begin{aligned}(5.18)\quad \int_\Omega |g(u+v) - g(v)| dx &\leq C_\epsilon \int_\Omega (g(u+v) - g(v)) u\, dx \\ &\quad + \epsilon \left[1 + \int_\Omega (g(u+v)(u+v) + g(v)v) dx\right],\end{aligned}$$

for any $\epsilon > 0$ and $u, v \in H^1(\Omega)$. To prove it we split the region of integration into the following subregions:

$$\begin{aligned}\Omega_A &\equiv \{x \in \Omega : |u(x)| \geq 1\}, \\ \Omega_B &\equiv \{x \in \Omega : |u(x)| \leq 1, |v(x)| \leq M\}, \\ \Omega_C &\equiv \{x \in \Omega : |u(x)| \leq 1, |v(x)| \geq M\},\end{aligned}$$

where $M \geq 2$ is a constant. It is clear that
$$\int_{\Omega_A} |g(u+v) - g(v)| dx \leq \int_\Omega (g(u+v) - g(v)) u\, dx.$$

We also have that
$$\begin{aligned}&\int_{\Omega_B} |g(u+v) - g(v)| dx \\ &\leq \int_{\Omega_B} |g(u+v) - g(v)|^{1/2} |u|^{1/2} \left(\frac{|g(u+v) - g(v)|}{|u|}\right)^{1/2} dx \\ &\leq \left[\int_{\Omega_B} (g(u+v) - g(v)) u\, dx\right]^{1/2} \cdot \left[\int_{\Omega_B} \left|\int_0^1 g'(v + \lambda u) d\lambda\right| dx\right]^{1/2} \\ &\leq C \cdot (1+M)^{(p-1)/2} \cdot \left[\int_{\Omega_B} (g(u+v) - g(v)) u\, dx\right]^{1/2} \\ &\leq \frac{2}{M} + C_M \cdot \int_{\Omega_B} (g(u+v) - g(v)) u\, dx.\end{aligned}$$

At last, since $|v| \geq M$ and $|u+v| \geq M/2$ on Ω_C, we have that
$$\begin{aligned}\int_{\Omega_C} |g(u+v) - g(v)| dx &\leq \int_{\Omega_C} (|g(u+v)| + |g(v)|)\, dx \\ &\leq \frac{2}{M} \int_{\Omega_C} (g(u+v)(u+v) + g(v)v) dx.\end{aligned}$$

Thus, if we choose $M = 2/\epsilon$ in the inequalities above, then we easily arrive to (5.18).

Thus, the conclusion of Theorem 3.47 applies under condition (5.16).

5.3.4. Proof of Theorem 5.5 - Regularity of attractor. We start with parts [i] and [ii]. In the case $n = 2$ (which is always subcritical) and $n = 3, q < 2$ this is straightforward consequence of Theorem 4.19. Indeed, it suffices to notice that the condition $g'(s) > 0$ along with growth condition (5.9) (resp. (5.8)) imply (4.54). In the critical case $n = 3, q \leq 2$, but with subcritical damping $0 \leq p \leq 3$ we apply Theorem 4.27. Here, again, (4.54) follows from (5.8). Restriction imposed on p along with Sobolev's embeddings imply validity of the first hypothesis imposed on the damping D in Theorem 4.27. Conditions (4.81) and (4.83) imposed on the nonlinear term F are straightforward. As for (4.82), this follows from the fact that the set of stationary points is bounded in $H^2(\Omega)$ (where the latter statement results from standard elliptic theory).

The most demanding case is the "double critical case", $n = 3, q = 2$ and $p = 5$, or more generally $3 \leq p \leq 5$ corresponding to the statement in part [iii] of Theorem 5.5. For this we shall resort to Theorem 4.25. We begin with the verification of Assumption 4.11. For the case considered $\Pi(u) = \int_\Omega \hat{f}(u)dx$, where $\hat{f}$ denotes the antiderivative of $f(u) + \mu_0 u$. Therefore the second and the third Frechet derivatives of $\Pi(u)$ has the form

(5.19)
$$\langle \Pi^{(2)}(u); v_1, v_2 \rangle = \int_\Omega f'(u)v_1 v_2 dx, \quad \langle \Pi^{(3)}(u); v_1, v_2, v_3 \rangle = \int_\Omega f''(u)v_1 v_2 v_3 dx,$$

for every $u, v_i \in \mathcal{V} \subset H^1(\Omega)$. Consequently, by (5.5) using the embedding $H^s(\Omega) \subset L_r(\Omega)$ for $s = 3(1/2 - 1/r)$ we easily obtain the estimates

(5.20)
$$\begin{aligned} \left| \langle \Pi^{(2)}(u); v, v \rangle \right| &\leq C \int_\Omega (1 + |u|^2) |v|^2 dx \leq C \left(1 + |u|_{L_4}^2 \right) |v|_{L_4}^2 \\ &\leq C \left(1 + \|u\|_{3/4}^2 \right) \|v\|_{3/4}^2 \leq C(\rho) \|v\|_{3/4}^2 \end{aligned}$$

for all $u, v \in H^1(\Omega)$ with $\|u\|_1 \leq \rho$, which implies (4.31) with (at least) $\sigma = 3/8$. Similar argument applies to $\langle \Pi^{(3)}(u); v_1, v_2, v_3 \rangle$:

(5.21)
$$\begin{aligned} \left| \langle \Pi^{(3)}(u); v_1, v_2, v_3 \rangle \right| &\leq C \int_\Omega (1 + |u|) |v_1 v_2 v_3| dx \\ &\leq C \left[\int_\Omega (1 + |u|^2) |v_1 v_2|^2 dx \right]^{1/2} \|v_3\|_{L_2} \\ &\leq C (1 + \|u\|_{L_6}) \|v_1\|_{L_6} \|v_2\|_{L_6} \|v_3\|_{L_2} \\ &\leq C (1 + \|u\|_1) \|v_1\|_1 \|v_2\|_1 \|v_3\|_{L_2} \end{aligned}$$

which implies condition (4.32).

The linear growth of the damping at the origin implies that we can take $H_0(s) = m^{-1}s$, as desired. In line with Theorem 4.1 we need to verify validity of (3.60) with $\kappa = 2$ (rather than $\kappa = 1$ as for Theorem 5.3). This additional restriction on the value of κ forced us to impose more stringent assumptions on g. We turn next to Assumption 3.21, in particular to the most demanding condition (3.60). We need to verify validity of (3.60) with $\kappa = 2$ (rather than $\kappa = 1$ as for Theorem 5.3). To prove inequality (3.60), we use the relation

$$\left| \int_\Omega (g(u+v) - g(u))w \, dx \right| \leq \epsilon \int_\Omega |g(u+v) - g(u)| \frac{|w|^2}{|v|} dx$$
(5.22)
$$+ C_\epsilon \int_\Omega (g(u+v) - g(u)) v \, dx$$

for any $\epsilon > 0$ and carry the following computations.

Let δ be an arbitrary small positive constant and $r, \bar{r}$ conjugate Holder's exponents such that $\beta r \leq 1$. Since (5.8) implies that

$$|g(u) - g(v)| \leq C|u - v|[1 + ug(u) + vg(v)]^\beta,$$

we obtain

$$\int_\Omega (g(u+v) - g(u)) \frac{|w|^2}{|v|} dx$$

$$\leq C\|w\|^2_{L^{2\bar{r}}(\Omega)} \left[\int_\Omega (1 + ug(u) + (u+v)g(u+v))^{\beta r} dx \right]^{1/r}$$

$$\leq C\|w\|^2_{L^{2\bar{r}}(\Omega)} \left[1 + \int_\Omega (g(u+v)(u+v) + g(u)u) \, dx \right]$$

(5.23)
$$\leq C|w|^2_{1-2\delta} \left[1 + \int_\Omega (g(u+v)(u+v) + g(u)u) dx \right],$$

where we have selected appropriate values of r in line with Sobolev's embeddings and restrictions on the growth of g implied by (5.9). In the case $n = 3$, $\beta = \frac{2}{3}$, so we take $r = 3/2$ and $\bar{r} = 3$. Thus (5.23) holds with $\delta = 0$ and, therefore, by (5.22) inequality in (3.60) is satisfied. Indeed

$$\left| \int_\Omega (g(u+v) - g(u))w \, dx \right| \leq C_\epsilon \int_\Omega (g(u+v) - g(u)) v \, dx$$
(5.24)
$$+ \epsilon |w|^2_1 \left[1 + \int_\Omega (g(u+v)(u+v) + g(u)u) dx \right],$$

which is condition (3.60) in Assumption 3.21 (this fact has been also used in the proof of Theorem 5.4). Conclusion of Theorem 4.25 stated in (4.78) implies that A is bounded in $\mathcal{V} \times \mathcal{V} \subset H^1(\Omega) \times H^1(\Omega)$ with additional spatial regularity

$$\Delta u \in L_2(\Omega) + D(H^1(\Omega)) \subset L_2(\Omega) + L_{6/p}(\Omega) = L_{\min\{2, 6/p\}}(\Omega),$$

where $(u, v) \in A$ and u satisfies homogenous (Robin or Dirichlet) boundary conditions. The above formula and standard L_p elliptic theory yield additional regularity of the variable u in line with the statement in Theorem 5.5. Proof of Theorem 5.5 is thus completed.

5.3.5. Proof of Theorem 5.8 -finite dimensionality. Let $g^* = 0$. For subcritical cases, $n = 2$ and $n = 3, q < 2$ we can just refer to Theorem 4.1, whose assumptions have been verified above. Indeed, (3.60) with $\kappa = 2$ follows from (5.23) where for $n = 2$ we take $r > 1$ close to one, so that $\beta r \leq 1$. This leads to (3.60) without the term $\epsilon |\mathcal{A}^{1/2} w|^2$. For $n = 3$ we have (5.24), which implies (3.60). Thus, due to subcriticality (either $n = 2$ or $q < 2$), Theorem 4.1 applies and yields finite dimensionality. This is not the case for $n = 3$ and critical exponent $q = 2$. We need to resort, again, to Theorem 4.25. Assumptions of this Theorem have been verified above (see (5.20) and (5.21)), hence the final conclusion of finite dimensionality has been established in all the cases considered with $g^* = 0$.

As for the last statement in Theorem 5.8, dealing with the case $g^* \neq 0$, we apply the second part of Theorem 4.1. Property (4.2) with $l = 1$ follows from the assumption $p \leq 2$.

As we recall, the proof of finite dimensionality of the attractor, under the hypotheses of Theorem 4.25 depends on the additional smoothness of the attractor (in the case of wave equation this is Theorem 5.5). We note that an additional smoothness of attractors was used critically in [81] for the purpose of proving finite dimensionality of attractors with a nonlinear damping. However the conditions imposed on the dissipation in [81] are much more restrictive than in the present work.

5.3.6. Proof of Theorem 5.10. Decay to equilibrium. Theorem 5.10 follows from Theorem 4.31. Critical role is played by the requirement (4.94) which, by Remark 4.32, follows from (3.60) with $\kappa = 2$. However, (3.60) with $\kappa = 2$ was shown already owing the growth condition imposed on g. Verification of (4.95) and (4.96) is straightforward. Indeed, to verify (4.96) in the case $n = 3$, $q = 2$ we note that by the imbedding $H^1(\Omega) \subset L_6(\Omega)$ we have that

$$|\langle F'(u) - F'(v), w \rangle|_{[\mathcal{D}(\mathcal{A}^{1/2})]'} \leq C \left(\int_\Omega \left|(1 + |u|^2 + |v|^2)(u - v)w\right|^{6/5} dx \right)^{5/6}$$

$$\leq C \left(\int_\Omega \left[(1 + |u|^2 + |v|^2)|u - v|\right]^{3/2} dx \right)^{2/3} \|w\|_{L_6(\Omega)}$$

$$\leq C \left(\int_\Omega (1 + |u|^6 + |v|^6) dx \right)^{1/3} \left(\int_\Omega |u - v|^3 dx \right)^{1/2} \|w\|_{L_6(\Omega)}.$$

This implies (4.96) in the case $n = 3$, $q = 2$. A similar argument applies for (4.95) and for other values of n and q.

The validity of energy *inequality* for generalized solutions can be obtained in the following way. From energy equality (5.7) we have that

$$(5.25) \qquad \mathcal{E}(t) + \int_s^t \int_\Omega \psi_N \left(g(w_t(\tau, x)) w_t(\tau, x) \right) dx d\tau \leq \mathcal{E}(s), \quad N = 1, 2, \ldots$$

for any strong solution, where $\psi_N(\sigma) = \sigma$ for $0 \leq \sigma \leq N$ and $\psi_N(\sigma) = N$ for $\sigma > N$. Since the function $s \mapsto \psi_N(g(s)s)$ is globally Lipschits, relation (5.25) remains true after the limit transition to generalized solutions. Now, since $\psi_N(\sigma) \leq \psi_{N+1}(\sigma)$ we can apply Levi-Lebesgue theorem on monotone convergence to obtain the energy inequality for generalized solutions. $\square$

5.3.7. Proof of Theorem 5.13 on determining functionals. It easily follows from Corollary 4.40. $\square$

CHAPTER 6

Von Karman evolutions with a nonlinear dissipation

6.1. The model

In what follows we consider a nonlinear system of dynamic elasticity described by von Karman evolution with a *nonlinear dissipation*.

Let $\Omega \subset \mathbb{R}^2$ be a bounded domain with a sufficiently smooth boundary Γ. We denote by $\nu = (\nu_1, \nu_2)$ the outer normal to Γ. Consider the following von Karman model with clamped boundary conditions:

(6.1)
$$u_{tt} - \alpha \Delta u_{tt} + k\left[g_0(u_t) - \alpha \mathrm{div} g(\nabla u_t)\right] + \Delta^2 u$$
$$= [v(u) + F_0, u] + p + L(u) + g^*(u_t) \quad \text{in} \quad \Omega \times (0, \infty),$$
$$u = \frac{\partial}{\partial \nu} u = 0 \quad \text{on} \quad \Gamma \times (0, \infty).$$

The Airy stress function $v(u)$ satisfies the following elliptic problem

(6.2) $\quad \Delta^2 v(u) + [u, u] = 0, \text{ in } \Omega, \quad \dfrac{\partial}{\partial \nu} v(u) = v(u) = 0 \text{ on } \Gamma.$

The von Karman bracket $[u, v]$ is given by

(6.3) $\quad [u, v] = \partial_{x_1}^2 u \cdot \partial_{x_2}^2 v + \partial_{x_2}^2 u \cdot \partial_{x_1}^2 v - 2 \cdot \partial_{x_1 x_2}^2 u \cdot \partial_{x_1 x_2}^2 v.$

The functions g_0 and g represent dissipation in the model. They are assumed monotone increasing and continuous. The damping parameter k is positive. The parameter $\alpha \geq 0$ represents rotational forces. The functions $F \in H^2(\Omega)$ and $p \in L_2(\Omega)$ are given. They describe in-plane and transverse forces applied to the plate. The operator L is a linear first order differential operator. In typical applications (see, e.g., [38]) one has $Lu = -\rho u_{x_1}$, where ρ depends on the velocity of the gas. The term $g^*(u_t)$ with $g^* \in W_\infty^1(\mathbb{R})$ represents an anti-damping. The presence of non-conservative terms Lu and g^* is responsible for the fact that energy is not decreasing.

Equations of von Karman (6.1) and (6.2) are well known in nonlinear elasticity and constitute a basic model describing nonlinear oscillations of a plate accounting for large displacements, see e.g., [83] and references therein. The model which accounts for moments of inertia, i.e. $\alpha > 0$ is often referred as *modified von Karman equations*. The rotational case when $\alpha > 0$ represents a purely hyperbolic dynamics with finite speed of propagation. Instead, when $\alpha = 0$ we have an infinite speed of propagation. Thus, the mathematical properties of these two models are very different necessitating different treatments and analysis in each case.

We note that the energy function associated with (6.1) takes the following form

$$\mathcal{E}(t) = \int_\Omega \left[\frac{1}{2}|u_t(x,t)|^2 + \frac{\alpha}{2}|\nabla u_t(x,t)|^2 + \frac{1}{2}|\Delta u(x,t)|^2 + \frac{1}{4}|\Delta v(u)(x,t)|^2 \right.$$
(6.4)
$$\left. - \frac{1}{2}[F_0(x), u(x,t)]u(x,t) - p(x)u(x,t) \right] dx$$

and the corresponding energy relation (which is valid for strong solutions) has the form

$$\mathcal{E}(t) + k \int_0^t \int_\Omega [g_0(u_t(x,\tau))u_t(x,\tau) + \alpha g(\nabla u_t(x,\tau))\nabla u_t(x,\tau)] \, dx d\tau$$
(6.5)
$$= \mathcal{E}(0) + \int_0^t \int_\Omega [g^*(u_t(x,\tau))u_t(x,\tau) + Lu(x,\tau)u_t(x,\tau)] \, dx d\tau.$$

Two properties of the energy function that should be emphessized are: (i) it is not obvious that $\mathcal{E}(t)$ is bounded from below, (ii) $\mathcal{E}(t)$ may not be decreasing. Indeed, the last two terms in (6.5) have uncontrolled signs. This is the effect of non-conservative forces.

6.2. Properties of von Karman bracket

This preliminary section contains some properties of bracket (6.3) and nonlinear forcing terms from equation (6.1) which we need to check the hypotheses of Chapters 3 and 4 in the both cases $\alpha > 0$ and $\alpha = 0$. The following notations will be used:

$$\| u \|_s \equiv \| u \|_{H^s(\Omega)}, \quad \| u \| \equiv \| u \|_{L^2(\Omega)} \quad \text{and} \quad (u, v) \equiv (u, v)_{L^2(\Omega)}.$$

PROPOSITION 6.1. *The mapping $\{u, v\} \mapsto [u, v]$ is a symmetric bilinear mapping from $H^2(\Omega) \times H^2(\Omega)$ into $L_1(\Omega)$. The trilinear form $([u, v], w)$ is symmetric on $H^2(\Omega)$ if either at least one of elements u, v or w belongs to $H_0^2(\Omega)$. Furthermore:*

- *The following estimates are valid*

(6.6)
$$\| [u, v] \|_{-j} \leq C \| u \|_{2-\beta} \cdot \| v \|_{3-j+\beta},$$

where $j = 1, 2$ and $0 \leq \beta < 1$ and

(6.7)
$$|[u, v]|_{-1-2\delta} \leq C|u|_{2-\delta}|v|_{2-\delta}, \quad 0 < \delta < 1/2.$$

- *For any $u \in H^2(\Omega)$ problem (6.2) has a unique solution $v = v(u)$ in $H_0^2(\Omega) \cap W_\infty^2(\Omega)$. This solution has the following properties*

(6.8)
$$\| v(u_1) - v(u_2) \|_{W_\infty^2(\Omega)} \leq C \cdot \| u_1 + u_2 \|_2 \cdot \| u_1 - u_2 \|_2$$

and

(6.9)
$$\| v(u_1) - v(u_2) \|_{3-\delta} \leq C_\delta \| u_1 + u_2 \|_2 \cdot \| u_1 - u_2 \|_{2-\delta}$$

for every $0 \leq \delta \leq 1$. We also have that

(6.10)
$$\left\| (\Delta_D^2)^{-1} [u_1, u_2] \right\|_{W_\infty^2(\Omega)} \leq C \cdot \| u_1 \|_2 \cdot \| u_2 \|_2,$$

where Δ_D^2 is the biharmonic operator with the Dirichlet boundary conditions and $(\Delta_D^2)^{-1}$ is the corresponding Green operator.

- *For any $\eta > 0$ there exists C_η such that*

(6.11)
$$\|u\|^2 \leq \eta(\|\Delta v(u)\|^2 + \|\Delta u\|^2) + C_\eta, \quad u \in (H_0^1 \cap H^2)(\Omega).$$

PROOF. The algebraic properties follow from standard and well known considerations for von Karman brackets and Airy stress function [83]. Particular cases of estimates (6.6) and (6.9) can be found in [83]. The proof of (6.6) and (6.7) for the all cases described can be found in [17]. Estimates (6.8) and (6.10) follow from recent developments on sharp regularity of Airy stress function (see [47] and also [26] and the references therein). Estimate (6.9) follows from elliptic regularity theory and from (6.6). The proof of (6.11) relies on the uniqueness property of Monge-Ampere equation and has been provided in [26] (see also [13] and [77]). □

6.3. Abstract setting of the model

In order to put von Karman model into the abstract setting of Chapter 3, we define

(6.12) $\quad F(u,w) \equiv [v(u) + F_0, u] + p + L(u) + g^*(w), \quad u \in H_0^2(\Omega), w \in L_2(\Omega),$

where $v = v(u)$ solves (6.2). The nonlinear term $F(u,v)$ can be represented in the form

$$F(u,v) = -\Pi'(u) + F^*(u,v)$$

with

(6.13) $\quad \Pi(u) = \frac{1}{4}\|\Delta v(u)\|^2 - \frac{1}{2}([u,u], F_0) - (p, u),$

where $v(u) \in H_0^2(\Omega)$ is defined by (6.2), and $F^*(u,v) = L(u) + g^*(v)$. Thus we set

(6.14) $\quad \Pi_0(u) = \frac{1}{4}\|\Delta v(u)\|^2 \quad \text{and} \quad \Pi_1(u) = -\frac{1}{2}([u,u], F_0) - (p, u).$

The following assertion makes it possible to prove that F satisfies the hypotheses from Chapters 3 and 4.

LEMMA 6.2. *For any $u \in H_0^2(\Omega)$ the following relations hold:*

(6.15) $\quad \|Lu\|_\sigma \leq C\|u\|_{1+\sigma} \text{ for } \sigma = 0, -1,$

(6.16) $\quad \|u\|^2 \leq \eta\left(\|\Delta u\|^2 + \Pi_0(u)\right) + C_\eta \text{ for any } \eta > 0,$

(6.17) $\quad |\Pi_1(u)| \leq \eta\left(\|\Delta u\|^2 + \Pi_0(u)\right) + C_\eta \text{ for any } \eta > 0,$

(6.18) $\quad (-\Pi'(u) + Lu, u) \leq -\frac{1}{2}\|\Delta v(u)\|^2 + \eta\|\Delta u\|^2 + C_\eta \text{ for any } \eta > 0.$

If $F_0 \in H^{2+\sigma}(\Omega)$ for some $\sigma \in (0,1)$, then

(6.19) $\quad \|\Pi'(u_1) - \Pi'(u_2)\|_{-1} \leq C_\rho\|u_1 - u_2\|_{2-\sigma}, \quad \sigma > 0,$

and if $F_0 \in W_\infty^2(\Omega)$, then

(6.20) $\quad \|\Pi'(u_1) - \Pi'(u_2)\| \leq C_\rho\|u_1 - u_2\|_2,$

for any $u_1, u_2 \in H_0^2(\Omega)$ such that $\|u_i\|_2 \leq \rho$, $i = 1,2$, where $\rho > 0$ is arbitrary and C_ρ is a non-decreasing function of ρ.

If $F_0 \in H^2(\Omega)$ and $p \in L_2(\Omega)$, then there exists $\delta > 0$ such that

(6.21) $\quad |\Pi(u_1) - \Pi(u_2)| \leq C\left(\|u_1\|_{2-\delta}^3 + \|u_2\|_{2-\delta}^3\right)\|u_1 - u_2\|_{2-\delta},$

for any $u_1, u_2 \in H_0^2(\Omega)$.

PROOF. Estimate (6.15) follows from the fact that L is a linear first order differential operator.

Relation (6.16) follows from (6.11).

Using (6.6) with $j = 2$ and $\beta = 0$ we obtain that

$$|\Pi_1(u)| \leq \frac{1}{2}\|F_0\|_2 \cdot \|u\|_1 \cdot \|u\|_2 + \|u\|_2 \cdot \|p\|_{-2}$$

$$\leq \eta\|\Delta u\|^2 + C_\eta^{(1)}\|u\|^2 + C_\eta^{(2)}$$

for any $\eta > 0$. Therefore (6.17) follows from (6.16).

A simple calculation using the symmetry of von Karman bracket (see Proposition 6.1) gives

$$(-\Pi'(u), u) = -\|\Delta v(u)\|^2 + ([u, F_0], u) + (p, u).$$

Therefore, as above, from (6.6) we obtain

$$(-\Pi'(u), u) \leq -\|\Delta v(u)\|^2 + \eta\|\Delta u\|^2 + C_\eta^{(1)}\|u\|^2 + C_\eta^{(2)}$$

for any positive η. Now using (6.11) we get (6.18).

To prove (6.19) we note that

$$-\Pi'(u_1) + \Pi'(u_2) = [u_1, v(u_1) - v(u_2)] + [u_1 - u_2, v(u_2) + F_0].$$

Using (6.6) with $j = 1$ and appropriate choice of β and also (6.9) with $\delta = 1$ we obtain that

$$\|\Pi'(u_1) - \Pi'(u_2)\|_{-1} \leq C(1 + \|u_1\|_2^2 + \|u_2\|_2^2)\|u_1 - u_2\|_{2-\sigma}$$

Therefore we get (6.19) from (6.15). In a similar way (6.20) follows from (6.8).

To prove (6.21) we note that $[u_1, u_1] - [u_2, u_2] = [u_1 - u_2, u_1 + u_2]$ and

$$\left|\|\Delta v(u_1)\|^2 - \|\Delta v(u_2)\|^2\right| \leq \|\Delta(v(u_1) - v(u_2))\| \left(\|\Delta v(u_1)\| + \|\Delta v(u_2)\|\right)$$

$$\leq C\|[u_1 - u_2, u_1 + u_2]\|_{-2} \left(\|[u_1, u_1]\|_{-2} + \|[u_2, u_2]\|_{-2}\right).$$

Therefore (6.21) follows from (6.6) with $j = 2$. □

As it was shown in Chapter 1 (see Example 1.3), problem (6.1) and (6.2) can be written in the form

(6.22) $$\begin{cases} Mu_{tt}(t) + \mathcal{A}u(t) + k \cdot D(u_t(t)) = F(u(t), u_t(t)), \\ u|_{t=0} = u_0 \in \mathcal{D}(\mathcal{A}^{1/2}), \ u_t|_{t=0} = u_1 \in V = \mathcal{D}(M^{1/2}), \end{cases}$$

with the following notation the following spaces and operators:

- $\mathcal{H} \equiv L_2(\Omega)$, $V \equiv V_\alpha$, where $V_\alpha = H_0^1(\Omega)$ for $\alpha > 0$ and $V_\alpha = L_2(\Omega)$ for $\alpha = 0$.
- $\mathcal{A}u \equiv \Delta^2 u$, $u \in \mathcal{D}(\mathcal{A})$: $\mathcal{D}(\mathcal{A}) \equiv H_0^2(\Omega) \cap H^4(\Omega)$.
- $Mu \equiv u - \alpha\Delta u$, $u \in \mathcal{D}(M)$, where $\mathcal{D}(M) \equiv H_0^1(\Omega) \cap H^2(\Omega)$ in the case $\alpha > 0$ and $\mathcal{D}(M) \equiv L_2(\Omega)$ when $\alpha = 0$.
- $F(u, u_t)$ is given by (6.12).
- $D(u) \equiv g_0(u) - \alpha \operatorname{div}(g(\nabla u))$.

It is clear that $\mathcal{A}$ and M satisfy the conditions introduced at the beginning of Sect. 1.2. The difference between the two cases $\alpha > 0$ and $\alpha = 0$ is, of course, different topology of V'. This difference of topology will result in very different properties and results for the corresponding two cases. The model with $\alpha > 0$ is hyperbolic with a finite speed of propagation, while the case $\alpha = 0$ corresponds to infinite speed of propagation. As we shall see below, this difference will produce different theories for the two cases. We shall begin with a more regular case of rotational model when $\alpha > 0$. Here, the analysis and the results are reminiscent to the wave equation case, treated in the previous chapter.

6.4. Model with rotational forces: $\alpha > 0$

6.4.1. Main results. Our main standing assumption is

ASSUMPTION 6.3.
- $g_0 \in C^1(\mathbb{R})$ is a monotone nondecreasing function such that $g_0(0) = 0$.
- The function g has the form $g(s_1, s_2) = (g_1(s_1), g_2(s_2))$, where $(s_1, s_2) \in \mathbb{R}^2$ and $g_i \in C^1(\mathbb{R})$ is monotone increasing function such that $g_i(0) = 0$, $i = 1, 2$. Moreover, we assume that g_i are of polynomial growth at infinity, i.e. we assume that

(6.23) $$0 < m \leq g_i'(s) \leq M|s|^{p-1}, \quad |s| \geq 1, \ i = 1, 2,$$

with some constants m, M and $p \geq 1$. In the case when $L \not\equiv 0$ (non-conservative case), we assume that this inequality holds with $p = 1$ and also that g_0 possesses the same property.
- $p(x) \in L_2(\Omega)$ and $F_0(x) \in H^{2+\sigma}(\Omega)$ for some $\sigma > 0$.
- L is a linear first order differential operator with smooth coefficients.

Below we consider the following two case only: (a) $L \not\equiv 0$, $g^* \equiv 0$ and (b) $L \equiv 0$, $g^* \not\equiv 0$. One could also consider a simultaneous effect of "static" and "dynamic" non-conservative forces L and g^*. However, this would require additional assumptions synchronizing these two effects (see Subsection 3.1.3 and Theorem 3.15). Similarly one could consider dynamic force acting on Δu_t as well as *nonlinear* non-conservative forces in line with setup in Theorem 3.15. We have chosen not to complicate matters too much and rather to adhere to a simpler framework that is relevant in applications.

Our main result in the case $g^* \equiv 0$ is the following assertion.

THEOREM 6.4. *Let $g^* \equiv 0$ Under Assumption 6.3 equations (6.1) and (6.2) with $\alpha > 0$ generates a continuous semiflow S_t in the space $H \equiv H_0^2(\Omega) \times H_0^1(\Omega)$ which possesses a global compact attractor A. Moreover*

- *If, in addition, we assume that (i) $g_0(s)$ is of polynomial growth at infinity, i.e. there exist $p_0 \geq 1$ and $M_0 > 0$ such that*

(6.24) $$0 \leq g_0'(s) \leq M_0[1 + |s|^{p_0 - 1}], \quad s \in \mathbb{R},$$

and (ii) there exists $0 \leq \beta < 1$ such that the functions g_i satisfy the inequality

(6.25) $$0 < m \leq g_i'(s) \leq M[1 + sg_i(s)]^\beta, \quad s \in \mathbb{R}, \ i = 1, 2,$$

then the attractor A is a bounded closed set in $(H^3 \cap H_0^2)(\Omega) \times H_0^2(\Omega)$ and its fractal dimension is finite.
- *If $L \equiv 0$, then $A = \mathcal{M}^u(\mathcal{N})$, where $\mathcal{M}^u(\mathcal{N})$ is the unstable manifold emanating from the set $\mathcal{N}$ (see the definition in Sect. 2.4) and $\mathcal{N}$ is the set of equilibria for the semiflow S_t. Moreover, the statements of Theorem 2.31 and Corollary 2.32 are in force.*

- Let us assume that the set of stationary solutions consists of finitely many isolated equilibrium points which are hyperbolic and that $L \equiv 0$. Then, under conditions (6.24) and (6.25) for any initial condition $y_0 \in H$ there exists an equilibrium point $e = (w, 0)$, $w \in H_0^2(\Omega)$, such that

(6.26) $$|S_t y_0 - e|_H \leq Ce^{-\omega t}, \quad C, \omega > 0.$$

- Under conditions (6.24) and (6.25) the system possesses a fractal exponential attractor (see Definition 4.41) whose dimension is finite in the space $H_0^1(\Omega) \times W$, where W is a completion of $H_0^1(\Omega)$ with respect to the norm $\|\cdot\|_W = \|(1 - \alpha\Delta)\cdot\|_{-2}$.

REMARK 6.5. We note that the property $g_i'(s) \leq M(1 + sg_i(s))^\beta$ for some $\beta < 1$ holds true, if we assume that either $p < 3$ or else $p \geq 3$ and $g_i(s)s \geq m_1|s|^l$ for all $|s| \geq 1$ and for some $m_1 > 0$ and $l > p - 1$.

We can also suggest another version of the stabilization result which covers the case when $m = \min_{i=1,2} g_i'(0)$ can be zero. To formulate this result we first prove the following assertion.

LEMMA 6.6. Let Assumption 6.3 hold. Then the form $D(v, u)$ given by the relation

(6.27) $$D(v, u) \equiv \int_\Omega g_0(v) u \, dx + \alpha \sum_{i=1,2} \int_\Omega g_i(v_{x_i}) u_{x_i} \, dx, \quad v, u \in H^1(\Omega),$$

possesses the property

(6.28) $$H_0(D(u + v, v) - D(u, v)) \geq \|v\|^2 + \alpha\|\nabla v\|^2, \quad v \in H_0^1(\Omega), \ u \in H^1(\Omega),$$

where $H_0 : \mathbb{R}_+ \mapsto \mathbb{R}_+$ is a strictly increasing continuous concave function. If $g_i'(s) \geq m > 0$ for all $s \in \mathbb{R}$, $i = 1, 2$, then (6.28) holds with $H_0(s) = h_0 \cdot s$.

PROOF. As in the proof Theorem 5.3 (see relation (5.14)) we can construct a strictly increasing continuous concave function $\widetilde{H}_0(s)$ such that

$$2s^2 \leq \widetilde{H}_0(sg_i^a(s)), \ i = 1, 2, \quad \text{for all} \ a \in \mathbb{R}, \ g_i^a(s) \equiv g_i(s + a) - g_i(a).$$

Hence

(6.29) $$s_1^2 + s_2^2 \leq \widetilde{H}_0(s_1 g_i^a(s_1) + s_2 g_i^a(s_2)) \quad \text{for all} \ a, s_1, s_2 \in \mathbb{R}.$$

Therefore the application Jensen's inequality leads to relation (6.28) with the function $H_0(s) = c_0 Vol(\Omega) \widetilde{H}_0(Vol(\Omega)^{-1} s)$. If $g_i'(s) \geq m > 0$ for all $s \in \mathbb{R}$, then it is clear that (6.29) holds with $\widetilde{H}_0(s) = \tilde{h}_0 \cdot s$. $\square$

Now we are in position to formulate the following stabilization theorem.

THEOREM 6.7. Let $g^* \equiv 0$ and $L \equiv 0$. In addition to Assumption 6.3 assume that (6.24) holds and there exists $0 \leq \beta < 1$ such that the functions g_i satisfy the inequality

(6.30) $$0 < g_i'(s) \leq M[1 + sg_i(s)]^\beta, \quad s \in \mathbb{R}, \ s \neq 0, \ i = 1, 2.$$

If the the set of stationary solutions consists of finitely many isolated equilibrium points which are hyperbolic, then for any initial condition $y_0 \in H$ there exists an equilibrium point $e = (w, 0)$, $w \in H_0^2(\Omega)$, such that

(6.31) $$|S_t y - e|_H^2 \leq C \cdot \sigma\left([tT^{-1}]\right), \ t > 0,$$

where C and T are positive constants, $[a]$ denotes the integer part of a and $\sigma(t)$ satisfies the following ODE:

(6.32) $$\frac{d\sigma}{dt} + Q(\sigma) = 0, \ t > 0, \ \sigma(0) = C(y,e).$$

Here $C(y,e)$ is a constant depending on y and e, $Q(s) = s - (I + G_0)^{-1}(s)$ with $G_0(s) = c_1 (I + H_0)^{-1}(c_2 s)$, where c_1 and c_2 are positive numbers and H_0 is defined in Lemma 6.6.

Few words about the relation of the result given in Theorem 6.4 with respect to the literature. Existence and finite-dimensionality of global attractor, for the case of *linear damping*, was proved earlier in [13] (see also [14]) where the structure of the attractor was established). The case of *nonlinear* but *linearly bounded* damping was considered in [79]. The result on exponential stabilization is new even in the case of linear damping. Consequently Theorem 6.4 extends the results known in the literature in two directions: (i) it allows to treat strongly nonlinear dissipation, and (ii) it provides estimates on the convergence to equilibria.

In the case when $L \equiv 0$ and $g^* \not\equiv 0$ we have the following result.

THEOREM 6.8. *Let $L \equiv 0$. Assume that $g^* \in C^1(\mathbb{R})$ possesses the properties*

(6.33) $$|[g^*]'(s)| \leq C[1 + |s|^{p_* - 1}], \quad s \in \mathbb{R},$$

for some $p_* \geq 1$ and $C > 0$, and

(6.34) $$\limsup_{|s| \to \infty} \frac{|g^*(s)|}{g_0(s)} < k.$$

Then under Assumption 6.3 equations (6.1) and (6.2) with $\alpha > 0$ generates a continuous semiflow S_t in the space $H \equiv H_0^2(\Omega) \times H_0^1(\Omega)$ which possesses a global compact attractor A. If, in addition, we assume that (i) $g_0(s)$ is of polynomial growth at infinity, i.e. (6.24) holds, and (ii) the functions g_i satisfy the inequality

$$0 < m \leq g_i'(s) \leq M[1 + |s|], \quad s \in \mathbb{R}, \ i = 1, 2,$$

then the attractor A is a bounded closed set in $(H^3 \cap H_0^2)(\Omega) \times H_0^2(\Omega)$ and the fractal dimension of the attractor A is finite.

We note that long-time behaviour of von Karman system with an anti-damping term was not considered before.

6.4.2. Proof of Theorem 6.4. Since (6.1) is presented in the form (6.22) with $V = H_0^1(\Omega)$ and

$$F(u, u_t) \equiv F(u) = [v(u) + F_0, u] + p + L(u), \quad u \in H_0^2(\Omega),$$

we conclude from Lemma 6.2 that the nonlinearity F in this case satisfies all hypotheses concerning F postulated in Chapters 1 and 3 (see Assumptions 1.1(F), 3.1(F) 3.7(F) and 3.21(F) with $n_V \equiv 0$). Indeed, (6.16) coincides with (3.18), relation (6.17) is identical with (1.5), (6.18) coincides with (3.17) and (6.19) is the same as (3.61) in the framework considered. Moreover, Assumption 3.21(F) holds with $\widetilde{\eta} > 0$ and, hence, in the case $\alpha > 0$ the nonlinearity F is subcritical. Therefore, in order to establish existence of global attractors, we shall appeal to Corollary 3.28.

Thus, to prove the theorem we need only to check the corresponding assumptions concerning the damping D.

Verifications of Assumptions imposed on the damping operator. The argument given in Example 1.3 shows that Assumption 1.1(D) holds and, therefore, by Theorem 1.5 problem (6.1) with $g^* \equiv 0$ generates a dynamical system in the space $H = H_0^2(\Omega) \times H_0^1(\Omega)$.

LEMMA 6.9. *Let Assumption 6.3 hold and $D(v, u)$ be given by (6.27). Then*

-

(6.35) $$D(v,v) \geq c_0 \left(\|v\|^2 + \alpha \|\nabla v\|^2 \right) - c_1, \quad v \in H_0^1(\Omega),$$

where c_0 and c_1 are positive constants;

- *If $p = 1$ in (6.23) and $g_0'(s)$ is bounded (assumed when $L \neq 0$), then*

(6.36) $$\|D(v)\|_{-1} \leq c_2 \|u\|_1, \quad v \in H_0^1(\Omega),$$

and

(6.37) $$|D(u+v,w) - D(u,w)| \leq c_3 \|w\|_1 \left[1 + D(u+v, u+v)^{1/2} + D(u,u)^{1/2} \right]$$

for any $v, u \in H_0^1(\Omega)$ and $w \in H^1(\Omega)$, where $c_2 > 0$ and $c_3 > 0$ are constants;

- *If $1 \leq p < \infty$ (relevant when $L = 0$) then*

(6.38) $$|D(v,u)| \leq C_1 \|u\|_s \cdot D(v,v) + C_2(1 + \|u\|_1^2), \quad v \in H^1(\Omega), \; u \in H^2(\Omega),$$

and

(6.39) $$|D(u+v,w) - D(u,w)| \leq C_3 \|w\|_s \left[1 + D(u+v, u+v) + D(u,u) \right],$$

for any $v, u \in H^1(\Omega)$ and $w \in H^2(\Omega)$, where $s = \frac{2p}{1+p}$ if $p > 1$ and $s > 1$ is arbitrary in the case $p = 1$, the constants C_i are positive.

PROOF. By the Friedrichs inequality we have

(6.40) $$\|v\|^2 + \alpha \|\nabla v\|^2 \leq C(1+\alpha) \|\nabla v\|^2, \quad v \in H_0^1(\Omega).$$

It follows from (6.23) that
$$sg_i(s) \geq ms^2 - c, \quad s \in \mathbb{R}, \quad \text{for some} \quad c > 0.$$
We also have that $sg_0(s) \geq 0$. Therefore
$$D(v,v) \geq \alpha \sum_{i=1,2} \int_\Omega g_i(v_{x_i}) v_{x_i} dx \geq \alpha m \|\nabla v\|^2 - \alpha c |\Omega|.$$
Thus (6.35) follows from (6.40).

In the case $L \neq 0$ we have that

(6.41) $$|g_0(v)| \leq a_1 |v| \quad \text{and} \quad |g_1(v_1)|^2 + |g_2(v_2)|^2 \leq a_1^2 (v_1^2 + v_2^2).$$

This implies that
$$|D(v,u)| \leq C \cdot (\|v\|\|u\| + \alpha \|\nabla v\|\|\nabla u\|) \leq C \cdot \|v\|_1 \|u\|_1.$$
Therefore we obtain (6.36).

Similarly, from (6.41) we have
$$|D(u+v,w) - D(u,w)|$$
$$\leq C \int_\Omega (|u+v| + |u|)|w|dx + C\alpha \int_\Omega (|\nabla(u+v)| + |\nabla u|)|\nabla w|dx.$$
$$\leq C \cdot \left[\left(\|u+v\|^2 + \alpha^2 \|\nabla(u+v)\|^2 \right)^{1/2} + \left(\|u\|^2 + \alpha^2 \|\nabla u\|^2 \right)^{1/2} \right] \|w\|_1.$$

As above this implies (6.37).

Now we consider the case of an arbitrary finite p (of relevance when $L = 0$). We will use the diagonal structure of the mapping $g = [g_1, g_2]^T$.

Denoting $\Omega_1 \equiv \{x \in \Omega : |v_{x_i}| \leq 1\}$ and $\Omega_2 \equiv \{x \in \Omega : |v_{x_i}| \geq 1\}$ for each $i = 1, 2$ we have

$$\left| \int_\Omega g_i(v_{x_i}) u_{x_i} dx \right| \leq \int_{\Omega_1} |g_i(v_{x_i}) u_{x_i}| dx + \int_{\Omega_2} |g_i(v_{x_i}) u_{x_i}| dx. \tag{6.42}$$

By the Cauchy-Schwarz inequality we obtain

$$\int_{\Omega_1} |g_i(v_{x_i}) u_{x_i}| dx \leq \frac{1}{4} \int_{\Omega_1} |g_i(v_{x_i})|^2 dx + \int_{\Omega_1} |u_{x_i}|^2 dx \leq C_\Omega + \|u\|_1^2. \tag{6.43}$$

In the region Ω_2 we use Hölder's inequality applied with some $1 < r \leq 2$:

$$\int_{\Omega_2} |g_i(v_{x_i}) u_{x_i}| dx \leq \left(\int_{\Omega_2} |g_i(v_{x_i})|^r dx \right)^{1/r} \left(\int_{\Omega_2} |u_{x_i}|^{\bar{r}} dx \right)^{1/\bar{r}},$$

where $r^{-1} + \bar{r}^{-1} = 1$. Since $H^{1-\frac{2}{\bar{r}}}(\Omega) \subset L_{\bar{r}}(\Omega)$ (see, e.g., [**102**, Chap.4]), from polynomial growth condition (6.23) assumed on g_i we have

$$\int_{\Omega_2} |g_i(v_{x_i}) u_{x_i}| dx \leq C \left(\int_{\Omega_2} |g_i(v_{x_i})| |v_{x_i}|^{p(r-1)} dx \right)^{1/r} \|u\|_{2-2/\bar{r}}.$$

If we choose $r = 1 + \frac{1}{p}$, we obtain

$$\int_{\Omega_2} |g_i(v_{x_i}) u_{x_i}| dx \leq C \left(\int_{\Omega_2} |g_i(v_{x_i})| |v_{x_i}| dx \right)^{\frac{p}{1+p}} \|u\|_{2-2/(1+p)}$$

$$\leq C \int_{\Omega_2} |g_i(v_{x_i})| |v_{x_i}| dx \cdot \|u\|_{2-2/(1+p)}, \tag{6.44}$$

where in the last inequality we use the fact that on Ω_2 we have

$$\left(\int_{\Omega_2} |g_i(v_{x_i})| |v_{x_i}| dx \right)^{\frac{p}{1+p}} \leq C \int_{\Omega_2} |g_i(v_{x_i})| |v_{x_i}| dx.$$

Combining (6.42), (6.43) and (6.44) yields the inequality

$$\left| \int_\Omega g_i(v_{x_i}) u_{x_i} dx \right| \leq C_1 \|u\|_{\frac{2p}{1+p}} \cdot \int_\Omega g_i(v_{x_i}) v_{x_i} dx + C_2(1 + \|u\|_1^2). \tag{6.45}$$

Similar, but simpler argument applies to the term g_0. Due to Sobolev's embedding $H^{1+\delta}(\Omega) \subset L_\infty(\Omega)$ (see, e.g., [**102**, Chap.4]), we need no growth conditions on $g_0(v)$. Indeed, on the set

$$\widetilde{\Omega}_2 \equiv \{x \in \Omega : |v(x)| \geq 1\}$$

we have

$$\left| \int_{\widetilde{\Omega}_2} g_0(v) u dx \right| \leq C \max_\Omega |u| \cdot \int_{\widetilde{\Omega}_2} |g_0(v)| dx \leq C \|u\|_{1+\delta} \int_{\widetilde{\Omega}_2} g_0(v) v dx \tag{6.46}$$

for every positive δ. Thus from (6.45) and (6.46) we we can easily obtain (6.38).

Now we shall verify relation (6.39). We shall carry the computations for terms g_i which are more complicated. In the same way as above using the embedding $H^{1-\frac{2}{\bar{r}}}(\Omega) \subset L_{\bar{r}}(\Omega)$ for $i = 1, 2$ we write

$$\int_\Omega |g_i((u+v)_{x_i}) - g_i(u_{x_i})||w_{x_i}|dx$$

(6.47) $$\leq C\|w\|_{2-2/\bar{r}}\left[\left(\int_\Omega |g_i((u+v)_{x_i})|^r dx\right)^{1/r} + \left(\int_\Omega |g_i(u_{x_i})|^r dx\right)^{1/r}\right],$$

where we have applied Hölder's inequality with exponent $1 < r \leq 2$ to be determined later, $\bar{r}^{-1} = 1 - r^{-1}$. We split the region of integration into two domains

$$\Omega_1 \equiv \{x \in \Omega : |u_{x_i}(x) + v_{x_i}(x)| \leq 1\}; \quad \Omega_2 \equiv \{x \in \Omega : |u_{x_i}(x) + v_{x_i}(x)| \geq 1\}.$$

Using the assumptions that $g_i(s)s \geq 0$ and the growth condition on g_i we obtain

$$\int_\Omega |g_i((u+v)_{x_i})|^r dx = \left[\int_{\Omega_1} + \int_{\Omega_2}\right]|g_i((u+v)_{x_i})|^{r-1}|g_i((u+v)_{x_i})|dx$$
$$\leq C_\Omega + \int_{\Omega_2} |(u+v)_{x_i}|^{p(r-1)}|g_i((u+v)_{x_i})|dx.$$

From here after selecting $r = 1 + 1/p$ we get

$$\left(\int_\Omega |g((u+v)_{x_i})|^r dx\right)^{1/r} \leq C_\Omega\left(1 + \int_\Omega g((u+v)_{x_i})(u+v)_{x_i}dx\right)^{1/r}$$

(6.48) $$\leq C_\Omega\left(1 + \int_\Omega g_i((u+v)_{x_i})(u+v)_{x_i}dx\right).$$

The same argument applies to the term $g_i(u_{x_i})$ giving

(6.49) $$\left(\int_\Omega |g_i(u_{x_i})|^r dx\right)^{1/r} \leq C_\Omega\left[1 + \int_\Omega g_i(u_{x_i})u_{x_i}dx\right].$$

As above it is also easy to find that

$$\int_\Omega |[g_0(u+v) - g_0(u)]w|dx$$
$$\leq C\max_\Omega |w| \cdot \int_\Omega [|g_0(u+v)| + |g_0(v)|]dx$$

(6.50) $$\leq C_\Omega\|w\|_{1+\delta}\left(1 + \int_\Omega g_0(u+v)(u+v)dx + \int_\Omega g_0(v)vdx\right)$$

for every $\delta > 0$. Therefore combining with (6.47), (6.48), (6.49) we obtain (6.39). $\square$

Now we are in position to prove existence of global attractor for the system (H, S_t) generated by (6.1) and (6.2).

We first we note that (H, S_t) is ultimately dissipative. Indeed, in the case $L \neq 0$ it follows from (6.35) and (6.36) that Assumption 3.7(D) holds and, hence, we can apply Theorem 3.10. If $L \equiv 0$, then by (6.35) and (6.38) we are in position to apply Theorem 3.4 because (6.38) implies (3.12). Indeed, the substitution $u := \delta u$ in (6.38) gives that

$$|D(v, u)| \leq C_1\|u\|_s \cdot D(v, v) + C_2(1/\delta + \delta\|u\|_1^2), \; v \in H^1(\Omega), \; u \in H^2(\Omega),$$

which implies (3.12) with $\delta > 0$ as small as we want. Thus (H, S_t) is dissipative and it possesses a forward invariant absorbing set.

Now we note that by (6.37), (6.39) and (6.28) Assumption 3.21(D) holds (with $C^\epsilon(r) = 0$, $\kappa = 1$ and without the term $\epsilon|\mathcal{A}^{1/2}w|^2$). Therefore, the existence of a compact global attractor $\mathcal{A}$ follows from Corollary 3.28.

To prove the finiteness of fractal dimension of $\mathcal{A}$ and the stabilization properties we need the following Lemma.

LEMMA 6.10. *Under Assumption 6.3 and hypotheses (6.24) and (6.30) the form $D(v, u)$ given by (6.27) possesses the properties*
- *if $p = 1$ and $g_0'(s)$ is bounded (when $L \ne 0$), then*

(6.51) $\quad |D(u + v, w) - D(u, w)| \le C_1[D(u + v, v) - D(u, v)] + C_2\|w\|_1^2$

  for any $v \in H_0^1(\Omega)$, $u \in H^1(\Omega)$ $w \in H^1(\Omega)$;
- *if $1 \le p < \infty$ (when $L = 0$), then*

$$|D(u+v,w) - D(u,w)| \le C_1[D(u+v,v) - D(u,v)]$$
(6.52) $\quad + C_2\|w\|_s^2 \left[1 + \|u\|_1^{p_0-1} + \|v\|_1^{p_0-1} + D(u+v, u+v) + D(u,u)\right]$

  for any $v \in H_0^1(\Omega)$, $u \in H^1(\Omega)$ $w \in H^2(\Omega)$, where $s \in (1, 2)$.

PROOF. Property (6.30) implies the relation

$$0 \le \frac{g_i(s_2) - g_i(s_1)}{s_2 - s_1} \le C\left(1 + s_1 g_i(s_1) + s_2 g_i(s_2)\right)^\beta, \quad \forall s_1 < s_2, \; i = 1, 2,$$

where $0 \le \beta < 1$ (in the first case $\beta = 0$). Therefore, as in the proof of Theorem 5.8 (cf. (5.22)), for arbitrary $\delta > 0$ we have

$$\int_\Omega |g_i((u+v)_{x_i}) - g_i(u_{x_i})||w_{x_i}|dx \le \delta \int_\Omega [g_i((u+v)_{x_i}) - g_i(u_{x_i})] \cdot |v_{x_i}|dx$$

$$+ C_\delta \int_\Omega \left(1 + (u+v)_{x_i} g_i((u+v)_{x_i}) + u_{x_i} g_i((u)_{x_i})\right)^\beta \cdot |w_{x_i}|^2 dx$$

$$\le \delta[D(u+v,v) - D(u,v)] + C_\delta \|w\|_s^2 [1 + a_\beta D(u+v, u+v) + D(u,u)]$$

with some $1 < s < 2$, where $a_\beta = 0$ if $\beta = 0$. Since $g_0(s)$ is of polynomial growth (see (6.24)), in a similar way we also have

$$\int_\Omega |g_0(u+v) - g_0(u)||w|dx \le \delta \int_\Omega (g_0(u+v) - g_0(u))v dx$$

$$+ C_\delta \|w\|_s^2 \cdot \int_\Omega \left(1 + |u|^{p_0-1} + |v|^{p_0-1}\right) dx, \quad s > 1,$$

where $p_0 = 1$ in the first case. Therefore (6.51) and (6.52) easily follow. $\square$

Completion of the proof of Theorem 6.4. Lemma 6.10 implies (3.60) with $\kappa = 2$. Thus the hypotheses of Theorem 4.1 hold and, hence, we are in position to conclude that the global attractor A of the system (H, S_t) generated by (6.1) and (6.2) has a finite fractal dimension. The smoothness of the attractor follows from Theorem 4.17.

The last statement of the Theorem 6.4 - existence of exponential fractal attractors - follows from Theorem 4.42.

The stabilization property (6.26) (in the case $L \equiv 0$) follows from Theorem 6.7.

6.4.3. Proof of Theorem 6.7. We apply Theorem 4.31. By Remark 4.32 property (4.94) follows from (3.60) which is true in our case under conditions (6.24) and (6.30) (see the argument above). Since for $L \equiv 0$ and $g^* \equiv 0$ we have
$$\langle F'(u), w\rangle = [v(u) + F_0, w] + [v(u, w), u],$$
where $v = v(u)$ satisfies (6.2) and $\widetilde{v} = v(u, w)$ solves the problem
$$\Delta^2 \widetilde{v} + 2[w, u] = 0, \text{ in } \Omega, \quad \frac{\partial}{\partial \nu}\widetilde{v} = \widetilde{v} = 0 \text{ on } \Gamma,$$
properties (4.95) and (4.96) easily follows from (6.6) and (6.9). To prove the energy inequality for this case we can use the same approximation method as for the wave equation in Chapter 5 (see the proof of Theorem 5.10). Thus the proof of Theorem 6.7 is complete.

REMARK 6.11. Under the conditions of Theorem 6.4 we can also apply Corollary 4.40 on the existence of a finite number of determining functionals. We also note that this question for problem (6.1) and (6.2) with $\alpha > 0$ was studied in [**25**] by another method.

6.4.4. Proof of Theorem 6.8. We first prove well-posedness of system (6.1) and (6.2) with $L \equiv 0$ and $g^* \neq 0$. For this we need to check Assumption 1.1.

Since $L_q(\Omega) \subset H^{-1}(\Omega)$ for every $q > 1$, it follows from (6.33) that
$$\|g^*(v_1) - g^*(v_2)\|_{-1} \leq C\|g^*(v_1) - g^*(v_2)\|_{L_{3/2}(\Omega)}$$
$$\leq C\left[\int_\Omega \left(1 + |v_1|^{p_*-1} + |v_2|^{p_*-1}\right)^{3/2} |v_1 - v_2|^{3/2} dx\right]^{2/3}$$
$$\leq C\left[\int_\Omega \left(1 + |v_1|^{p_*-1} + |v_2|^{p_*-1}\right)^6 dx\right]^{1/6} \|v_1 - v_2\|.$$

Using the embedding $H^1(\Omega) \subset L_{6(p_*-1)}(\Omega)$, we obtain that

(6.53) $$\|g^*(v_1) - g^*(v_2)\|_{-1} \leq C\left(1 + r^{p_*-1}\right)\|v_1 - v_2\|$$

for any $v_i \in H_0^1(\Omega)$, $\|v_i\|_1 \leq r$. From (6.34) we also have that
$$(g^*(v), u) \leq k_0(g_0(v), v) + C$$
for some $k_0 < k$ and $C \geq 0$. Thus (1.3) and (1.6) hold for the model considered. Other requirements from Assumption 1.1 are also true by the argument given in proof of Theorem 6.4. Thus the existence of semiflow follows from Theorem 1.5.

To prove dissipativity of this semiflow we apply Corollary 3.17. To conclude, note that conditions (3.52) follows from (6.34) and (3.51) in Assumption 3.16 is satisfied with $\gamma = 1/2$. This latter assessment results from (6.38). See also Example 3.20 for some details.

To prove the existence of a compact global attractor we use Corollary 3.28. Indeed, Assumption 3.21(D) was checked in the proof of Theorem 6.4. As for Assumption 3.21(F) it follows from (6.19) and (6.53) that relation (3.61) holds with $n_V(z) \equiv \|z\|_{L_2(\Omega)}$. Consequently we can apply Corollary 3.28 and obtain the existence of a compact global attractor A.

To prove finiteness of fractal dimension of A we need to check (4.2) in Theorem 4.1. Our hypotheses concerning g_0 and g_i in the last part of Theorem 6.8 guarantee (4.2) with $l = 1$. The smoothness of the attractor follows from Theorem 4.17. Thus the proof of Theorem 6.8 is complete.

6.5. Non-rotational case $\alpha = 0$

In the "non-rotational case", the situation is more delicate due to *less regularity of the phase space*. In this case, the nonlinear term in the equation is no longer *compact*.

We consider the following von Karman model with clamped boundary conditions:

(6.54)
$$u_{tt} + k\, g(u_t) + \Delta^2 u = [v(u) + F_0, u] + L(u) + p \text{ in } \Omega \times (0, \infty),$$
$$u = \frac{\partial}{\partial \nu} u = 0, \text{ on } \Gamma \times (0, \infty).$$

As above, the Airy stress function $v(u)$ satisfies the elliptic problem (6.2) The von Karman bracket $[u, v]$ is given by (6.3) and the damping parameter k is positive. This model also accounts for non-conservative force modeled by operator L.

The feature of the model in comparison with the case $\alpha > 0$ that we wish to emphasize is that equation (6.54) does not account for rotational inertia. As a consequence, well-posedness (uniqueness) of weak solutions has been, until recently, an open problem. In fact, only recently due to discovery of sharp regularity of Airy's stress function, the uniqueness of finite energy solutions could be shown [**47**]. This implies that one has a well-posed semiflow on $H_0^2(\Omega) \times L_2(\Omega)$. However, the effect of nonlinear term is *not compact* with respect to the topology of phase space. So, once more we deal with a situation of *noncompact* nonlinearity. Unlike the case of semilinear wave equation, the nonlinearity is also non-local. This prevents the usual "splitting" of nonlinearity into a "compact" part and "dissipative" part - applicable to wave equation -as seen before (see [**4, 48, 60, 101**] and also the discussion in Chapter 3).

Our standing assumption is the following:

ASSUMPTION 6.12.
- The increasing function $g \in C^1(\mathbb{R})$ is assumed to satisfy $g(0) = 0$ along with the following bound: there exists a positive constant m such that $m \leq g'(s)$ for all $|s| \geq 1$. If $L \not\equiv 0$ we assume that $g'(s)$ is bounded from above.
- $p(x) \in L_2(\Omega)$ and $F_0(x) \in W_\infty^2(\Omega)$.
- L is a linear first order differential operator with smooth coefficients.

Our main result reads as follows:

THEOREM 6.13.
- **Generation of the flow.** *Under Assumption 6.12 equations (6.54) and (6.2) generates a continuous semiflow S_t in the space $H \equiv H_0^2(\Omega) \times L_2(\Omega)$ with the following properties.*
- **Global attractor.** *There exists a compact global attractor A. In the case $L \equiv 0$ this attractor has the form $A = \mathcal{M}^u(\mathcal{N})$, where $\mathcal{M}^u(\mathcal{N})$ is the unstable manifold emanating from the set $\mathcal{N}$ (see the definition in Sect. 2.4) and $\mathcal{N}$ is the set of equilibria for the semiflow S_t. Moreover, the statements of Theorem 2.31 and Corollary 2.32 are in force.*
- **Smoothness of attractor.** *Assuming that $L \equiv 0$ and that there exist constants $m > 0$ and $M < \infty$ such that*

(6.55)
$$m \leq g'(s) \leq M\left(1 + s g(s)\right) \text{ for all } s \in \mathbb{R},$$

the attractor A is a bounded set in $(H^4 \cap H_0^2)(\Omega) \times H_0^2(\Omega)$. Assuming, in addition that F_0 and p are $C^\infty(\Omega)$, $g \in C^\infty(\mathbb{R})$, the attractor belongs to $C^\infty(\Omega) \times C^\infty(\Omega)$.

- **Finite dimension of the attractor.** Assuming (6.55) with $m > 0$ and $L = 0$ the fractal dimension of the attractor is finite. If $L \neq 0$, the same conclusion holds under additional assumption that the damping parameter k is sufficiently large.
- **Uniform decay rates to equilibria** Let (6.55) be valid with $m = 0$. If the the set of stationary solutions consists of finitely many isolated equilibrium points which are hyperbolic, then for any initial condition $y_0 \in H$ there exists an equilibrium point $e = (w, 0)$, $w \in H_0^2(\Omega)$, such that

(6.56) $$|S_t y - e|^2 \leq C \cdot \sigma\left([tT^{-1}]\right), \ t > 0,$$

where C and T are positive constants, $[a]$ denotes the integer part of a and $\sigma(t)$ solves (6.32) with $Q(s) = s - (I + G_0)^{-1}(s)$ and $G_0(s) = c_1 (I + H_0)^{-1}(c_2 s)$. Here $H_0(s) = c_3 h(c_4 s)$, where c_i are positive numbers and $h : \mathbb{R}_+ \mapsto \mathbb{R}_+$ is a concave, strictly increasing, continuous function with the properties

(6.57) $$h(0) = 0 \quad and \quad s^2 + g^2(s) \leq h(sg(s)) \quad for \quad |s| \leq 1.$$

If (6.55) holds with $m > 0$, then (6.56) holds with $\sigma(t) = C(y, e)e^{-\omega t}$, where $\omega > 0$.

REMARK 6.14. Similarly to Remark 5.6 and Remark 6.5 we note that the property $g'(s) \leq M(1 + sg(s))$ holds true, if we assume that there exist $p \geq 1$ and $M_1 > 0$ such that $g'(s) \leq M_1[1 + |s|^{p-1}]$, and when $p > 3$, $g(s)s \geq m_1|s|^{p-1}$ for all $|s| \geq 1$ and for some $m_1 > 0$. However our assumption (6.55) allows also exponential behavior for $g(s)$, e.g., $|g(s)| \sim e^{\alpha|s|}$ as $|s| \to \infty$ for some $\alpha > 0$.

REMARK 6.15. One could also add another non-conservative term $g^*(u_t)$ to equation (6.54). However, as for the wave equation we need some compactness assumptions imposed on the mapping $v \mapsto g^*(v)$ of $L_2(\Omega)$ into itself. In this case by adapting arguments given in Chapter 5, we could easily include the term $g^*(u_t)$ to our analysis. We are not pursuing this, mostly because we prefer to keep the exposition as focused as possible by avoiding redundant arguments.

Few words about the literature relevant to the problem. First study of attractors in the context of dynamics described by (6.54) has been carried out in [**15**] in the case of *linear* damping. Later, in [**77**], existence and finite dimensionality of attractors was shown in the case of *nonlinear* dissipation subject to rather severe growth restrictions -including linear growth condition required for finite dimensionality of attractor and large size of the damping parameter. The main contribution of of Theorem 6.13 is that it allows for rather unrestricted, possibly superlinear growth of dissipation. In fact, one of the main challenges of the present paper was to relax linear growth assumption imposed on the damping in [**77**]. It turned out that this was possible due to novel criteria for asymptotic smoothness and finite-dimensionality introduced in Chapter 2. These are generalizations of criterions known before and applicable mostly to hyperbolic like problems with linear (almost linear) damping only. Some of these criterions (e.g. Theorem 4.4) were used in recent publications [**66, 67**] in order to obtain sharp results on existence and finite dimensionality of attractor for a simple - source free - von Karman equation with $F_0 = 0, L = 0$. Theorem 6.13 generalizes these results to the case of full von Karman model that accounts for full sources $F_0 \neq 0$ and non-conservative terms $L \neq 0$. In addition, the last assertion in Theorem 6.13 provides quantitative information on decay rates of solutions to equilibria points. The result on stabilization to equilibrium is, to our best knowledge, completely new in the literature for this problem even in the case of linear damping.

We also note that the result given in Theorem 6.13 can be obtained with other boundary conditions such as hinged or free. See for some details in [**24**]. We can also consider the systems with a mixed boundary-interior damping [**31**].

REMARK 6.16. Although the result of Theorem 6.13 is a consequence of several abstract results, the main wheel behind the proof (say- in the case when $g'(s) > 0$) is the following "stabilizability" estimate

(6.58) $$E(z(t)) \leq C_1 e^{-\omega t} E(z(0)) + C_2 \sup_{0 \leq \tau \leq t} \left\{ \|z(\tau)\|_{H^{2-\epsilon}(\Omega)}^2 \right\}.$$

where $z(t) \equiv u(t) - w(t)$, $(u(t), u'(t))$ and $(w(t), w'(t))$ are two trajectories of the original problem which belong to the attractor $\mathcal{A}$, and the (free) energy E is given by

$$E(z(t)) = \int_\Omega [|z_t|^2 + |\Delta z|^2] dx.$$

This particular estimate, known in control theory as observability/stabilizability estimate, provided a motivation and impetus to search for more general abstract criteria for compactness and finite dimensionality of attractors, which can be applicable to a large class of systems. Theorems presented in Chapter 3 result from this effort.

REMARK 6.17. One question that remains open is whether finite dimensionality of attractor can be proved for *non-conservative case - $L \neq 0$* without assuming large damping parameter.

Proof of Theorem 6.13. Step 1: Generation of the flow and dissipativity. We shall recast system (6.54) as an abstract evolution dynamics described by (6.22) with $V = L_2(\Omega)$ and $M = I$. As in the case $\alpha > 0$ we also have that $\mathcal{H} \equiv L_2(\Omega)$, $\mathcal{A}u = \Delta^2 u, u \in \mathcal{D}(A) \equiv H_0^2(\Omega) \cap H^4(\Omega)$. We have the identification $\mathcal{D}(\mathcal{A}^{1/2}) \sim H_0^2(\Omega)$. The nonlinear operators $D(u)$ and $F(u, u_t) \equiv F(u)$ are defined as:
$$D(u) \equiv kg(u), \quad F(u) \equiv [v(u) + F_0, u] + L(u) + p.$$
So $F(u) = -\Pi'(u) + F^*(u)$ with
(6.59) $$\Pi(u) = \Pi_0(u) + \Pi_1(u),$$
where
$$\Pi_0(u) \equiv \frac{1}{4} \|\Delta v(u)\|^2, \quad \Pi_1(u) \equiv -\frac{1}{2}([u, u], F_0) - (p, u)$$
and $F^*(u) = L(u)$.

As above we can conclude from Lemma 6.2 that the nonlinearity F in this case satisfies all hypotheses concerning F which appear in Chapters 1 and 3 (see Assumptions 1.1(F), 3.1(F) 3.7(F) and 3.21(F)). Thus generation of the flow is guaranteed by Theorem 1.5. Ultimate dissipativity follows from Theorem 3.4 (in the non-conservative case) and Theorem 3.10 in the conservative case.

Step 2: Compactness of attractor. Assumption 3.21(F) holds with $\widetilde{\eta} = 0$ and, hence, in the case $\alpha = 0$ the nonlinearity F is critical. However, due to (6.19) and (6.21), Assumption 3.33(F) holds and we can appeal to Theorem 3.34 and Proposition 3.36.

Since we have that

$$\left|\int_\Omega (g(u+v) - g(u))w\, dx\right| \leq \|w\|_{L_\infty(\Omega)} \int_\Omega (|g(u+v)| + |g(u)|)dx$$

$$\leq C\|w\|_{2-2\delta}\left(\int_\Omega (g(u+v)(u+v) + g(u)u)dx + 1\right)$$

(6.60) $$\leq C|\mathcal{A}^{1/2-\delta}w|[1 + (D(u+v), u+v) + (Du, u)],$$

relation (3.60) postulated by Assumption 3.33(D) holds with $\kappa = 1$. Therefore by Proposition 3.36 the system (H, S_t) is asymptotically smooth. Consequently, in the case $L \equiv 0$, we use Corollary 2.29 along with already established dissipativity to guarantee the existence of a compact global attractor. In the case $L \not\equiv 0$ we use the first part of Theorem 3.10 to claim dissipativity of the system (with an absorbing set which may depend on the damping parameter k). Hence by Theorem 3.34 a compact global attractor exists. The proof of existence of global attractors is thus completed.

Step 3: Independence of the size of attractor on the damping parameter k. In the case when $m = \inf_{s\in\mathbb{R}} g'(s) > 0$ and the damping parameter k is large, we can also appeal to Theorem 3.4 in the case $L \equiv 0$ and to the second part of Theorem 3.10 when $L \not\equiv 0$ to guarantee independence of k for the size of an absorbing set. Indeed, since $g'(s) \geq m$ for all $s \in \mathbb{R}$ and $g(0) = 0$ we have that $ms^2 \leq sg(s)$ for $s \in \mathbb{R}$. Thus

$$m\|v\|^2 \leq \int_\Omega v(x)g(v(x))dx \quad \text{for all} \quad v \in H_0^2(\Omega).$$

Thus (3.2) holds with $c_0 = 0$. In the case $L \not\equiv 0$ we obviously have that $|g(s)| \leq C \cdot |s|$ for all s and therefore

$$\|g(v)\|_{-2} \leq C\left(\int_\Omega |g(v(x))|^2 dx\right)^{1/2} \leq C\|v\| \quad \text{for all} \quad v \in L_2(\Omega).$$

Hence the operator D satisfies (3.24). Consequently, by Theorem 3.10 the system possesses a bounded forward invariant absorbing set with the size independent of $k \geq k_0 > 0$. If $L \equiv 0$, we use the relation $|g(s)| \leq \delta + C_\delta sg(s)$ for every $\delta > 0$ and for all $s \in \mathbb{R}$ to obtain that

$$\left|\int g(v)u\, dx\right| \leq \max_\Omega |u|\left(|\Omega| \cdot \delta + C_\delta \int vg(v)dx\right)$$

$$\leq \widetilde{C}_\delta \cdot [E_0(u,v)]^{1/2} \int vg(v)dx + C \cdot \delta(1 + E_0(u,v)),$$

where $E_0(u,v) = \frac{1}{2}\left(\|v\|^2 + \|\Delta u\|^2\right)$. This implies (3.3) and, hence, by Theorem 3.4 we obtain the same conclusion as for the case $L \not\equiv 0$.

Step 4: Smoothness of the attractor. To obtain the third assertion in Theorem 6.13 we use Corollary 4.26. For this we need to verify Assumption 3.21(D) with $H_0(s) = m^{-1}s$ and $\kappa = 2$ and Assumption 4.11.

Verification of Assumption 3.21(D) is accomplished on the strength of the growth restriction (6.55) imposed on g. Computations are almost the same as

in (5.22) and (5.23) where we can take $r = 1$ (and $\bar{r} = \infty$) and use the fact that (6.55) implies the relation
$$m \le \frac{g(s_2) - g(s_1)}{s_2 - s_1} \le M\left(1 + s_1 g(s_1) + s_2 g(s_2)\right), \ \forall s_1 < s_2.$$
Additional hypotheses on g imposed in Corollary 4.26 are also obvious.

We turn to Assumption 4.11. Since $\Pi(u)$ is given by (6.59), after a simple calculation we have that the second $\Pi^{(2)}(u)$ and the third $\Pi^{(3)}(u)$ Frechet derivatives of $\Pi(u)$ have the form
$$\begin{aligned}\langle \Pi^{(2)}(u); w_1, w_2\rangle &= -([v(u) + F_0, w_1], w_2) - 2([v(u, w_1), u], w_2),\\ \langle \Pi^{(3)}(u); w_1, w_2, w_3\rangle &= -2([v(w_1, w_2), u], w_3) - 2([v(w_1, u), w_2], w_3)\\ &\quad -2([v(w_2, u), w_1], w_3),\end{aligned} \quad (6.61)$$
for every $u, w_i \in H_0^2(\Omega)$, where we denote by $v(w_1, w_2) \in H_0^2(\Omega)$ the solution to the problem
$$(6.62) \quad \Delta^2 v(w_1, w_2) = -[w_1, w_2] \text{ in } \Omega, \quad \frac{\partial}{\partial \nu} v(w_1, w_2) = v(w_1, w_2) = 0 \text{ on } \Gamma.$$

We need to check (4.31) and (4.32) for these $\Pi^{(k)}(u)$.

Let $\|u\|_2 \le \rho$. Using (6.6) with $j = 2$, $\beta = 1/2$ and (6.9) with $\delta = 1$, $u_1 = u$, $u_2 = 0$ we get that
$$\begin{aligned}|([v(u) + F_0, w_1], w_2)| &= |(v(u) + F_0, [w_1, w_2])| \le C\|v(u) + F_0\|_2 \|[w_1, w_2]\|_{-2}\\ &\le C_\rho \|w_1\|_{3/2} \|w_2\|_{3/2}.\end{aligned}$$

Similarly, from (6.6) with $j = 2$, $\beta = 0$ and from elliptic regularity of Δ^2 we have that
$$|([v(u, w_1), u], w_2)| = |(v(u, w_1), [u, w_2])| \le C_\rho \|w_1\|_1 \|w_2\|_1.$$
Therefore we have the estimate
$$(6.63) \quad \left|\langle \Pi^{(2)}(u); w_1, w_2\rangle\right| \le C_\rho \|w_1\|_{3/2} \|w_2\|_{3/2},$$
which implies (4.31). We also obviously have that
$$\left|\langle \Pi^{(3)}(u); w_1, w_2, w_3\rangle\right| = 2\Phi(u; w_1, w_2) \cdot \|w_3\|,$$
where
$$\Phi(u; w_1, w_2) = \|[v(w_1, w_2), u]\| + \|[v(w_1, u), w_2]\| + \|[v(w_2, u), w_1]\|.$$
Using (6.10) we obtain that
$$\Phi(u; w_1, w_2) \le C\|u\|_2 \|w_1\|_2 \|w_2\|_2 \le C_\rho \|w_1\|_2 \|w_2\|_2,$$
which implies (4.32). Thus, Assumption 4.11 holds and by Corollary 4.26, we can conclude that the attractor A is a bounded subset in $(H^4 \cap H_0^2)(\Omega) \times H_0^2(\Omega)$.

Higher level smoothness $C^\infty(\Omega) \times C^\infty(\Omega)$ for smooth data g, F_0 and p follows by applying the argument of Theorem 4.19 (see also Corollary 4.23) to equation satisfied for time derivatives of solutions. This is the usual boot-strap type of argument (see [**81**] and also [**16**] where a similar result was proved for von Karman equation (6.54) in the case of linear damping). In the present case, nonlinear operators involved become algebra on the respective Sobolev spaces (since $H^2(\Omega$ is embedded in $C(\overline{\Omega}))$, so that reiteration of the argument poses no difficulties.

Step 5: Finite dimensionality of attractor. In order to establish finite dimensionality of the attractor, in the case $L \equiv 0$, we appeal to the third statement in Theorem 4.25, whose assumptions have been verified above. In the case $L \neq 0$ we apply Theorem 4.1 which relies on Theorem 3.58 (with $\kappa = 2$) and requires large size of the damping parameter. Since the size of attractor is independent on the large size of the damping parameter k (Step 3 in the proof of Theorem 6.13), application of that theorem (note that other assumptions like (3.60) and (3.21) have been already verified) leads to a desired conclusion.

Step 6: Uniform decay rates. Theorem 4.31 implies the conclusions of the fifth part of Theorem 6.13. Indeed, the additional requirements posted in (4.95) and (4.96) follow from standard regularity properties of von Karman brackets and Airy's stress functions (see Proposition 6.1 and also the argument given in the previous section). Validity of energy *inequality* for generalized solutions follows from the same argument as in the case of wave equation. □

REMARK 6.18. If Assumption 6.12 and relation (6.55) hold, we can apply Corollary 4.40 on determining functionals and also Theorem 4.42 and Theorem 4.44 on the existence of exponential fractal attractors. We also note that determining functionals for problem (6.54) was considered before in [**17**] for linear damping and in [**25**] for nonlinear function g and $L \equiv 0$ by another method. The existence of exponential fractal attractors for this problem has not been yet addressed.

CHAPTER 7

Other models from continuum mechanics

In this chapter we consider several other PDE models arising in continuum mechanics. Our aim is to demonstrate that the methods developed in Chapters 3 and 4 provide a description of long-time dynamics also for these specific dynamical systems.

7.1. Berger's plate model

This model was suggested in [7] as a simplification of von Karman plate equations which describes large deflections of plates. In the case of hinged plate the Berger equation takes the form

$$u_{tt} + \Delta^2 u + \left(Q - \int_\Omega |\nabla u(x,t)|^2 dx\right) \cdot \Delta u = p(u, u_t, x) \quad \text{in} \quad \Omega \times (0, \infty),$$

$$u = \Delta u = 0 \quad \text{on} \quad \Gamma \times (0, \infty),$$

(7.1)
$$u|_{t=0} = u_0(x), \ u_t|_{t=0} = u_1(x).$$

Here $\Omega \subset \mathbb{R}^n$ is a bounded domain with a sufficiently smooth boundary Γ, $n = 1, 2$. The function $u(x,t)$ is the displacement of the plate, the parameter Q describes in-plane forces applied to the plate and the function p represents transverse loads which may depend on the displacement u and the velocity u_t. The case $n = 1$ corresponds to the plate which is infinite in one direction (infinite panel). In this case equation (7.1) becomes a well-known *beam equation* which was treated by many authors (see, e.g., the monographs [19, 59], the survey [40] and also the references therein). In the abstract form (which covers the both dimensions $n = 1$ and $n = 2$) long-time behaviour of linearly damped problem (7.1) was treated in [19, Chap.4] with all details. Our main emphasis is placed on nonlinear dissipation and transwersal loads that are represented by non-compact operators $p(u, u_t)$.

The unforced energy function associated with Berger's model takes the form:

$$\mathcal{E}(u, u_t) = \frac{1}{2}\left[\|u_t\|^2_{L_2(\Omega)} + \|\Delta u\|^2_{L_2(\Omega)} - Q\|\nabla u\|^2_{L_2(\Omega)} + \frac{1}{2}\|\nabla u\|^4_{L_2(\Omega)}\right].$$

The energy may be negative, however due to the presence of nonlocal superlinear term in the equation, it is bounded from below,

$$\mathcal{E}(u, u_t) \geq \frac{1}{2}\left[\|u_t\|^2_{L_2(\Omega)} + \|\Delta u\|^2_{L_2(\Omega)} + \frac{1}{4}\|\nabla u\|^4_{L_2(\Omega)}\right] - Q^2.$$

In what follows we shall consider effects of aeroelastic forces applied to the plate. Our concern is long time behaviour of dynamics governed by the corresponding models. We shall begin with a simplest case of aeroelastic force that is conservative. Later in the section we shall consider non-conservative forces which constitute typical models of aeroelastic forces. For this latter case, we shall see that the developments in Chapter 3 that accounts for presence of non-conservative forces is going to play critical role.

7.1.1. Damped system with conservative forces. We assume that the transverse loads in the right hand side of (7.1) has the form

(7.2) $$p(u_t, x) \equiv -k \cdot g(u_t(x,t)) + p_0(x),$$

where $p_0(x) \in L_2(\Omega)$, k is a positive parameter, and $g \in C^1(\mathbb{R})$ is a monotone increasing function such that $g(0) = 0$.

It is clear that problem (7.1) and (7.2) can be written in the form

(7.3) $$\begin{cases} u_{tt}(t) + \mathcal{A}u(t) + k \cdot D(u_t(t)) = F(u(t)), \\ u|_{t=0} = u_0 \in \mathcal{D}(\mathcal{A}^{1/2}), \ u_t|_{t=0} = u_1 \in \mathcal{H}, \end{cases}$$

with the following notation:
- $\mathcal{H} \equiv L_2(\Omega)$.
- $\mathcal{A}u \equiv \Delta^2 u$, $u \in \mathcal{D}(\mathcal{A}) \equiv \{u \in H_0^1(\Omega) \cap H^4(\Omega) : \Delta u = 0 \text{ on } \Gamma\}$.
- $F(u, u_t) \equiv F(u) \equiv \left(Q - |\mathcal{A}^{1/4} u(t)|^2\right) \mathcal{A}^{1/2} u + p_0$.
- $D(v) \equiv g(v)$.

It is obvious that $\mathcal{A}$ satisfies the conditions introduced at the beginning of Sect. 1.2. Moreover, we have that $\mathcal{A}^{1/2} u = -\Delta u$ on the domain $\mathcal{D}(\mathcal{A}^{1/2}) \equiv H_0^1(\Omega) \cap H^2(\Omega)$ and $|\mathcal{A}^{1/4} u| = \|\nabla u\|_{L^2(\Omega)}$ for any $u \in \mathcal{D}(\mathcal{A}^{1/4}) \equiv H_0^1(\Omega)$.

The nonlinear term $F(u)$ has the form

$$F(u) = -\Pi'(u)$$

with

(7.4) $$\Pi(u) = \frac{1}{4}|\mathcal{A}^{1/4} u|^4 - \frac{Q}{2}|\mathcal{A}^{1/4} u|^2 - (p_0, u) \equiv \Pi_0(u) + \Pi_1(u),$$

where

(7.5) $$\Pi_0(u) = \frac{1}{4}|\mathcal{A}^{1/4} u|^4 \quad \text{and} \quad \Pi_1(u) = -\frac{Q}{2}|\mathcal{A}^{1/4} u|^2 - (p_0, u).$$

A simple calculation shows that Assumption 1.1 holds in this case and, hence, by Theorem 1.5 equations (7.1) and (7.2) generates a semiflow S_t in the space $H = (H_0^1 \cap H^2)(\Omega) \times L_2(\Omega)$. Since the nonlinear term F satisfies (3.61) from Assumption 3.21 with $\tilde{\eta} = 0$, we are in the *critical* case. This suggests that we could apply the same argument as in the case of von Karman plate with $\alpha = 0$ (see Chapter 6). In fact, we shall see that the analysis in this case is simpler. Under the same assumptions (as in von Karman case with $\alpha = 0$) imposed on the damping we shall obtain existence of global attractor. Our main results for the Berger's model read as follows.

THEOREM 7.1. *With the reference to problem (7.1) and (7.2) we assume that $g(s) \in C^1(\mathbb{R})$ is increasing, $g(0) = 0$ and*

(7.6) $$g'(s) \geq m \quad \text{for all} \quad |s| \geq 1,$$

where m is a positive constant. Then
- *there exists compact, global attractor A of the form $A = \mathcal{M}^u(\mathcal{N})$, where $\mathcal{M}^u(\mathcal{N})$ is the unstable manifold emanating from the set $\mathcal{N}$ (see the definition in Sect. 2.4) and $\mathcal{N}$ is the set of equilibria for S_t. Moreover, the statements of Theorem 2.31 and Corollary 2.32 are in force;*

- *assuming in addition that*

(7.7) $\quad\quad\quad 0 < m \leq g'(s) \leq M \cdot (1 + sg(s))$ *for all $s \in \mathbb{R}$,*

the fractal dimension of the attractor is finite. Moreover, in this case the attractor is a closed bounded set in $\mathcal{D}(\mathcal{A}) \times (H_0^1 \cap H^2)(\Omega)$.

PROOF. In the same as in the Karman case with $\alpha = 0$ (see Theorem 6.13) we can use Theorem 3.34. Assumption 3.33(D) was checked in the proof of Theorem 6.13. The requirements listed in Assumption 3.33(F) are obvious. The critical assumption on compactness of potential energy follows from trivial compact embedding $D(\mathcal{A}^{1/2}) \subset D(\mathcal{A}^{1/4})$. Therefore Theorem 3.34 implies the existence of a compact global attractor A. By Proposition 4.30 we have that $A = \mathcal{M}^u(\mathcal{N})$.

To prove the finiteness of fractal dimension and the smoothness of the global attractor we must resort (due to criticality of F) to either Theorem 4.27 or more involved Theorem 4.25. However, as in the case of von Karman equation, subcritical conditions imposed on the set of stationary points (4.82) is *not satisfied* (regardless the smoothness of equilibria points). Therefore, we must rely on Theorem 4.25 and on its Corollary 4.26. Indeed, conditions on the damping operator given in Theorem 4.25 and Corollary 4.26 was verified in Section 6.5. Thus we need to check Assumption 4.11 only. Since $\Pi(u)$ is given by (7.4), simple computations give that the second $\Pi^{(2)}(u)$ and the third $\Pi^{(3)}(u)$ Frechet derivatives of $\Pi(u)$ have the form

$$\langle \Pi^{(2)}(u); w_1, w_2 \rangle = \left[|\mathcal{A}^{1/4} u|^2 - Q \right] (\mathcal{A}^{1/4} w_1, \mathcal{A}^{1/4} w_2) \\ + 2(\mathcal{A}^{1/4} u, \mathcal{A}^{1/4} w_1)(\mathcal{A}^{1/4} u, \mathcal{A}^{1/4} w_2)$$

and

$$\langle \Pi^{(3)}(u); w_1, w_2, w_3 \rangle = 2(\mathcal{A}^{1/4} u, \mathcal{A}^{1/4} w_3)(\mathcal{A}^{1/4} w_1, \mathcal{A}^{1/4} w_2) \\ + 2(\mathcal{A}^{1/4} w_3, \mathcal{A}^{1/4} w_1)(\mathcal{A}^{1/4} u, \mathcal{A}^{1/4} w_2) \\ + 2(\mathcal{A}^{1/4} u, \mathcal{A}^{1/4} w_1)(\mathcal{A}^{1/4} w_3, \mathcal{A}^{1/4} w_2)$$

for every $u, w_i \in (H^2 \cap H_0^1)(\Omega)$. Therefore

$$\left| \langle \Pi^{(2)}(u); w_1, w_2 \rangle \right| \leq C_\rho |\mathcal{A}^{1/4} w_1| |\mathcal{A}^{1/4} w_2|$$

and

$$\left| \langle \Pi^{(3)}(u); w_1, w_2, w_3 \rangle \right| \leq C_\rho |w_3| \left(|\mathcal{A}^{1/2} w_1| |w_2| + |w_1| |\mathcal{A}^{1/2} w_2| \right)$$

for every $u, w_i \in (H^2 \cap H_0^1)(\Omega)$ such that $|\mathcal{A}^{1/2} u| \leq \rho$. This implies relations (4.31) and (4.32) and thus Assumption 4.11 holds for the case considered.

Therefore we can apply Corollary 4.26 to establish the smoothness of the attractor and the third statement of Theorem 4.25 to prove finiteness of its fractal dimension. □

We note that the statement of Theorem 7.1 can be also obtained as a corollary of the corresponding result proved in [9] concerning a coupled plate–wave system.

We also note that, as in Chapters 5 and 6 we can obtain results on the rate of stabilization to equilibria, on determining functionals and on exponential attractor. We do not provide details here because the corresponding arguments are quite similar to the ones given in Chapters 5 and 6.

REMARK 7.2. Problem (7.1) and (7.2) belongs to the class of (abstract) equations of the form

(7.8) $$\begin{cases} Mu_{tt}(t) + D(u_t(t)) + \mathcal{A}u(t) + Q\left(\|\mathcal{A}^{1/4}u(t)\|^2\right)\mathcal{A}^{1/2}u(t) = F^*(u), \\ u|_{t=0} = u_0,\ u_t|_{t=0} = u_1, \end{cases}$$

where the function $Q(z) \in C^1(\mathbb{R}_+)$ possesses the properties

(7.9) $$\int_0^z Q(\xi)d\xi \geq -a_0,\ zG(z) - a_1 \int_0^z Q(\xi)d\xi \geq a_1 z^{1+\eta} - a_3,\ z > 0,$$

with positive constants a_i and η. In the Berger model (7.1) we have $Q(z) = z - Q$. The long time behaviour of problem (7.8) with *linear* damping operator D and linear non-conservative force $F^*(u)$ was studied with details in [**19**, Chap.4].

7.1.2. Non-conservative forces: a case study. If we assume that the plate is located in a supersonic flow of gas, then according to "piston" theory (the semi-empirical plane sections law [**63**]) the load p in the right hand side of (7.1) has the form

$$p(u, u_t, x) \equiv p(u, u_t) = -\sum_{k=1}^{\infty} a_k(u_t + \rho \cdot u_{x_1})^k,$$

for the flow which moves along x_1-axis. Here a_k are a positive parameters which depend on the properties of the gas, and $\rho > 0$ is determined from the gas velocity. It was suggested in [**8**] to consider not more than 3 first terms in this representation for p in the case of plate equations because of consistency with other (plate) hypotheses. If we will follow this recommendation keeping *two* first terms and the *only* purely damping item from the third-order term, we obtain the following structure of aerodynamical load

(7.10) $$p(u, u_t) = -\left(a_1 u_t + a_2 u_t^2 + a_3 u_t^3\right) - \rho \cdot (a_1 + 2a_2 u_t) u_{x_1} - \rho^2 a_2 u_{x_1}^2,$$

where a_k and ρ are a positive parameters. As we have seen, the presence of aeroelastic nonlinear force contributes to: (i) superlinear non-conservative forces and, in addition, to (ii) non-monotone damping. In fact, nonconservative forces take the form:

$$F^*(u, v) = a_0 v - \rho(a_1 + 2a_2 v)u_{x_1} + a_2 u_{x_1}^2.$$

It is immediate to see that $F^*(u, v)$ is not locally Lipschitz with respect to the phase space topology $H^2(\Omega) \times L_2(\Omega) \to L_2(\Omega)$. Indeed, term vu_{x_1} is not in $L_2(\Omega)$ with $(u; v) \in H = H^2(\Omega) \times L_2(\Omega)$. This, of course raises the problem of the well-posedness of the flow. While existence of weak solutions can be shown (see below), the uniqueness of weak solutions is problematic. Since the study of non-unique flows is beyond the scope of this paper, we consider this model within the context of rotational inertia. Indeed, instead of (7.1) we consider equation with rotational forces added to the model, i.e.

(7.11) $$\begin{aligned}(1 - \alpha\Delta)u_{tt} - \alpha\Delta u_t + \Delta^2 u + \left(Q - \int_\Omega |\nabla u(x,t)|^2 dx\right) \cdot \Delta u \\ = p(u, u_t)\ \text{in}\ \Omega \times (0, \infty), \\ u = \Delta u = 0\ \text{on}\ \Gamma \times (0, \infty),\ u|_{t=0} = u_0(x),\ u_t|_{t=0} = u_1(x).\end{aligned}$$

where $p(u, u_t)$ is given above in (7.10). In that case we have $V = H_0^1(\Omega)$, and consequently all the assumptions of Theorem 1.5 are satisfied. By appealing to

Theorem 3.15 we show that this model is dissipative with respect to all finite energy solutions.

THEOREM 7.3. *Problem (7.11) and (7.10) generates a dissipative dynamical system (H, S_t) with the phase space $H = (H_0^1 \cap H^2)(\Omega) \times H_0^1(\Omega)$. Moreover, there exists a compact global attractor of finite fractal dimension.*

PROOF. We first put the problem in the form (1.1) with $k = 1$, i.e. we in the form

(7.12) $\quad \begin{cases} Mu_{tt}(t) + \mathcal{A}u(t) + D(u_t(t)) = F(u(t), u_t(t)), \\ u|_{t=0} = u_0 \in \mathcal{D}(\mathcal{A}^{1/2}), \ u_t|_{t=0} = u_1 \in V = \mathcal{D}(M^{1/2}), \end{cases}$

with $\mathcal{H} \equiv L_2(\Omega)$ and $V \equiv H_0^1(\Omega)$. The operator $\mathcal{A}$ is defined as in the previous subsection and $Mu \equiv u - \alpha \Delta u$ for $u \in \mathcal{D}(M) \equiv H_0^1(\Omega) \cap H^2(\Omega)$. The damping operator D has the form

$$D(v) \equiv -\alpha \Delta v + D_0(v), \quad v \in \mathcal{D}(\mathcal{A}^{1/2}) = (H^2 \cap H_0^1)(\Omega),$$

where

(7.13) $\quad D_0(v) = g(v) \equiv (a_0 + a_1)v + a_2 v^2 + a_3 v^3, \quad v \in (H_0^1 \cap H^2)(\Omega),$

and $F(u, v) = -\Pi'(u) + F^*(u, v)$, where $\Pi(u)$ is given by (7.4) with $p_0 = 0$ and

(7.14) $\quad F^*(u, v) = a_0 v - \rho \cdot (a_1 + 2a_2 v) u_{x_1} - \rho^2 a_2 u_{x_1}^2, \quad (u; v) \in H.$

We choose the parameter $a_0 > 0$ such that $g'(s) \geq 1$ for all $s \in \mathbb{R}$.

By Theorem 1.5 problem (7.12) generates semiflow in H and the main new issue (with respect to the analysis of Section 6.4) is proving ultimate dissipativity. In order to accomplish this we shall use Theorem 3.15. To this end we verify the corresponding assumptions. We start with Assumption 3.14.

Relation (3.38) is obvious.

Let us prove (3.39). From (7.13) we obviously have that

(7.15) $\quad \begin{aligned} -\int_\Omega g(v) u \, dx &\leq C \left[\int_\Omega (1 + v^4) dx \right]^{3/4} \cdot \left[\int_\Omega |u|^4 dx \right]^{1/4} \\ &\leq \varepsilon \|u\|_{L_4}^4 + C_\varepsilon \left(1 + \int_\Omega g(v) v \, dx \right) \end{aligned}$

for any $\varepsilon > 0$. Since $\|u\|_{L_4} \leq C \|\nabla u\|$ and

$$\alpha(\Delta v, u) \leq \alpha \|\nabla v\| \cdot \|\nabla u\| \leq \varepsilon \|\nabla u\|^2 + C_\varepsilon \|\nabla v\|^2,$$

(here and below $\|\cdot\|$ denotes the norm in $L_2(\Omega)$), we conclude from (7.15) that for any $\varepsilon > 0$ there is C_ε such that

(7.16) $\quad -(Dv, u) \leq \varepsilon \|\nabla u\|^4 + C_\varepsilon \left[1 + (D(v), v) \right].$

Using (7.14) and the Hölder inequality we obtain that

$$\begin{aligned} (F^*(u, v), u) &\leq C \left(\|u\| \cdot \|v\| + \int_\Omega (1 + |v|) \cdot |u| \cdot |u_{x_1}| dx \right) + a_2 \rho^2 \int_\Omega |u| \cdot u_{x_1}^2 dx \\ &\leq \varepsilon \|\nabla u\|^4 + C_\varepsilon \left(1 + \int_\Omega g(v) v \, dx \right) + a_2 \rho^2 \int_\Omega |u| \cdot u_{x_1}^2 dx \end{aligned}$$

for any $\varepsilon > 0$. Using embedding $H^{1+\delta}(\Omega) \subset L_\infty(\Omega)$ and interpolation we have that

$$a_2 \rho^2 \int_\Omega |u| \cdot u_{x_1}^2 dx \leq C \|u\|_{H^{1+\delta}} \|\nabla u\|^2 \leq C \|\nabla u\|^{3-\delta} \|\Delta u\|^\delta.$$

By the Hölder inequality we obtain that
$$a_2\rho^2 \int_\Omega |u| \cdot u_{x_1}^2 dx \leq \frac{1}{2}\|\Delta u\|^2 + C\|\nabla u\|^{2\frac{3-\delta}{2-\delta}} \leq \frac{1}{2}\|\Delta u\|^2 + \varepsilon\|\nabla u\|^4 + C_\varepsilon$$
for any $\varepsilon > 0$. Therefore
$$(7.17) \quad (F^*(u,v), u) \leq 2\varepsilon\|\nabla u\|^4 + \frac{1}{2}\|\Delta u\|^2 + C_\varepsilon \left(1 + \int_\Omega g(v)v dx\right)$$
for any $\varepsilon > 0$. We also have that
$$(7.18) \quad -(\Pi'(u), u) = -\|\nabla u\|^4 + Q\|\nabla u\|^2 \leq -\frac{1}{2}\|\nabla u\|^4 + \frac{Q^2}{2}.$$

Choosing ε small enough, from (7.16)–(7.18) we obtain (3.39) with $\gamma = 0$.

To prove relation (3.41) from Assumption 3.14 we need the following assertion.

LEMMA 7.4. *Let $2 \leq p < 3$. Then there exists positive constants c_1 and c_2 such that*
$$(7.19) \quad \int_\Omega |\nabla u|^p dx \leq c_1\|\nabla u\|^2 \|\Delta u\|^{p-2} \leq \delta \left(\|\nabla u\|^4 + \|\Delta u\|^2\right) + c_2 \left(\frac{1}{\delta}\right)^{\frac{p-1}{3-p}}$$
for any $u \in H_0^1(\Omega) \cap H^2(\Omega)$ and for any $\delta \in (0,1]$.

PROOF. We consider the case $2 < p < 3$ only. Since $H^{1-2/p}(\Omega) \subset L_p(\Omega)$ (see, e.g., [**102**, Chap.4]), using an interpolation inequality we have that
$$\|w\|_{L_p} \leq c\|w\|_{H^{1-2/p}} \leq c\|w\|^{2/p} \cdot \|w\|_{H^1}^{1-2/p}, \quad w \in H^1(\Omega),$$
which implies the first inequality in (7.19). To prove the second one we use Young's inequality:
$$\begin{aligned} c_1\|\nabla u\|^2 \|\Delta u\|^{p-2} &\leq \delta\|\nabla u\|^4 + C\delta^{-1}\|\Delta u\|^{2(p-2)} \\ &\leq \delta\|\nabla u\|^4 + C\delta^{-1}\left[\|\Delta u\|^{2q(p-2)}\epsilon^q + \epsilon^{-q/(q-1)}\right] \end{aligned}$$
Selecting $q = (p-2)^{-1}$ gives
$$c_1\|\nabla u\|^2\|\Delta u\|^{p-2} \leq \delta\left(\|\nabla u\|^4 + C_1\delta^{-2}\epsilon^{1/(p-2)}\cdot\|\Delta u\|^2\right) + C_2\delta^{-1}\epsilon^{-1/(3-p)}.$$
The final estimate follows by taking $\epsilon = C_1^{-1}\delta^{2(p-2)}$. □

Finally we verify (3.41). From (7.14) we find that
$$\begin{aligned} (F^*(u,v), v) &\leq C\left(\|v\|^2 + \int_\Omega (1+|v|^2)\cdot|u_{x_1}|dx + \int_\Omega |v|\cdot|u_{x_1}|^2 dx\right) \\ &\leq \frac{1}{2}\int_\Omega g(v)v dx + C_1 + C_2 \int_\Omega |u_{x_1}|^{8/3} dx. \end{aligned}$$
Therefore Lemma 7.4 implies that
$$(F^*(u,v), v) \leq \frac{1}{2}\int_\Omega g(v)v dx + \delta\left(\|\nabla u\|^4 + \|\Delta u\|^2\right) + C_1 + C_2\left(\frac{1}{\delta}\right)^5.$$
for any $\delta \in (0,1]$. This implies (3.41).

Ultimate dissipativity follows now from Theorem 3.15.

Proof of the remaining statements in the theorem (compactness and finite dimensionality of an attractor) is an easy consequence of the fact that the semilinear terms are compact when $\alpha > 0$ (with respect to the of the topology generated by

$V = H_0^1(\Omega)$). The arguments rely on Corollary 3.28 and Theorem 4.4 and are the same as in the von Karman case with rotational inertia. □

REMARK 7.5. We note that the proof of ultimate dissipativity does not depend on the topology in $V = H_0^1(\Omega)$. In fact, the estimates are valid for all solutions of the original model (7.1) (with $\alpha = 0$) which satisfy the energy inequality. Of course, strong solutions do satisfy energy identity. However lack of uniqueness of weak solutions prevents us from inferring the same conclusion as being valid for all weak solutions. Existence of weak solutions is a consequence of an ad-hoc argument which takes advantage of the presence of superlinear damping in the equation. Indeed, though the nonconservative term is not locally Lipschitz from $\mathcal{D}(\mathcal{A}^{1/2}) \times L_2(\Omega)$ into $L_2(\Omega)$, as required for the well-posedness of the flow with $\alpha = 0$ by Theorem 1.5, we can circumvent the difficulty by noting that such well-posedness is indeed guaranteed due to the presence of the cubic velocity term in the damping operator $D(v)$. Indeed, the contribution of the non-Lipschitz term vu_{x_1} in energy inequality leads to the term

$$\int_\Omega |v|^2 |u_{x_1}| dx \le \epsilon \int_\Omega |v|^4 dx + C_\epsilon \int_\Omega |u_{x_1}|^2 dx \le \epsilon \int_\Omega |v|^4 dx + C_\epsilon |\mathcal{A}^{1/2} u|^2.$$

The above inequality allows to obtain a priori bound for solutions, hence the existence of solutions in a finite energy space. The above observations lead us to the conclusion that in the case $\alpha = 0$ weak solutions of system (7.1) with p given by (7.10) are ultimately dissipative, provided they satisfy energy inequality. The question on the existence of a compact global attractor for problem (7.1) and (7.10) is open.

7.2. Mindlin-Timoshenko plates and beams

Let $\Omega \subset \mathbb{R}^2$ be a bounded domain with a sufficiently smooth boundary Γ. Let $v(x,t) = (v_1(x,t), v_2(x,t))^T$ be a vector function and $w(x,t)$ be a scalar function on $\Omega \times \mathbb{R}_+$. The system of Mindlin-Timoshenko equations describes dynamics of a plate taking into account transverse shear effects (see, e.g., [**75**, Chap.1] and the references therein). This system has the form

(7.20) $$\alpha v_{tt} + k \cdot g(v_t) - Av + \kappa \cdot (v + \nabla w) = -f_0(v) + \nabla_x [f_1(w)],$$

(7.21) $$w_{tt} + k \cdot g_0(w_t) - \kappa \cdot \text{div}(v + \nabla w) = -f_2(w).$$

Here the functions $v_1(x,t)$ and $v_2(x,t)$ are the angles of deflection of a filament (they are measures of transverse shear effects) and $w(x,t)$ is the bending component (transverse displacement). The vector $f_0(v) = (f_{01}(v_1, v_2), f_{02}(v_1, v_2))^T$ and scalar f_1 and f_2 functions represents (nonlinear) feedback forces, $g(v_1, v_2) = (g_1(v_1), g_2(v_2))^T$ and g_0 are monotone damping functions describing resistance forces (with the intensity $k > 0$). The parameter $\alpha > 0$ describes rotational inertia of filaments. The factor $\kappa > 0$ is the so-called shear modulus (from mechanical point of view the limiting situation $\kappa \to +0$ corresponds to plane strain and the case $\kappa \to +\infty$ corresponds to absence of transverse shear). The operator A has the form

$$A = \begin{bmatrix} \partial_{x_1}^2 + \frac{1-\nu}{2} \partial_{x_2}^2 & \frac{1+\nu}{2} \partial_{x_1 x_2}^2 \\ \frac{1+\nu}{2} \partial_{x_1 x_2}^2 & \frac{1-\nu}{2} \partial_{x_1}^2 + \partial_{x_2}^2 \end{bmatrix},$$

where $0 < \nu < 1$ is the Poissons ratio. We supplement problem (7.20) and (7.21) with boundary conditions

(7.22) $\quad v_1(x,t) = v_2(x,t) = 0, \; w(x,t) = 0 \quad \text{on} \quad \Gamma \times \mathbb{R}_+.$

In 1D case ($n = \dim \Omega = 1$) the corresponding problem looks like

(7.23) $\quad \alpha v_{tt} + k \cdot g(v_t) - v_{xx} + \kappa \cdot (v + w_x) = -f_0(v) + \partial_x \left[f_1(w) \right], \quad x \in (0,1), \; t > 0,$

(7.24) $\quad w_{tt} + k \cdot g_0(w_t) - \kappa \cdot \partial_x(v + w_x) = -f_2(w), \quad x \in (0,1), \; t > 0,$

(7.25) $\quad v(0,t) = v(1,t) = 0, \; w(0,t) = w(1,t) = 0 \quad \text{for} \quad t > 0.$

Here $v(x,t)$ and f_0 are scalar functions. Equations (7.23)–(7.25) models dynamics of *beams* under the Mindlin-Timoshenko hypotheses. For details concerning the Mindlin-Timoshenko hypotheses and governing equations see, e.g. [74] and [75]. We also refer to the recent paper [30] for an analysis of long time behaviour of the Mindlin-Timoshenko plate under another set of assumptions concerning nonlinear feedback forces.

We rewrite problem (7.20)–(7.22) in the Hilbert space $\mathcal{H} = L_2(\Omega) \times L_2(\Omega) \times L_2(\Omega)$ in the following form

(7.26) $\quad \begin{cases} M u_{tt}(t) + \mathcal{A} u(t) + k \cdot D(u_t(t)) = F(u(t)), \\ u|_{t=0} = u_0 \in \mathcal{D}(\mathcal{A}^{1/2}), \; u_t|_{t=0} = u_1 \in \mathcal{H}. \end{cases}$

Here $u(t) = (v_1(t), v_2(t), w(t))^T$, and

$$\mathcal{A} = \begin{bmatrix} -A + \kappa I & \kappa \cdot \nabla \\ -\kappa \cdot \mathrm{div} & -\kappa \cdot \Delta \end{bmatrix}$$

with the domain

$$\mathcal{D}(\mathcal{A}) = \left\{ (u_1, u_2, w) \in (H^2 \cap H_0^1)(\Omega) \times (H^2 \cap H_0^1)(\Omega) \times (H^2 \cap H_0^1)(\Omega) \right\}.$$

It is clear that $\mathcal{A}$ is positive self-adjoint operator and

$$\mathcal{D}(\mathcal{A}^{1/2}) = H_0^1(\Omega) \times H_0^1(\Omega) \times H_0^1(\Omega).$$

We define the damping operator by the formula

$$D(u) = (g_1(v_1), g_2(v_2), (g_0(w))^T, \quad u = (v_1, v_2, w)^T.$$

We also have that $V = \mathcal{H}$ and the operator M is a bounded operator in $\mathcal{H}$ of the form

$$M = \begin{bmatrix} \alpha I & 0 \\ 0 & 1 \end{bmatrix}.$$

The nonlinear term F is given by

(7.27) $\quad F(u) = \begin{bmatrix} -f_{01}(v_1, v_2) + f_1'(w) \partial_{x_1} w \\ -f_{02}(v_1, v_2) + f_1'(w) \partial_{x_2} w \\ -f_2(w) \end{bmatrix}, \quad u = \begin{bmatrix} v_1 \\ v_2 \\ w \end{bmatrix}.$

Obviously, a representation similar to (7.26) can be also written for one-dimensional version (7.23)–(7.25) of the Mindlin-Timoshenko plate equations.

As an example of an application of the results from Chapters 3 and 4 we prove the following assertion.

THEOREM 7.6. *Assume the following hypotheses:*

- $f_0 \equiv (f_{01}(v_1, v_2), f_{02}(v_1, v_2))$ is C^1-function of the polynomial growth possessing the representation

$$f_{0i}(v) = \frac{\partial \Psi_0(v_1, v_2)}{\partial v_i} + f^*_{0i}(v_1, v_2), \quad i = 1, 2,$$

where $\Psi_0 : \mathbb{R}^2 \mapsto \mathbb{R}$ is a C^2-function which is bounded from below and the functions $f^*_{0i} : \mathbb{R}^2 \mapsto \mathbb{R}$ belong to C^1 and have bounded derivatives. Moreover we assume the relation

(7.28) $\quad v_1 f_{01}(v_1, v_2) + s_2 f_{02}(v_1, v_2) \geq c_0 \Psi_0(v_1, v_2) - c_1, \quad (v_1; v_2) \in \mathbb{R}^2,$

where $c_0 > 0$ and $c_1 \geq 0$ are constants.
- $f_1(w) \equiv 0$.
- $f_2(w)$ is C^1-function of the polynomial growth such that

(7.29) $$\liminf_{|w| \to \infty} \frac{f_2(w)}{w} \geq 0.$$

- $g_i(s)$ are monotone functions such that there exist two positive constants m_1 and m_2 such that

(7.30) $\quad m_1 \leq g'(s) \leq m_2(1 + |s|^{p-1}) \quad \text{for all} \quad |s| \geq 1,$

where $1 \leq p < \infty$ when $f^*_{0i}(v_1, v_2) \equiv 0$ and $p = 1$ in the case $f^*_{0i} \neq 0$.

Then

- the dynamical system (H, S_t) generated by (7.20) and (7.21) with the boundary conditions (7.22) in the energy space $H = \mathcal{D}(\mathcal{A}^{1/2}) \times \mathcal{H}$ possesses a compact global attractor.
- If in addition we assume that (7.30) holds for all $s \in \mathbb{R}$ and for $p \geq 3$ we have $sg_i(s) \geq m|s|^l$ for all $|s| \geq 1$, and for some $l > p - 1$, then the attractor has a finite fractal dimension.

PROOF. Since $f_0(v_1, v_2)$ is of polynomial growth and $\dim \Omega = 2$, one can find from (7.27) that

(7.31) $\quad |F(u) - F(u^*)| \leq C_r |\mathcal{A}^{1/2 - \tilde{\eta}}(u - u^*)|, \quad 0 < \tilde{\eta} < 1/2,$

for any $u, u^* \in \mathcal{D}(\mathcal{A}^{1/2})$ such that $|\mathcal{A}^{1/2} u| \leq r$ and $|\mathcal{A}^{1/2} u^*| \leq r$. Thus we are in the *subcritical* case. We also have the representation

$$F(u) = -\Pi'(u) + F^*(u), \quad u = (v_1; v_2, ; w) \in \mathcal{D}(\mathcal{A}^{1/2}),$$

where we denote

$$\Pi(u) = \int_\Omega \left(\Psi_0(v_1(x), v_2(x)) + \int_0^{w(x)} f_2(\xi) d\xi \right) dx$$

and $F^*(u) = (f^*_{0i}(v_1, v_2); f^*_{0i}(v_1, v_2); 0)^T$. This observations allow us to apply the same argument as in the case of wave equation in *two-dimensional* domain (see Chapter 5) and conclude the proof. □

Under some conditions we can also establish analogs to other assertions from Chapter 5 (e.g., decay rates when $f^*_{0i} \equiv 0$) and consider the case $f_1 \neq 0$.

7.3. Kirchhoff limit in Mindlin-Timoshenko plates and beams

By [74] in the limit $\kappa \to +\infty$, the Mindlin-Timoshenko-model (7.20) and (7.21) can be approximated by a nonlinear Kirchhoff plate model. *Formally*, this Kirchhoff model can be obtain by the following procedure: (i) apply the divergence operator to (7.20) and add the result to (7.21), then (ii) put the relation $v = -\nabla w$ in the sum obtained. This formal procedure, after some calculations, leads to the following equation

$$(7.32) \quad (1 - \alpha\Delta)w_{tt} + k \cdot [g_0(w_t) - \alpha\mathrm{div}\,\{\widetilde{g}(\nabla w_t)\}] + \Delta^2 w = \mathrm{div}\left[\widetilde{f}_0(\nabla w)\right] + \Delta\left[f_1(w)\right] - f_2(w),$$

where $\widetilde{g}(s) = -g(-s)$ and $\widetilde{f}_0(s) = -f_0(-s)$. Boundary conditions (7.22) in this limit are transformed in the form

$$(7.33) \quad w(x,t) = 0, \ \nabla w(x,t) = 0 \quad \text{on} \quad \Gamma \times \mathbb{R}_+.$$

We note that in the theory of plates and shells the relation $v = -\nabla w$ corresponds to absence of transverse shear and it is one of the Kirchhoff hypotheses in the shells theory. Thus it is natural to call the limiting procedure $\kappa \to +\infty$ as the *Kirchhoff limit* (for details we refer to [74]). We also refer to [30] for the justification of the Kirchhoff limit for the case of linear damping and for another choice of *nonlinear forces*.

Special cases of equation (7.32) (with $n = 1$) have appeared in some models of 1D viscoelasticity:

$$(7.34) \quad w_{tt} - k \cdot w_{xxt} + w_{xxxx} = \partial_x\left[\sigma(w_x)\right],$$

and 1D compressible gas dynamics:

$$(7.35) \quad w_{tt} - k \cdot w_{xxt} + w_{xxxx} = \partial_{xx}\left[\sigma(w)\right].$$

We refer to [**42, 87, 97, 98**] and to the literature quoted therein for the detailed discussion of problems (7.34) and (7.35). We also mention that two-dimensional version of equation (7.35) with $\sigma(w) = w + w^2$ has the form

$$(7.36) \quad w_{tt} - k \cdot \Delta w_t + \Delta^2 w - \Delta w = \Delta\left[w^2\right], \quad \Omega \subset \mathbb{R}^2, \ t > 0,$$

and it is known as the "good" Boussinesq equation (see, e.g., [**103**] and the references therein).

In order to focus our considerations and explain the main ideas in a clear way we first consider the problem

$$(7.37) \quad \begin{aligned}(1 - \alpha\Delta)w_{tt} + k_0 \cdot g_0(w_t) - \alpha \cdot k_1 \cdot \mathrm{div}\,\{g(\nabla w_t)\} \\ = -\Delta^2 w + \mathrm{div}\left[|\nabla w|^2 \nabla w\right] + \Delta\left[w^2\right] - \rho|u|^{l-1}u, \quad \Omega \subset \mathbb{R}^2, \ t > 0,\end{aligned}$$

which we call Kirchhoff-Boussinesq equation. Here $\alpha > 0$, $\rho \geq 0$ and $m \geq 1$ are constants. The functions g_0 and g and the parameters k_0 and k_1 will be specified later on. In comparison with (7.36) the Kirchhoff-Boussinesq equation (7.37) takes into account rotational inertia of filaments ($\alpha > 0$), damping effects and contains additional (potential) nonlinearities $\mathrm{div}\left[|\nabla w|^2 \nabla w\right]$ and $\rho|u|^{l-1}u$ (however, the case $\rho = 0$ is also allowed). We also note that in the case $\alpha = 0$ problem (7.37) was studied in [**29**] for linear damping function g_0.

With (7.37) we can associate any set of boundary conditions that are well posed for the plate. To be specific, we consider the clamped boundary conditions (7.33).

We represent problem (7.37) and (7.33) in the form

(7.38)
$$\begin{cases} Mu_{tt}(t) + \mathcal{A}u(t) + k \cdot D(u_t(t)) = F(u(t)), \\ u|_{t=0} = u_0 \in \mathcal{D}(\mathcal{A}^{1/2}),\; u_t|_{t=0} = u_1 \in V = \mathcal{D}(M^{1/2}), \end{cases}$$

with the following notation the spaces and operators:
- $\mathcal{H} \equiv L_2(\Omega)$, $V = H_0^1(\Omega)$.
- $\mathcal{A}u \equiv \Delta^2 u$, $u \in \mathcal{D}(\mathcal{A})$: $\mathcal{D}(\mathcal{A}) \equiv H_0^2(\Omega) \cap H^4(\Omega)$.
- $Mu \equiv u - \alpha \Delta u$, $u \in \mathcal{D}(M)$, where $\mathcal{D}(M) \equiv H_0^1(\Omega) \cap H^2(\Omega)$.
- $F(u) = \operatorname{div}\left[|\nabla u|^2 \nabla u\right] + \Delta\left[u^2\right] - \rho|u|^{l-1}u$.
- $D(u) \equiv \frac{k_0}{k} \cdot g_0(u) - \alpha \cdot \frac{k_1}{k} \cdot \operatorname{div}\left[g(\nabla u)\right]$ with $k = \max\{k_0, k_1\}$.

As in Chapter 6, $\mathcal{A}$ and M satisfy the conditions introduced at the beginning of Sect. 1.2. Simple calculations shows that the nonlinear term F is *subcritical* in the case $\alpha > 0$, i.e. relation (7.31) holds for the case considered.

REMARK 7.7. We emphasize that the results below are stated for the case $\alpha > 0$ only. In the case $\alpha = 0$, $n = 2$ the nonlinear term F in equation (7.38) is no longer locally Lipschitz and it is not obvious that problem (7.38) generates a semi-flow which is well defined on a finite energy space. This is the main reason why we consider $\alpha > 0$ when $n = 2$. The analysis of the case $\alpha = 0$ requires different technicalities (see [**29**] where the case of linear damping is considered).

7.3.1. Model 1: Kirchhoff-Boussinesq plate with $\rho = 0$. The following assumptions are imposed on the damping function:

ASSUMPTION 7.8.
- $g_0 \in C^1(\mathbb{R})$ is a monotone nondecreasing function such that $g_0(0) = 0$ and $0 < m \leq g_0'(s) \leq M|s|^{q_0}$ for $|s| \geq 1$, where $0 \leq q_0 \leq 2$.
- The function $g : \mathbb{R}^2 \mapsto \mathbb{R}^2$ has the form $g(s_1, s_2) = (g_1(s_1), g_2(s_2))$, where $(s_1, s_2) \in \mathbb{R}^2$ and $g_i \in C^1(\mathbb{R})$ is monotone nondecreasing function such that $g_i(0) = 0$, $i = 1, 2$. Moreover, we assume that g_i are of a cubic growth at infinity, i.e. $0 < m \leq g_i'(s) \leq M|s|^{q_1}$ for $i = 1, 2$ and $|s| \geq 1$, where $0 \leq q_1 \leq 2$.

Our main result for this model reads as follows:

THEOREM 7.9. *Under the Assumption 7.8 the equations (7.37) and (7.33) with $\alpha > 0$ and $\rho = 0$ generate a continuous semiflow S_t in the space $H \equiv H_0^2(\Omega) \times H_0^1(\Omega)$. Assuming that $q_0 < 2$, $q_1 < 2$, $k_1 > 0$ and $k = \max\{k_0, k_1\}$ is sufficiently large, the semiflow possesses a global compact attractor A. Moreover*

- *If, in addition, we assume that*

(7.39)
$$g_i'(s) \geq m > 0, \quad \text{for all}\quad s \in \mathbb{R},\; i = 1, 2,$$

  then the fractal dimension of the attractor A is finite.
- *Under condition (7.39) the system possesses a fractal exponential attractor (see Definition 4.41) whose dimension is finite in the space $H_0^1(\Omega) \times W$, where W is a completion of $H_0^1(\Omega)$ with respect to the norm $\|\cdot\|_W = \|(1-\alpha\Delta)\cdot\|_{-2}$.*

PROOF. In the case considered the nonlinear force F admits the representation

(7.40)
$$F(u) = -\Pi'(u) + F^*(u),$$

where
(7.41)
$$\Pi(u) = \Pi_0(u) + \Pi_1(u), \quad F^*(u) = |\nabla u|^2,$$
with
(7.42)
$$\Pi_0(u) = \frac{1}{4} \int_\Omega |\nabla u(x)|^4 dx, \quad \Pi_1(u) = \int_\Omega u(x)|\nabla u(x)|^2 dx.$$

Existence of the semiflow follows now from Theorem 1.5.

The further arguments are divided in several steps.

Step 1: Dissipativity. To establish the ultimate dissipativity we use Theorem 3.11. To this end we need to verify Assumption 3.7 with the modifications stated in Theorem 3.11.

It is clear that
$$(u, F(u)) = -\int_\Omega |\nabla u|^4 dx + \int_\Omega u\Delta\left[u^2\right] dx = -4\Pi_0(u) - 2\int_\Omega u|\nabla u|^2 dx.$$
Since
$$\int_\Omega u|\nabla u|^2 dx \leq C \cdot \|u\|_{L_2(\Omega)} \cdot \|\nabla u\|_{L_4(\Omega)}^2 \leq C \cdot \|\nabla u\|_{L_2(\Omega)} \cdot \|\nabla u\|_{L_4(\Omega)}^2$$
$$\leq C \cdot \|\nabla u\|_{L_4(\Omega)}^3 \leq C \cdot [\Pi_0(u)]^{3/4},$$
we obtain that
$$(u, F(u)) \leq -4\Pi_0(u) + C \cdot [\Pi_0(u)]^{3/4} \leq -\Pi_0(u) + C.$$
which proves (3.17). As for (3.18) we have
$$|u|^2 \leq C \int_\Omega |\nabla u|^2 dx \leq \eta \int_\Omega |\nabla u|^4 dx + C_\eta,$$
where η can be arbitrary small. To obtain (3.34) we use the estimate
$$|F^*(u)|_{V'}^2 \leq C|F^*(u)|^2 \leq C \int_\Omega |\nabla u|^4 dx \leq C\Pi_0(u).$$
Thus the hypotheses of Theorem 3.11 concerning F hold.

We now verify hypotheses of Theorem 3.11 imposed on the damping operator D. To satisfy (3.15) we clearly need $k_1 > 0$. As for (3.33) we have that
$$\frac{k_0}{k} \cdot (g_0(v), u) \leq C_k \|u\|_{L_{q_0+2}(\Omega)}^{q_0+2} + \frac{k_0}{k^2} \int_\Omega |g_0(v)|^{\frac{q_0+2}{q_0+1}} dx.$$
Since $|g(v)|^{\frac{q_0+2}{q_0+1}} \leq C g_0(v)v$ for $|v| \geq 1$ and $q_0 < 2$, we can conclude that
(7.43)
$$\frac{k_0}{k} \cdot (g_0(v), u) \leq \frac{\delta}{2k} \cdot \Pi_0(u) + \frac{c_0 k_0}{k^2}(g_0(v), v) + C_{\delta,k},$$
where δ can be taken arbitrary small and c_0 does not depend on k and k_0. In a similar way we have that
$$\frac{k_1}{k} \cdot (g_i(v_{x_i}), u_{x_i}) \leq C_k \|\nabla u\|_{L_{q_1+2}(\Omega)}^{q_1+2} + \frac{k_1}{k^2} \int_\Omega |g_i(v_{x_i})|^{\frac{q_1+2}{q_1+1}} dx$$
(7.44)
$$\leq \frac{\delta}{4k} \cdot \Pi_0(u) + \frac{c_1 k_1}{k^2}(g_i(v_{x_i}), v_{x_i}) + C_{\delta,k}$$
for any $\delta > 0$ and $i = 1, 2$, where c_1 does not depend on k and k_1. Thus from (7.43) and (7.44) we obtain
$$k \cdot (D(v), u) \leq \delta \cdot \Pi_0(u) + C_0(D(v), v) + C_{\delta,k}$$

for any $\delta > 0$, where C_0 does not depend on k. Hence (3.33) holds. There we can apply Theorem 3.11 to obtain dissipativity.

REMARK 7.10. The fact that $\alpha > 0$ has no bearing on the existence of absorbing ball. If the problem (7.37) and (7.33) with $\alpha = 0$ and $\rho = 0$ possesses a generalized solution satisfying the corresponding energy inequality, then the same argument as above allows us to check the hypotheses of Theorem 3.11 and to prove dissipativity of semiflow when $k_0 > 0$ is large enough and g_0 is of subcubic growth. Thus ultimate dissipativity holds for generalized solutions to problem (7.37) and (7.33) regardless whether $\alpha > 0$ or not.

The following steps conclude the proof of Theorem 7.9.

Step 2: Existence of a global attractor. Lemma 6.6 implies (3.59). By (6.39) in Lemma 6.9 the damping operator satisfies (3.60) with $\kappa = 1$ and $\epsilon = 0$. Thus by Corollary 3.28 the dynamical system generated by (7.37) and (7.33) with $\alpha > 0$ and $\rho = 0$ possesses a compact global attractor.

Step 3: Finite dimension and exponential attractor. Since $q_1 < 2$, It follows from (7.39) that there exists $0 \leq \beta < 1$ such that the functions g_i satisfy the inequality

$$(7.45) \qquad 0 < m \leq g_i'(s) \leq M[1 + sg_i(s)]^\beta, \quad s \in \mathbb{R}, \ i = 1, 2.$$

Therefore it follows from relation (6.52) in Lemma 6.10 that the damping operator D satisfies (3.60) with $\kappa = 2$ and $\epsilon = 0$. Thus we can apply Theorem 4.1 to prove finite dimensionality of the global attractor. The existence of a fractal exponential attractor follows from Theorem 4.42. □

7.3.2. Model 2: Kirchhoff-Boussinesq plate with $\rho > 0$. In this case the additional potential term $\rho|u|^{l-1}u$ will allow us to dispense with a necessity of assuming large values for the damping parameter. The following assumptions are imposed on the damping function:

ASSUMPTION 7.11.
- $g_0 \in C^1(\mathbb{R})$ is a monotone increasing function such that $g_0(0) = 0$ and $0 < m \leq g_0'(s) \leq M|s|^q$ for $|s| \geq 1$, where $0 \leq q \leq l - 1$.
- The function g has the form $g(s_1, s_2) = (g_1(s_1), g_2(s_2))$, where $s_i \in \mathbb{R}$ and $g_i \in C^1(\mathbb{R})$ is monotone increasing function such that $g_i(0) = 0$, $i = 1, 2$. Moreover, we assume that $0 < m \leq g_i'(s) \leq M|s|^2$ for $i = 1, 2$ and $|s| \geq 1$.

Our main result reads as follows:

THEOREM 7.12. *Under the Assumption 7.11 the equations (7.37) and (7.33) with $\alpha > 0$, $\rho > 0$ and $l > 3$ generates a continuous semiflow S_t in the space $H \equiv H_0^2(\Omega) \times H_0^1(\Omega)$. The corresponding dynamical system possesses a global compact attractor A. If, in addition, we assume that there exists $0 \leq \beta < 1$ such that the functions g_i satisfy inequality (7.45), then the fractal dimension of the attractor A is finite. Under condition (7.45) the system possesses a fractal exponential attractor (see Definition 4.41) with finite dimension in the space $H_0^1(\Omega) \times W$, where W is the same as in Theorem 7.9.*

PROOF. In the case considered the nonlinear term F admits the representation

$$(7.46) \qquad F(u) = -\Pi'(u) + F^*(u),$$

where

$$(7.47) \quad \Pi(u) = \Pi_0(u) = \int_\Omega \left[\frac{1}{4} |\nabla u(x)|^4 + \frac{\rho}{l+1} |u(x)|^{l+1} \right] dx, \quad F^*(u) = \Delta\left[u^2\right].$$

Due to the compactness of $F(u)$ the conclusion of Theorem 7.12 follows by the same arguments as these given for von Karman model with $\alpha > 0$ (see Section 6.4) or in the proof of Theorem 7.9, provided we manage to prove ultimate dissipativity. For this we use Corollary 3.17 with $D^* \equiv 0$ and $G^*(u) = F^*(u) = \Delta\left[u^2\right]$.

We first show that relation (3.51) holds with $\gamma = 0$. By Hölder inequality

$$(g_0(v), u) \leq \eta |u|_{L_{l+1}(\Omega)}^{l+1} + c_\eta \int_\Omega |g_0(v)|^{1+1/l} dx.$$

Since $|g_0(v)|^{1/l} \leq C|v|$ for $|v| \geq 1$, this implies that

$$(g_0(v), u) \leq \eta \Pi_0(u) + C_\eta^1 (D(v), v) + C_\eta^2$$

for any positive η. As for the second damping, the argument is the same as in the proof of Theorem 7.9. Indeed, similar to (7.44) we have that

$$(g_i(v_{x_i}), u_{x_i}) \leq \eta \|\nabla u\|_{L_4(\Omega)}^4 + C_\eta^1 (g_i(v_{x_i}), v_{x_i}) + C_\eta^2$$

for any $\eta > 0$ and $i = 1, 2$. Summing up we obtain

$$-k(Dv, u) \leq \delta \Pi_0(u) + C_\delta^1 (D(v), u) + C_\delta^2$$

for any $\delta > 0$. This proves (3.51) in Assumption 3.16.

To prove (3.53) we note that

$$\begin{aligned} |F^*(u)|_{V'}^2 &\leq C \|\nabla [u^2]\|_{L_2(\Omega)}^2 \\ &\leq C \cdot \|u\|_{L_4(\Omega)}^2 \cdot \|\nabla u\|_{L_4(\Omega)}^2 \leq \epsilon \int_\Omega |\nabla u|^4 dx + C_\epsilon |u|_{L_4(\Omega)}^4 \end{aligned}$$

for every $\epsilon > 0$. Since $l > 3$ we have that

$$\|u\|_{L_4(\Omega)}^4 = \int_\Omega |u|^4 dx \leq \eta \int_\Omega |u|^{l+1} dx + C_\eta$$

for any $\eta > 0$. Therefore we obtain that

$$|F^*(u)|_{V'}^2 \leq \delta \Pi_0(u) + C_\delta$$

for any $\delta > 0$. Thus (3.53) holds and by Corollary 3.17 the system is dissipative.

To prove finite-dimensionality and existence of a fractal exponential attractor we use the same arguments as in the proof of Theorem 7.9. $\square$

7.3.3. Generalization of Model 2. Under appropriate (physically reasonable) assumptions concerning nonlinear functions, the general Kirchhoff model (7.32) and (7.33) is also covered by the theory developed in Chapters 3 and 4 in the cases when either $n = 1$ or else $n = 2$ and $\alpha > 0$. In this subsection we shall consider some generalizations of the canonical model considered above.

Here we impose the following set of assumptions.

ASSUMPTION 7.13.
- $\widetilde{f}_0 \equiv (\widetilde{f}_{01}(v_1, v_2), \widetilde{f}_{02}(v_1, v_2))$ is C^1-function of the polynomial growth possessing the representation

$$\widetilde{f}_0(v) = \left(\frac{\partial \Psi(v_1, v_2)}{\partial v_1} ; \frac{\partial \Psi(v_1, v_2)}{\partial v_2} \right),$$

where $\Psi_0 : \mathbb{R}^2 \mapsto \mathbb{R}$ is a C^2-function such that

(7.48) $$\Psi(v_1, v_2) \geq c_0 \left(v_1^2 + v_2^2\right)^{1+\delta} - c_1, \quad \text{for all} \quad (v_1; v_2) \in \mathbb{R}^2,$$

where $c_0 > 0$, $c_1 \geq 0$ and $\delta > 0$ are constants.
- $f_2(w))$ is C^1-function such that $\Phi(w) = \int_0^w f_2(\xi) d\xi$ is bounded from below.
- $f_1(w)$ is C^1-function possessing the property

(7.49) $$\limsup_{|s| \to \infty} \frac{|f_1'(s)|^{2+2/\delta}}{1 + \Phi(s) - \Phi_{inf}} = 0,$$

where δ is the parameter from relation (7.48) and $\Phi_{inf} = \inf_{s \in \mathbb{R}} \Phi(s)$.
- $g_0 \in C^1(\mathbb{R})$ is a monotone nondecreasing function such that $g_0(0) = 0$ and

(7.50) $$-u g_0(v) \leq C_1^\eta \cdot v g_0(v) + \eta \Phi(u) + C_2^\eta, \quad (v; u) \in \mathbb{R}^2, \; \forall \eta > 0.$$

- The function $\widetilde{g}$ has the form $\widetilde{g}(s_1, s_2) = (\widetilde{g}_1(s_1), \widetilde{g}_2(s_2))$, where $(s_1, s_2) \in \mathbb{R}^2$ and $\widetilde{g}_i \in C^1(\mathbb{R})$ is monotone nondecreasing function, $\widetilde{g}_i(0) = 0$, $i = 1, 2$. Moreover, we assume that $\widetilde{g}_i$ are of polynomial growth at infinity and

(7.51) $$-\sum_{i=1,2} u_i g_i(v_i) \leq C_1^\eta \cdot [v_1 g_1(v_1) + v_2 g_2(v_2)] + \eta \Psi(u_1, u_2) + C_2^\eta, \quad \forall \eta > 0,$$

for all $(v_1; v_2; u_1; u_2) \in \mathbb{R}^4$.

We note that the assumption above concerning nonlinear functions is satisfied with $\delta = 1$, for example, if

$$\Psi(v_1, v_2) = \left(v_1^2 + v_2^2\right)^2, \quad f_1(w) = w^2, \quad f_2(w) = |w|^{l-1} w, \; l > 3,$$

and the functions $g_0(v)$ and $g_i(v)$ are the same as in Assumption 7.11.

We represent problem (7.32) and (7.33) in the form (7.38) with
- $\mathcal{H} \equiv L_2(\Omega)$, $V = H_0^1(\Omega)$.
- $\mathcal{A} u \equiv \Delta^2 u$, $u \in \mathcal{D}(\mathcal{A})$: $\mathcal{D}(\mathcal{A}) \equiv H_0^2(\Omega) \cap H^4(\Omega)$.
- $M u \equiv u - \alpha \Delta u$, $u \in \mathcal{D}(M)$, where $\mathcal{D}(M) \equiv H_0^1(\Omega) \cap H^2(\Omega)$.
- $F(u) = \mathrm{div}\left[\widetilde{f}_0(\nabla u)\right] + \Delta\left[f_1(u)\right] - f_2(u)$.
- $D(u) \equiv g_0(u) - \mathrm{div}(\widetilde{g}(\nabla u))$.

As above, $\mathcal{A}$ and M satisfy the conditions introduced at the beginning of Sect. 1.2 and the nonlinear term F is *subcritical*, i.e. relation (7.31) holds for the case considered. We also have the representation

(7.52) $$F(u) = -\Pi'(u) + F^*(u),$$

where

(7.53) $$\Pi(u) = \int_\Omega \left[\Psi(\nabla u(x)) + \Phi(u(x))\right] dx, \quad F^*(u) = \Delta\left[f_1(u)\right].$$

REMARK 7.14. It is also possible to present F in the form (7.52) with

(7.54) $$\Pi(u) = \int_\Omega \left[\Psi(\nabla u) + \Phi(u) + \frac{1}{2} f_1'(u) |\nabla u|^2\right] dx, \quad F^*(u) = \frac{1}{2} f_1''(u) |\nabla u|^2.$$

On the one hand this representation requires the existence (and some estimate) of the second derivative $f_1''(u)$. On the the other hand, if $f_1''(u)$ is uniformly bounded and $\delta = 1$, we can avoid assumption (7.49). This allow us to include the case $f_2(w) \equiv 0$ into consideration relying on argument similar ones given in the proof of Theorem 7.9.

Our main result concerning problem ((7.32) and (7.33)) reads as follows:

THEOREM 7.15. *Under the Assumption 7.13 the equations (7.32) and (7.33) with $\alpha > 0$ generates a continuous semiflow S_t in the space $H \equiv H_0^2(\Omega) \times H_0^1(\Omega)$ which possesses a global compact attractor A. Moreover*

- *If, in addition, we assume that (i) $g_0(s)$ is of a polynomial growth at infinity, i.e. there exist $p_0 \geq 1$ and $M_0 > 0$ such that*

(7.55) $$0 \leq g_0'(s) \leq M_0[1 + |s|^{p_0-1}], \quad s \in \mathbb{R},$$

  and (ii) there exists $0 \leq \beta < 1$ such that the functions g_i satisfy inequality (7.45), then the fractal dimension of the attractor A is finite.

- *Under conditions (7.55) and (7.45) the system possesses a fractal exponential attractor whose dimension is finite in the space $H_0^1(\Omega) \times W$, where W is the same as in Theorem 7.9.*

PROOF. As in the proof of Theorem 7.12 to prove dissipativity we use Corollary 3.17 with $D^* \equiv 0$ and $G^*(u) = F^*(u) = \Delta[f_1(u)]$.

It follows from (7.50) and (7.51) that relation (3.51) holds with $\gamma = 0$.

To prove (3.53) we note that

$$\begin{aligned}|F^*(u)|_{V'}^2 &\leq C\|\nabla[f_1(u)]\|_{L_2}^2 \leq C\|f_1'(u) \cdot \nabla u\|_{L_2}^2 \\ &\leq C \cdot \|f_1'(u)\|_{L_{2(1+1/\delta)}}^2 \cdot \|\nabla u\|_{L_{2(1+\delta)}}^2 \\ &\leq C_\eta \cdot \|f_1'(u)\|_{L_{2(1+1/\delta)}}^{2(1+1/\delta)} + \eta \cdot \|\nabla u\|_{L_{2(1+\delta)}}^{2(1+\delta)}\end{aligned}$$

for any $\eta > 0$. By (7.49) we have that

$$C_\eta \|f_1'(u)\|_{L_{2(1+1/\delta)}}^{2(1+1/\delta)} = C_\eta \int_\Omega |f_1'(u)|^{2(1+1/\delta)} dx \leq \eta \int_\Omega \Phi(u) dx + \widetilde{C}_\eta$$

for any $\eta > 0$. Therefore using (7.48) we obtain that

$$|F^*(u)|_{V'}^2 \leq \eta \Pi_0(u) + C_\eta$$

for any $\eta > 0$, where $\Pi_0(u) = \Pi(u) + c_0$ with an appropriate constant $c_0 \geq 0$.

Thus by Corollary 3.17 the system is dissipative.

To prove finite-dimensionality we apply Theorem 4.1 and the argument given in the proof of Theorem 6.4 (see Lemma 6.10).

As in the proof of Theorem 6.4 the existence of a fractal exponential attractor follows from Theorem 4.42. □

7.4. Systems with strong damping

Wave and plate equations exhibiting strong damping are also covered by our abstract framework. To see this, let us consider the following problems

(7.56) $$w_{tt} + g_0(w_t) - \operatorname{div}[g(\nabla w_t)] - \Delta w = f(w, w_t),$$

(7.57) $$w_{tt} + g_0(w_t) - \operatorname{div}[g(\nabla w_t)] + \Delta^2 w = f(w, w_t),$$

and

(7.58) $$w_{tt} + \Delta[g(\Delta w_t)] + \Delta^2 w = f(w, w_t).$$

These models with linear damping functions and with $f(w, w_t) \equiv f(w)$ were considered by many authors (see, e.g., the monograph [59], the survey [40] and the references therein, and also the recent papers [11, 88]). In fact, in the linear case the strongly damped wave or plate equations are associated with analytic semigroups (see, e.g., [**82**, Chap.3]). This, in turn, induces strong regularizing effect on solutions. Thus, for this class of models both stability and regularity theory is indeed very rich. We will focus here on nonlinear damping. Since the major regularizing term in linear equation is the damping, it should be clear that this effect may be severely diminished in the presence of nonlinearity in the dissipation. For this reasons analyticity type of arguments will no longer be applicable. On the other hand we can show that that theory presented in Chapters 3 and 4 applies to these models as well.

As an example we consider the following version of problem (7.56) in a smooth bounded domain $\Omega \subset \mathbb{R}^n$, $n \leq 3$:

(7.59) $\quad w_{tt} - \text{div}\,[g(\nabla w_t)] - \Delta w + f(w) = h(w_t), \quad x \in \Omega,\ t > 0,$

subject to the Dirichlet boundary condition

(7.60) $\quad w = 0 \ \text{ on } \ \Sigma \equiv [0, \infty) \times \partial\Omega.$

We impose the following hypotheses.

ASSUMPTION 7.16.   • The function g has the form

$$g(s_1, s_2) = (g_1(s_1), g_2(s_2)), \quad (s_1, s_2) \in \mathbb{R}^2,$$

where $g_i \in C^1(\mathbb{R})$ is monotone nondecreasing function such that $g_i(0) = 0$ and $0 \leq g'_i(s) \leq M_0$ for all $s \in \mathbb{R}$ and $i = 1, 2$ with some constants $M_0 > 0$. Moreover we assume that

(7.61) $\quad \omega_i \equiv \liminf_{|s| \to \infty} g'_i(s) > 0, \quad i = 1, 2.$

• The function $h \in C^1(\mathbb{R})$ is such that (i) $h(0) = 0$, (ii) there exists a positive constant M_1 such that

(7.62) $\quad |h'(s)| \leq M_1 \cdot (1 + |s|^{p-1}) \quad \text{for all} \quad s \in \mathbb{R},$

where $1 \leq p \leq 5$ when $n = 3$ and $1 \leq p < \infty$ when $n \leq 2$, and (iii) the relation

(7.63) $\quad \kappa \equiv \limsup_{|s| \to \infty} h'(s) < \lambda_1 \min\{\omega_1, \omega_2\}$

holds, where ω_i is defined by (7.61) and $\lambda_1 > 0$ is the first eigenvalue of the operator $-\Delta$ equipped with the Dirichlet boundary conditions.

• Function $f \in C^1(\mathbb{R})$ is of the following polynomial growth condition: there exists a positive constant $M_2 > 0$ such that

$$|f'(s)| \leq M_2 |s|^q,\ |s| \geq 1,$$

where $q < 2$ when $n = 3$ and $q < \infty$ when $n \leq 2$. Moreover, the dissipativity condition

(7.64) $\quad \liminf_{|s| \to \infty} \frac{f(s)}{s} \equiv \mu > -\lambda_1$

is assumed to hold, where λ_1 is the same as in 7.63).

REMARK 7.17. As an example of a function $h(s)$ with the properties described in Assumption 7.16 we can consider, for instance, a function of the form
$$h(s) = -\epsilon \cdot |s|^{p-1} s + h_*(s),$$
where $\epsilon > 0$ and the function $h_* \in C^1(\mathbb{R})$ satisfies the conditions
$$h_*(0) = 0 \quad \text{and} \quad \limsup_{|s| \to \infty} \frac{|h'_*(s)|}{|s|^{p-1}} = 0.$$
In this case $\kappa = -\infty$ if $p > 1$ and $\kappa = -\epsilon$ for $p = 1$. We can also avoid the asymptotically linear growth assumption given in (7.61) by imposing other (structural) hypotheses on $h(s)$ (see Example 3.19 in Chapter 3).

THEOREM 7.18. *Under Assumption 7.16 problem (7.59) and (7.60) generates a continuous ultimately dissipative semiflow S_t in the space $H \equiv H_0^1(\Omega) \times L_2(\Omega)$. Moreover,*
- *if, in addition, we assume that (i) $\inf_{s \in \mathbb{R}} g'_i(s) > 0$, $i = 1, 2$, and (ii) $p < 5$ in the case $n = 3$, where p is the parameter from (7.62), then S_t possesses a global compact attractor A;*
- *this attractor has finite fractal dimension provided $p \leq 2$ and $g(s)$ is a linear function.*

PROOF. We first split the function h in two functions:
$$h(s) = -g_0(s) + h_*(s).$$
Here the function $h_* \in C^1(\mathbb{R})$ has a bounded derivative and possesses the property
(7.65) $$\kappa_* \equiv \limsup_{|s| \to \infty} |h'_*(s)| < \lambda_1 \min\{\omega_1, \omega_2\} \equiv \omega$$
and the function $g_0 \in C^1(\mathbb{R})$ is a monotone nondecreasing function such that $g_0(0) = 0$, and there exists a positive constant m_1 and m_2 such that
(7.66) $$0 < m_1 \leq g'_0(s) \leq m_2 \left(1 + |s|^{p-1}\right) \quad \text{for all} \quad s \in \mathbb{R},$$
where $1 \leq p \leq 5$ when $n = 3$ and $1 \leq p < \infty$ when $n = 2$.

This splitting can be carried in the following way. Let $0 < \delta < (\omega - \kappa)/2$ and R_δ be chosen such that
$$0 < \delta < \omega \quad \text{and} \quad h'(s) \leq \omega - 2\delta \quad \text{for all} \ |s| \geq R_\delta.$$
We take a positive function $\psi_0 \in C^\infty(\mathbb{R})$ such that (i) $\psi_0(s) = \omega - \delta$ for $|s| \geq R_\delta + 1$; (ii) $\psi_0(s) \geq \omega - \delta$ for $R_\delta < |s| < R_\delta + 1$; and (iii) $\psi_0(s) = K + \delta$ for $|s| \leq R_\delta$, where K is a positive number such that $K \geq \sup_\mathbb{R} h'(s)$. Now we define
$$g_0(s) = -h(s) + h_*(s), \quad h_*(s) = \int_0^s \psi_0(\xi) d\xi.$$
Since $g'_0(s) = -h'(s) + \psi_0(s)$, it is clear that $g'_0(s) \geq \delta$ for all $s \in \mathbb{R}$. The other properties of g_0 and h_* stated above are also obvious.

Now we represent problem (7.59) and (7.60) in the form
(7.67) $$\begin{cases} u_{tt}(t) + \mathcal{A}u(t) + D(u_t(t)) = F(u(t)) + D^*(u_t(t)), \\ u|_{t=0} = u_0 \in \mathcal{D}(\mathcal{A}^{1/2}), \ u_t|_{t=0} = u_1 \in H, \end{cases}$$
with
- $\mathcal{H} \equiv L_2(\Omega)$.

- $\mathcal{A}u \equiv -(\Delta + \mu_0)u$, $u \in \mathcal{D}(\mathcal{A})$: $\mathcal{D}(\mathcal{A}) \equiv H_0^1(\Omega) \cap H^2(\Omega)$. The parameter μ_0 is chosen such that $-\mu < \mu_0 < \lambda_1$, where μ and λ_1 are the same as in (7.64).
- $F(u) = -f(u) - \mu_0 u$, $D^*(v) = h_*(v)$.
- $D(u) \equiv g_0(u) - \mathrm{div}\,[g(\nabla u)]$.

We note that in the case considered

$$F(u,v) = -\Pi'(u) + D^*(v) \quad \text{with} \quad \Pi(u) = \int_\Omega \hat{f}(u)dx,$$

where $\hat{f}$ denotes the antiderivative of $f(u) + \mu_0 u$.

As above, the existence of semiflow easily follows from Theorem 1.5.

The further arguments are divided in several steps.

Step 1: Dissipativity. As in Example 3.19, to prove dissipativity we use Corollary 3.17. Since $G^*(u) \equiv 0$ in our case, we need to check relations (3.51) and (3.52) only.

We obviously have that

$$(-D(v) + D^*(v), u) = -(g_0(v), u) - (g(\nabla v), \nabla u) + (h_*(v), u).$$

As in the proof of Theorem 5.3 one can see that

$$-(g_0(v), u) \leq \delta \|u\|_{H^1(\Omega)}^2 + C_1^\delta + C_2^\delta \cdot (1 + \|u\|_{H^1(\Omega)}) \cdot \int_\Omega g_0(v)v\,dx$$

for any $\delta > 0$ (cf. (5.12)). Since g_i' and h_*' are bounded, we obtain that

$$-(g(\nabla v), \nabla u) + (h_*(v), u) \leq \delta \|\nabla u\|_{L_2(\Omega)}^2 + C_\delta \|\nabla v\|_{L_2(\Omega)}^2$$

for any $\delta > 0$. From (7.61) we have that

$$\|\nabla v\|_{L_2(\Omega)}^2 \leq C_1(g(\nabla v), \nabla v) + C_2.$$

Therefore the relations above imply (3.51).

Now we check (3.52). From (7.65) we have that

$$\limsup_{|s| \to \infty} \frac{|h_*(s)|}{|s|} < \omega = \lambda_1 \min\{\omega_1, \omega_2\}.$$

Thus there exits $0 < \eta < 1$ and $C_\eta > 0$ such that

$$|s| \cdot |h_*(s)| \leq \eta^2 \omega \cdot |s|^2 + C_\eta, \quad s \in \mathbb{R}.$$

Consequently,

$$(7.68) \qquad (v, D^*(v)) = (v, h_*(v)) \leq \frac{\eta^2 \omega}{\lambda_1} \cdot \|\nabla v\|_{L_2(\Omega)}^2 + C.$$

By (7.61) we have that

$$s^2 \leq \frac{1}{\eta \omega_i} \cdot s g_i(s) + C_\eta, \quad s \in \mathbb{R},$$

for every $0 < \eta < 1$, which implies that

$$\|\nabla v\|_{L_2(\Omega)}^2 \leq \frac{1}{\eta} \cdot \max\left\{\frac{1}{\omega_1}, \frac{1}{\omega_2}\right\} \cdot (g(\nabla v), \nabla v) + C.$$

Therefore by (7.68) we obtain that

$$(v, D^*(v)) \leq \eta (g(\nabla v), \nabla v) + C.$$

Thus (3.52) holds and therefore the semiflow S_t is dissipative.

Step 2: Existence of a global attractor. We consider the case $n = 3$ only. The case $n \leq 2$ is much simpler. We rely on Remark 3.32 and therefore we need to to check Assumption 3.21(D) and property (3.89) in Remark 3.32.

Relation (3.59) can be established by the same method as in Section 6.4. As for (3.60) we note that, since $g_i'(s_i)$ are bounded, we have that

$$|(D(u+v) - D(u), w)| \leq \int_\Omega |g_0(u+v) - g_0(u)| \cdot |w| dx + C\|\nabla v\|_{L_2} \|\nabla w\|_{L_2}.$$

From (5.15) we have

$$\int_\Omega |g_0(u+v) - g_0(u)| \cdot |w| dx \leq \|w\|_{L_{1+1/p}} \left| \int_\Omega |(g(u+v) - g(u)|^{p+1} dx \right|^{1/(p+1)}$$

$$\leq C\|w\|_{H^{1-2\delta}} \left(1 + \int_\Omega [g_0(u+v)(u+v) + g_0(u)u] \, dx \right)$$

$$\leq C|\mathcal{A}^{1/2-\delta} w| \left(1 + (D(u+v), u+v) + (D(u), u) \right)$$

for some $\delta > 0$. Since $\inf_{s \in \mathbb{R}} g_i'(s) > 0$ for $i = 1, 2$, it is also easy to see that

$$C\|\nabla v\|_{L_2} \|\nabla w\|_{L_2} \leq \epsilon |\mathcal{A}^{1/2} w|^2 + C_\epsilon (D(u+v) - (D(u), v)$$

for every $\epsilon > 0$. Thus (3.60) holds with $\kappa = 1$.

Now we prove (3.89). Since $q < 2$ and $h_*(s)$ is globally Lipschits, we can see that

$$(7.69) \quad |(F(u,v) - F(\hat{u}, \hat{v}), v - \hat{v})| \leq C_1(r) |\mathcal{A}^{1/2-\eta}(u - \hat{u})|^2 + C_2 \|v - \hat{v}\|_{L_2}^2,$$

where $\eta > 0$, $C(r) > 0$ is non-decreasing function of r, $C_2 > 0$ does not depend on r, and $u, \hat{u} \in \mathcal{D}(\mathcal{A}^{1/2})$, $v, \hat{v} \in V$ satisfy the relations $|\mathcal{A}^{1/2} u| \leq r$ and $|\mathcal{A}^{1/2} \hat{u}| \leq r$. We obviously have

$$\|v - \hat{v}\|_{L_2}^2 = (\mathcal{A}^{1/2}(v - \hat{v}), \mathcal{A}^{-1/2}(v - \hat{v})) \leq \epsilon \|\nabla(v - \hat{v})\|_{L_2}^2 + C_\epsilon |\mathcal{A}^{-1/2}(v - \hat{v})|^2.$$

Therefore (7.69) implies that

$$(7.70) \quad \begin{aligned} |(F(u,v) - F(\hat{u}, \hat{v}), v - \hat{v})| &\leq \frac{1}{2} \int_\Omega g(\nabla(v - \hat{v})) \nabla(v - \hat{v}) dx \\ &\quad + C(r) \left(|\mathcal{A}^{1/2-\eta}(u - \hat{u})|^2 + |\mathcal{A}^{-1/2}(v - \hat{v})|^2 \right), \end{aligned}$$

and hence (3.89) holds.

Thus by Remark 3.32 the semiflow S_t generated by (7.59) and (7.60) possesses a global compact attractor A.

Step 3: Finite dimension. In the proof of finite dimensionality of the attractor A we rely on Theorem 4.4. To apply this theorem we need to establish relations (4.2) and (4.12).

We claim that (4.2) holds with $l = 1$. Indeed, since $g_i(s_i)$ are linear functions, using estimate (7.66) for g_0, we obtain that

$$|\mathcal{A}^{-1}(D(u+v) - D(v))| \leq \int_\Omega |g_0(u+v) - g_0(v)| dx + C\|u\|_{L_2}$$

$$\leq C_1 \int_\Omega \left(1 + |v|^{p-1} + |u|^{p-1} \right) \cdot |u| dx + C_1 \|u\|_{L_2}.$$

Since $p \leq 2$, it is easy to see that

$$|\mathcal{A}^{-1}(D(u+v) - D(v))| \leq C \left(1 + \|v\|_{L_2} + \|u\|_{L_2} \right) \cdot \|u\|_{L_2},$$

which implies (4.2) with $l = 1$.

To prove (4.12) we note that by Remark 3.32 this relation follows from estimate (7.70) and Assumption 3.21(D) with $\kappa = 2$ which can be easily checked under the condition $p \leq 2$. □

EXAMPLE 7.19. In a smooth bounded domain $\Omega \subset \mathbb{R}^n$, $n \leq 3$, we consider the following version of problem (7.59) and (7.60):

$$(7.71) \quad w_{tt} + (|w_t|^{p-1} - \lambda) \cdot w_t - \varepsilon \text{div}\,[g(\nabla w_t)] - \Delta w + f(w) = 0, \quad x \in \Omega, \, t > 0,$$

subject to the Dirichlet boundary condition (7.60). Here $1 < p < 5$, $\lambda \in \mathbb{R}$, $\varepsilon \geq 0$ (in the case $p = 1$ we assume that $\lambda \leq c_0 \varepsilon$, where $c_0 > 0$ is a constant determined by g). The function $f(w)$ satisfies the hypotheses in Assumption 7.16. The vector function $g(s) = (g_1(s_1), g_2(s_2))$ is continuously differentiable and possesses the properties $g_i(0) = 0$ and $0 < m \leq g_i'(s) \leq M < \infty$.

For all $\varepsilon \geq 0$ the system (H, S_t) generates by (7.71) and (7.60) is dissipative (see Example 3.19 and also [104] for $\varepsilon = 0$ and $p = 3$). It possesses a compact global attractor for *any* positive ε. If g is linear and $0 \leq p \leq 2$, then Theorem 7.18 asserts finite dimensionality of the attractor. Thus, introducing strong damping (even with a small intensity) in the wave equation with non-monotone damping stabilizes the system to a compact global attractor.

REMARK 7.20. In order to conclude finite dimensionality of the attractor in example 7.19, it is not necessary to assume linearity of g. The method of proof of Theorem 4.4 will give the same conclusion with a more general assumption $0 < m \leq g_i'(s) \leq M < \infty$. This can be achieved by exploiting additionally smoothing effects of strong damping which leads to a generalization of condition (4.2). Since the emphasis in this paper is on hyperbolic like dynamics, which do not exhibit smoothing effects, we did not strive in our presentation for the most general formulations accounting also for parabolic-like effects. Nevertheless, even in that case we are able to demonstrate that the theory presented yields new results in parabolic-like situations as well.

The results similar to Theorem 7.18 can be also stated for problems (7.57) and (7.58). We do not give details here, since it is more natural to study strongly damped problems directly by taking advantage of additional regularity provided by the strong damping terms.

Bibliography

[1] J. Arrietta and A.N. Carvalho and J.Hale, *A damped hyperbolic equations with critical exponents*, Commun. in Partial Diff. Eqs **17** (1992), 841–866.
[2] A. Babin and M. Vishik, *Regular attractors of semigroups and evolution equations*, J. Math. Pures et Appl. **62** (1983), 441–491.
[3] A. Babin and M. Vishik, *Maximal attractors of semigroups corresponding to evolution differential equations*, Math. USSR Sbornik **54** (1986), 387–408.
[4] A. Babin and M. Vishik, *Attractors of Evolution Equations*, North-Holland, Amsterdam, 1992.
[5] J. Ball, *Global attractors for semilinear wave equations*, Discr. Cont. Dyn. Sys. **10** (2004), 31–52.
[6] V. Barbu, *Nonlinear Semigroups and Differential Equations in Banach Spaces*. Noordhoff, 1976.
[7] M. Berger, *A new approach to the large deflection of plate*, J. Appl. Mech. **22** (1955), 465–472.
[8] V.V. Bolotin, *Nonconservative Problems of Elastic Stability*, Pergamon Press, Oxford, 1963.
[9] F. Bucci, I. Chueshov and I. Lasiecka, *Global attractor for a composite system of nonlinear wave and plate equations*, Commun. Pure Appl. Anal. **6** (2007), 113-140.
[10] H. Cartan, *Calculus Différentielles*, Hermann, Paris, 1967.
[11] A. Carvalho and J. Cholewa, *Attractors for strongly damped wave equations with critical nonlinearities*, Pacific J. Math. **207** (2002), 287-310.
[12] S. Ceron and O. Lopes, *α-contractions and attractors for dissipative semilinear hyperbolic equations and systems*, Ann. Math. Pura Appl. IV **160** (1991), 193–206.
[13] I. Chueshov, *Finite-dimensionality of the attractor in some problems of the nonlinear theory shells*, Math. USSR Sbornik **61** (1988), 411–420.
[14] I. Chueshov, *Structure of a maximal attractor of a modified system of von Karman equations*, J. of Soviet Mathematics **48** (1990), 692–696.
[15] I. Chueshov, *Strong solutions and attractor of a system of von Karman equations*, Math. USSR Sbornik **69** (1991), 25–36.
[16] I Chueshov, *Regularity of solutions and approximate inertial manifolds for von Karman evolution equations*, Math. Meth. in Appl. Sci., **17** (1994), 667–680.
[17] I. Chueshov, *On the finiteness of the number of determining elements for von Karman evolution equations*, Math. Meth. in the Appl. Sci., **20** (1997), 855–865.
[18] I.D Chueshov, *Theory of functionals that uniquely determine asymptotic dynamics of infinite-dimensional dissipative systems*, Russian Math. Surveys **53** (1998), 731–776.
[19] I.D. Chueshov, *Introduction to the Theory of Infinite-Dimensional Dissipative Systems*, Acta, Kharkov, 1999, in Russian; English translation: Acta, Kharkov, 2002; see also http://www.emis.de/monographs/Chueshov/
[20] I.D. Chueshov, M. Eller and I. Lasiecka, *On the attractor for a semilinear wave equation with critical exponent and nonlinear boundary dissipation*, Commun. in Partial Diff. Eqs **27** (2002), 1901–1951.
[21] I.D. Chueshov, M. Eller and I. Lasiecka, *Attractors and their structure for semilinear wave equations with nonlinear boundary dissipation*, Bol. Soc. Paran. Mat. **22** (2004), 38–57.
[22] I.D. Chueshov, M. Eller and I. Lasiecka, *Finite dimensionality of the attractor for a semilinear wave equation with nonlinear boundary dissipation*, Commun. in Partial Diff. Eqs., **29** (2004), 1847–1876.
[23] I. Chueshov and V. Kalantarov, *Determining functionals for nonlinear damped wave equations*, Matem. Fizika, Analyz, Geometriya **8** (2001), 215–227.

[24] I. Chueshov and I. Lasiecka, *Inertial manifolds for von Karman plate equations*, Appl. Math. Optim. (special issue dedicated to J. L. Lions) **46** (2002), 179–207.

[25] I. Chueshov and I. Lasiecka, *Determining functionals for a class of second order in time evolution equations with applications to von Karman equations*, In: Analysis and Optimization of Differential Systems, V.Barbu et al. (eds), Kluwer, Boston-Dordrecht-London, 2003, pp.109–122.

[26] I. Chueshov and I. Lasiecka, *Global attractors for von Karman evolutions with a nonlinear boundary dissipation*, J. Diff. Eqs. **198** (2004), 196–231.

[27] I. Chueshov and I. Lasiecka, *Attractors for second order evolution equations with a nonlinear damping*, J. Dyn. Diff. Eq., **16** (2004), 469–512.

[28] I. Chueshov and I. Lasiecka, *Kolmogorov's ε-entropy for a class of invariant sets and dimension of global attractors for second order in time evolution equations with nonlinear damping*, In: Control Theory of Partial Differential Equations, O.Imanuvilov et al. (eds). A Series of Lectures in Pure and Applied Mathematics, vol. 242, Chapman & Hall/CRC, Boca Raton, 2005, 51–69.

[29] I. Chueshov and I. Lasiecka, *Existence, uniqueness of weak solutions and global attractors for a class of nonlinear 2D Kirchhoff-Boussinesq models*, Discr. Cont. Dyn. Sys. **15** (2006), 777–809.

[30] I. Chueshov and I. Lasiecka, *Global attractors for Mindlin–Timoshenko plates and for their Kirchhoff limits*, Milan J. Math., **74** (2006), 117–138.

[31] I. Chueshov and I. Lasiecka, *Long-time dynamics of von Karman semi-flows with nonlinear boundary/interior damping*, J. Diff. Eqs. **233** (2007), 42-86.

[32] I. Chueshov and I. Lasiecka, *Long-time dynamics of wave equation with nonlinear interior/boundary damping and sources of critical exponents*, AMS Contemporary Mathematics, 2007, in press.

[33] I. Chueshov, I. Lasiecka and D. Toundukov, *Long-term dynamics of semilinear wave equation with nonlinear localized interior damping and a source term of critical exponent*, preprint, 2007.

[34] I. Chueshov and S. Siegmund, *On dimension and metric properties of trajectory attractors*, J. Dyn. Diff. Eq., **17** (2005), 621–641.

[35] B. Cockburn, D.A. Jones and E.S. Titi, *Determining degrees of freedom for nonlinear dissipative systems*, C.R. Acad. Sci. Paris Ser.I **321** (1995), 563–568.

[36] B. Cockburn, D. A. Jones and E. Titi, *Estimating the number of asymptotic degrees of freedom for nonlinear dissipative systems*, Math. Comp. **66** (1997), 1073–1087.

[37] P. Constantin, C. Doering and E. Titi, *Rigorous estimates of small scales in turbulent flows*, J. Math. Physics, **37** (1996), 6152–6156.

[38] E.H. Dowell, *Aeroelasticity of Plates and Shells*, Noordhoff International Publishing, Leyden, 1975.

[39] A. Eden, C. Foias, B. Nicolaenko and R. Temam, *Exponential Attractors for Dissipative Evolution Equations*, Research in Appl. Math. 37, Masson, Paris 1994.

[40] A. Eden and V. Kalantarov, *Finite dimensional attractors for a class of semilinear wave equations*, Turkish J. Math. **20** (1996), 425–450.

[41] A. Eden, A. Milani and B. Nicolaenko, *Finite dimensional exponential attractors for semilinear wave equations with damping*, J. Math. Anal. Appl. **169** (1992), 408–419.

[42] A. Eden, A. Milani and B. Nicolaenko, *Local exponential attractors for models of phase change for compressible gas dynamics*, Nonlinearity **6** (1993), 93–117.

[43] M. Efendiev, A. Miranville and S. Zelik, *Exponential attractors for nonlinear reaction-diffusion systems in $\mathbb{R}^n$*, C.R. Acad. Sci. Paris, Ser. I **330** (2000), 713–718.

[44] P. Fabrie, C. Galusinski and A. Miranville, *Uniform inertial sets for damped wave equations*, Discr. Cont. Dyn. Sys. **6** (2000), 393–418.

[45] P. Fabrie, C. Galusinski, A. Miranville and S. Zelik, *Uniform exponential attractors for singularly perturbed damped wave equation*, Discr. Cont. Dyn. Sys. **10** (2004), 211–238.

[46] K. Falconer, *Fractal Geometry: Mathematical Foundations and Applications*, Wiley, Chichester, 1990.

[47] A. Favini, M. Horn, I. Lasiecka and D. Tataru, *Global existence, uniqueness and regularity of solutions to a von Karman system with nonlinear boundary dissipation*, Diff. and Int. Eqs. **9** (1996), 287–294, and *Addendum to this paper* Diff. and Int. Eqs. **10** (1997), 197–201.

[48] E. Feireisl, *Attractors for wave equations with nonlinear dissipation and critical exponents*, C.R.Acad.Sc. Paris, Ser. I **315** (1992), 551–555.
[49] E. Feireisl, *Finite dimensional asymptotic behaviour of some semilinear damped hyperbolic problems*, J. Dyn. Diff. Eqs **6** (1994), 23–35.
[50] E. Feireisl, *Global attractors for damped wave equations with supercritical exponent*, J. Diff. Eqs **116** (1995), 431–447.
[51] E. Feireisl and E. Zuazua, *Global attractors for for semilinear wave equation with locally distributed nonlinear damping and critical exponent*, Commun. in Partial Diff. Eqs **18** (1993), 1538–1555.
[52] C. Foias, O. Manley, R.Temam, and Y.M. Treve, *Asymptotic analysis of the Navier-Stokes equations*, Physica D **9** (1983), 157–188.
[53] C. Foias and E. Olson, *Finite fractal dimension and Hölder-Lipschitz parametrization*, Indiana Univ. Math. J. **45** (1996), 603–616.
[54] C. Foias and G. Prodi, *Sur le comportement global des solutions nonstationnaires des équations de Navier-Stokes en dimension deux*, Rend. Sem. Mat. Univ. Padova **39** (1967), 1–34.
[55] C. Foias and R. Temam, *Determination of solutions of the Navier-Stokes equations by a set of nodal values*, Math. Comput. **43** (1984), 117–133.
[56] C. Foias and E. Titi, *Determining modes, finite difference schemes and inertial manifolds*, Nonlinearity, **4** (1991), 135–153.
[57] J.M. Ghidaglia and R. Temam, *Regularity of the solutions of second order evolution equations and their attractors*, Ann. della Scuola Norm. Sup. Pisa **14** (1987), 485–511.
[58] J.M. Ghidaglia and R. Temam, *Attractors of damped nonlinear hyperbolic equations*, J. Math. Pure et Appl. **66** (1987), 273–319.
[59] J.K. Hale, *Asymptotic Behavior of Dissipative Systems*. Amer. Math. Soc., Providence, RI, 1988.
[60] J. K. Hale and G. Raugel, *Attractors for Dissipative Evolutionary Equations*, International Conference on Differential Equations, River Edge NJ, vol.1, World Science Publishing, 1993, pp. 3–22.
[61] A Haraux, *Semilinear Hyperbolic Problems in Bounded Domains*, Mathematical Reports, vol.3, Harwood Gordon Breach, New York, 1987.
[62] A Haraux, *Two remarks on on dissipative hyperbolic problems*, Seminaire de College de France, J.L. Lions (ed.), Pitman, Boston, 1985.
[63] A.A. Il'ushin, *The plane sections law in aerodynamics of large supersonic speeds*, Prikladnaya Matem. i Mech. **20** (1956), 733–755, in Russian.
[64] D.A. Jones and E. Titi, *Determination of the solutions of the Navier-Stokes equations by finite volume elements*, Physica D **60** (1992), 165–174.
[65] D.A. Jones and E. Titi, *Upper bounds on the number of determining modes, nodes and volume elements for the Navier-Stokes equations*, Indiana Univ. Math. J. **42** (1993), 875–887.
[66] A. K. Khanmamedov, *Global attractors for von Karman equations with nonlinear dissipation*, J. Math. Anal. Appl. **318** (2006), 92–101.
[67] A. K. Khanmamedov, *Finite dimensionality of the global attractors to von Karman equations with nonlinear interior dissipation*, Nonlinear Analysis **66** (2007), 204–213.
[68] A. N. Kolmogorov and V.M. Tihomirov, ε-*entropy and* ε-*capacity of sets in functional spaces*, Uspehi Mat. Nauk **14** (1959), 3–86, in Russian.
[69] I. N. Kostin, *Rate of attraction to a non-hyperbolic attractor*, Asymptotic Analysis **16** (1998), 203–222.
[70] O. Ladyzhenskaya, *A dynamical system generated by the Navier–Stokes equations*, J. Soviet Math. **3** (1975), 458–479.
[71] O. Ladyzhenskaya, *Finite dimensionality of bounded invariant sets for Navier-Stokes systems and other dissipative systems*, J. Soviet Math. **28** (1985), 714–726.
[72] O. Ladyzhenskaya, *Estimates for the fractal dimension and number of deterministic modes for invariant sets of dynamical systems*, J. Soviet Math. **49** (1990), 1186–1201.
[73] O. Ladyzhenskaya, *Attractors for Semigroups and Evolution Equations*, Cambridge University Press, Cambridge, 1991.
[74] J. Lagnese, *Boundary Stabilization of Thin Plates*, SIAM, 1989.
[75] J. Lagnese and J.L. Lions, *Modeling, Analysis and Control of Thin Plates*, Collection RMA, Masson, Paris, 1988.

[76] I. Lasiecka, *Existence and uniqueness of solutions to second order nonlinear and nonmonotone boundary conditions*, Nonlin. Anal., TMA, **24** (1994), 797–823.

[77] I. Lasiecka, *Finite dimensionality and compactness of attractors for von Karman equations with nonlinear dissipation*, Nonlinear Differential Equations **6** (1999), 437–472.

[78] I. Lasiecka, *Mathematical Control Theory of Coupled PDE's*, CMBS-NSF Lecture Notes, SIAM Publications, 2001.

[79] I. Lasiecka and W. Heyman, *Asymptotic behaviour of solutions in nonlinear dynamic elasticity*, Discr. Cont. Dyn. Sys. **1** (1995), 237–252.

[80] I. Lasiecka and D. Tataru, *Uniform boundary stabilization of semilinear wave equation with nonlinear boundary dissipation*, Diff. Integral Eqs. **6** (1993), 507–533.

[81] I. Lasiecka and A. Ruzmaikina, *Finite dimensionality and regularity of attractors for a 2-D semilinear wave equation with nonlinear dissipation*, J. Math. Anal. Appl. **270** (2002), 16–50.

[82] I. Lasiecka and R. Triggiani, *Control Theory for Partial Differential Equations*, Cambridge University Press, Cambridge, 2000.

[83] J.L. Lions, *Quelques Méthodes de Résolution des Problèmes aux Limites Non Linéaires*, Dunod, Paris, 1969.

[84] J.Málek and J. Nečas, *A finite dimensional attractor for three dimensional flow of incompressible fluids*, J. Diff. Eqs. **127** (1996), 498–518.

[85] J.Málek and D. Pražak, *Large time behavior via the method of l-trajectories*, J. Diff. Eqs. **181** (2002), 243–279.

[86] V. Melnik and J. Valero, *On attractors of multivalued semi-flows and differential inclusions*, Set-Valued Anal. **6** (1998), 83–111.

[87] B. Nicolaenko and W. Qian, *Inertial manifolds for nonlinear viscoelasticity equations*, Nonlinearity **11** (1998), 1075–1093.

[88] V. Pata and M. Squassina, *On the strongly damped wave equation*, Commun. Math. Phys. **253** (2005), 511-533.

[89] D. Pražak, *On finite fractal dimension of the global attractor for the wave equation with nonlinear damping*, J. Dyn. Diff. Eqs. **14** (2002), 764–776.

[90] G. Raugel, *Une equation des ondes avec amortissment non lineaire dans le cas critique en dimensions trois*, C.R. Acad.Sci Paris, Ser.I **314** (1992), 177–182.

[91] G. Raugel, *Global attractors in partial differential equations*, In: Handbook of Dynamical Systems, Vol.2, B. Fiedler (ed.) Elsevier, Amsterdam, 2002, pp.885–982.

[92] G.R. Sell and Y. You, *Dynamics of Evolutionary Equations*, Springer, New York, 2002.

[93] M. Sermange and R. Temam, *Some mathematical questions related to MHD equations*, Commun. Pure Appl. Math. **36** (1983), 635–664.

[94] Z. Shengfan, *Dimension of the global attractor for damped nonlinear wave equation*, Proceedings AMS **127** (1999), 3623–3631.

[95] R. Showalter, *Monotone Operators in Banach Spaces and Nonlinear Partial Differential Equations*, AMS Providence, 1997.

[96] J. Simon, *Compact sets in the space $L^p(0,T;B)$*, Annali di Matematica Pura ed Applicata, Ser.4 **148** (1987), 65–96.

[97] M. Slimrod, *Dynamics of phase transition in a Van der Waals fluid*, J. Diff. Eqs **52** (1984), 1–23.

[98] M. Slimrod, *A limiting 'viscosity' approach to the Riemann problem for materials exhibiting change of phase*, Arch. Ration. Mech. Anal. **105** (1989), 327–365.

[99] C. Sun, M. Yang and C. Zhong, *Global attractors for the wave equation with nonlinear damping*, J. Diff. Eqs **227** (2006), 427-443.

[100] L. Tartar, *Interpolation non-lineaire et regularite*, J. Funct. Analysis **9** (1972), 469–489.

[101] R. Temam, *Infinite Dimensional Dynamical Systems in Mechanics and Physics*, Springer, Berlin-Heidelberg-New York, 1988.

[102] H. Triebel, *Interpolation Theory, Functional Spaces and Differential Operators*, North Holland, Amsterdam, 1978.

[103] V.V. Varlamov, *On the damped Boussinesq equation in a circle*, Nonlinear Analysis **38** (1999), 447–470.

[104] Y. You, *Global dynamics of nonlinear wave equations with cubic non-monotone damping*, Dynamics of Partial Diff. Eqs. **1** (2004), 65–87.

[105] S. Zelik, *Asymptotic regularity of solutions of singularly perturbed damped wave equtions with supercritical nonlinearities* Discr. Cont. Dyn. Sys. **11** (2004), 351–392.

Index

(ε, ϱ)-distinguishable subset, 22
ε-capacity, 23

absorbing set, 17
Airy stress function, 140
attractor
   fractal exponential, 122
   global, 17
   regular structure, 113
   smoothness, 103

beam equation, 158
Berger plate model, 158
Boussinesq equation, 167

compact seminorm, 23
completeness defect, 120
conservative force, 5

damped wave equation, 5, 125
determining functionals, 120
dynamical system, 17
   asymptotically smooth, 17
   dissipative, 17
   gradient, 32

energy inequality, 10
energy relation, 9
evolution semigroup, 17

fractal dimension, 22

generalized solution, 9

Hausdorff semidistance, 17

inertial set, 122

Karman bracket, 140
Karman evolution equations, 7, 140
Kirchhoff limit, 167
Kirchhoff plate model, 167
Kirchhoff-Boussinesq equation, 167
Kuratowski α-measure, 18

lower semicontinuous function, 13
Lyapunov function, 32

strict, 32

Mindlin-Timoshenko model, 164
modified von Karman equations, 7
Morse decomposition, 34

non-conservative force, 5

operator
   accretive, 10
   demicontinuous, 15
   hemicontinuous, 4
   locally Lipschitz, 4
   monotone, 4

phase space, 17
precompact pseudometric, 18, 20

radius of dissipativity, 17

spaces
   $C(a, b; H)$, 9
   $L_p(a, b; H)$, 9
   $W_p^1(a, b; H)$, 9
strong damping, 173
strong solution, 9

unstable manifold, 32

Editorial Information

To be published in the *Memoirs*, a paper must be correct, new, nontrivial, and significant. Further, it must be well written and of interest to a substantial number of mathematicians. Piecemeal results, such as an inconclusive step toward an unproved major theorem or a minor variation on a known result, are in general not acceptable for publication.

Papers appearing in *Memoirs* are generally at least 80 and not more than 200 published pages in length. Papers less than 80 or more than 200 published pages require the approval of the Managing Editor of the Transactions/Memoirs Editorial Board.

As of May 31, 2008, the backlog for this journal was approximately 17 volumes. This estimate is the result of dividing the number of manuscripts for this journal in the Providence office that have not yet gone to the printer on the above date by the average number of monographs per volume over the previous twelve months, reduced by the number of volumes published in four months (the time necessary for preparing a volume for the printer). (There are 6 volumes per year, each usually containing at least 4 numbers.)

A Consent to Publish and Copyright Agreement is required before a paper will be published in the *Memoirs*. After a paper is accepted for publication, the Providence office will send a Consent to Publish and Copyright Agreement to all authors of the paper. By submitting a paper to the *Memoirs*, authors certify that the results have not been submitted to nor are they under consideration for publication by another journal, conference proceedings, or similar publication.

Information for Authors

Memoirs are printed from camera copy fully prepared by the author. This means that the finished book will look exactly like the copy submitted.

Initial submission. The AMS uses Centralized Manuscript Processing for initial submissions. Authors should submit a PDF file using the Initial Manuscript Submission form found at www.ams.org/cgi-bin/peertrack/submission.pl, or send one copy of the manuscript to the following address: Centralized Manuscript Processing, MEMOIRS OF THE AMS, 201 Charles Street, Providence, RI 02904-2294 USA. If a paper copy is being forwarded to the AMS, indicate that it is for it Memoirs and include the name of the corresponding author, contact information such as email address or mailing address, and the name of an appropriate Editor to review the paper (see the list of Editors below).

The paper must contain a *descriptive title* and an *abstract* that summarizes the article in language suitable for workers in the general field (algebra, analysis, etc.). The *descriptive title* should be short, but informative; useless or vague phrases such as "some remarks about" or "concerning" should be avoided. The *abstract* should be at least one complete sentence, and at most 300 words. Included with the footnotes to the paper should be the 2000 *Mathematics Subject Classification* representing the primary and secondary subjects of the article. The classifications are accessible from www.ams.org/msc/. The list of classifications is also available in print starting with the 1999 annual index of *Mathematical Reviews*. The Mathematics Subject Classification footnote may be followed by a list of *key words and phrases* describing the subject matter of the article and taken from it. Journal abbreviations used in bibliographies are listed in the latest *Mathematical Reviews* annual index. The series abbreviations are also accessible from www.ams.org/publications/. To help in preparing and verifying references, the AMS offers MR Lookup, a Reference Tool for Linking, at www.ams.org/mrlookup/.

Electronically prepared manuscripts. The AMS encourages electronically prepared manuscripts, with a strong preference for $\mathcal{AMS}$-LaTeX. To this end, the Society has prepared $\mathcal{AMS}$-LaTeX author packages for each AMS publication. Author packages include instructions for preparing electronic manuscripts, samples, and a style file that generates the particular design specifications of that publication series. Though $\mathcal{AMS}$-LaTeX is the highly preferred format of TeX, author packages are also available in $\mathcal{AMS}$-TeX.

Authors may retrieve an author package from the AMS website starting from www.ams.org/tex/ or via FTP to ftp.ams.org (login as anonymous, enter username as password, and type cd pub/author-info). The *AMS Author Handbook* and the *Instruction Manual* are available in PDF format following the author packages link from www.ams.org/tex/. The author package can also be obtained free of charge by sending email to tech-support@ams.org (Internet) or from the Publication Division, American Mathematical Society, 201 Charles St., Providence, RI 02904-2294, USA. When requesting an author package, please specify $\mathcal{AMS}$-LaTeX or $\mathcal{AMS}$-TeX and the publication in which your paper will appear. Please be sure to include your complete mailing address.

After acceptance. The final version of the electronic file should be sent to the Providence office (this includes any TeX source file, any graphics files, and the DVI or PostScript file) immediately after the paper has been accepted for publication.

Before sending the source file, be sure you have proofread your paper carefully. The files you send must be the EXACT files used to generate the proof copy that was accepted for publication. For all publications, authors are required to send a printed copy of their paper, which exactly matches the copy approved for publication, along with any graphics that will appear in the paper.

Accepted electronically prepared files can be submitted via the web at www.ams.org/submit-book-journal/, sent via FTP, or sent on CD-Rom or diskette to the Electronic Prepress Department, American Mathematical Society, 201 Charles Street, Providence, RI 02904-2294 USA. TeX source files, DVI files, and PostScript files can be transferred over the Internet by FTP to the Internet node ftp.ams.org (130.44.1.100). When sending a manuscript electronically via CD-Rom or diskette, please be sure to include a message identifying the paper as a Memoir.

Electronically prepared manuscripts can also be sent via email to pub-submit@ams.org (Internet). In order to send files via email, they must be encoded properly. (DVI files are binary and PostScript files tend to be very large.)

Electronic graphics. Comprehensive instructions on preparing graphics are available at www.ams.org/jourhtml/. A few of the major requirements are given here.

Submit files for graphics as EPS (Encapsulated PostScript) files. This includes graphics originated via a graphics application as well as scanned photographs or other computer-generated images. If this is not possible, TIFF files are acceptable as long as they can be opened in Adobe Photoshop or Illustrator. No matter what method was used to produce the graphic, it is necessary to provide a paper copy to the AMS.

Authors using graphics packages for the creation of electronic art should also avoid the use of any lines thinner than 0.5 points in width. Many graphics packages allow the user to specify a "hairline" for a very thin line. Hairlines often look acceptable when proofed on a typical laser printer. However, when produced on a high-resolution laser imagesetter, hairlines become nearly invisible and will be lost entirely in the final printing process.

Screens should be set to values between 15% and 85%. Screens which fall outside of this range are too light or too dark to print correctly. Variations of screens within a graphic should be no less than 10%.

Inquiries. Any inquiries concerning a paper that has been accepted for publication should be sent to memo-query@ams.org or directly to the Electronic Prepress Department, American Mathematical Society, 201 Charles St., Providence, RI 02904-2294 USA.

Editors

This journal is designed particularly for long research papers, normally at least 80 pages in length, and groups of cognate papers in pure and applied mathematics. Papers intended for publication in the *Memoirs* should be addressed to one of the following editors. The AMS uses Centralized Manuscript Processing for initial submissions to AMS journals. Authors should follow instructions listed on the Initial Submission page found at www.ams.org/memo/memosubmit.html.

Algebra to ALEXANDER KLESHCHEV, Department of Mathematics, University of Oregon, Eugene, OR 97403-1222; email: ams@noether.uoregon.edu

Algebraic geometry and its application to MINA TEICHER, Emmy Noether Research Institute for Mathematics, Bar-Ilan University, Ramat-Gan 52900, Israel; email: teicher@macs.biu.ac.il

Algebraic geometry to DAN ABRAMOVICH, Department of Mathematics, Brown University, Box 1917, Providence, RI 02912; email: amsedit@math.brown.edu

Algebraic topology to ALEJANDRO ADEM, Department of Mathematics, University of British Columbia, Room 121, 1984 Mathematics Road, Vancouver, British Columbia, Canada V6T 1Z2; email: adem@math.ubc.ca

Combinatorics to JOHN R. STEMBRIDGE, Department of Mathematics, University of Michigan, Ann Arbor, Michigan 48109-1109; email: FRS@umich.edu

Complex analysis and harmonic analysis to ALEXANDER NAGEL, Department of Mathematics, University of Wisconsin, 480 Lincoln Drive, Madison, WI 53706-1313; email: nagel@math.wisc.edu

Differential geometry and global analysis to LISA C. JEFFREY, Department of Mathematics, University of Toronto, 100 St. George St., Toronto, ON Canada M5S 3G3; email: jeffrey@math.toronto.edu

Dynamical systems and ergodic theory and complex anaysis to YUNPING JIANG, Department of Mathematics, CUNY Queens College and Graduate Center, 65-30 Kissena Blvd., Flushing, NY 11367; email: Yunping.Jiang@qc.cuny.edu

Functional analysis and operator algebras to DIMITRI SHLYAKHTENKO, Department of Mathematics, University of California, Los Angeles, CA 90095; email: shlyakht@math.ucla.edu

Geometric analysis to WILLIAM P. MINICOZZI II, Department of Mathematics, Johns Hopkins University, 3400 N. Charles St., Baltimore, MD 21218; email: trans@math.jhu.edu

Geometric analysis to MARK FEIGHN, Math Department, Rutgers University, Newark, NJ 07102; email: feighn@andromeda.rutgers.edu

Harmonic analysis, representation theory, and Lie theory to ROBERT J. STANTON, Department of Mathematics, The Ohio State University, 231 West 18th Avenue, Columbus, OH 43210-1174; email: stanton@math.ohio-state.edu

Logic to STEFFEN LEMPP, Department of Mathematics, University of Wisconsin, 480 Lincoln Drive, Madison, Wisconsin 53706-1388; email: lempp@math.wisc.edu

Number theory to JONATHAN ROGAWSKI, Department of Mathematics, University of California, Los Angeles, CA 90095; email: jonr@math.ucla.edu

Partial differential equations to GUSTAVO PONCE, Department of Mathematics, South Hall, Room 6607, University of California, Santa Barbara, CA 93106; email: ponce@math.ucsb.edu

Partial differential equations and dynamical systems to PETER POLACIK, School of Mathematics, University of Minnesota, Minneapolis, MN 55455; email: polacik@math.umn.edu

Probability and statistics to RICHARD BASS, Department of Mathematics, University of Connecticut, Storrs, CT 06269-3009; email: bass@math.uconn.edu

Real analysis and partial differential equations to DANIEL TATARU, Department of Mathematics, University of California, Berkeley, Berkeley, CA 94720; email: tataru@math.berkeley.edu

All other communications to the editors should be addressed to the Managing Editor, ROBERT GURALNICK, Department of Mathematics, University of Southern California, Los Angeles, CA 90089-1113; email: guralnic@math.usc.edu.

Titles in This Series

913 **Ethan Akin, Joseph Auslander, and Eli Glasner,** The topological dynamics of Ellis actions, 2008

912 **Igor Chueshov and Irena Lasiecka,** Long-time behavior of second order evolution equations with nonlinear damping, 2008

911 **John Locker,** Eigenvalues and completeness for regular and simply irregular two-point differential operators, 2008

910 **Joel Friedman,** A proof of Alon's second eigenvalue conjecture and related problems, 2008

909 **Cameron McA. Gordon and Ying-Qing Wu,** Toroidal Dehn fillings on hyperbolic 3-manifolds, 2008

908 **J.-L. Waldspurger,** L'endoscopie tordue n'est pas si tordue, 2008

907 **Yuanhua Wang and Fei Xu,** Spinor genera in characteristic 2, 2008

906 **Raphaël S. Ponge,** Heisenberg calculus and spectral theory of hypoelliptic operators on Heisenberg manifolds, 2008

905 **Dominic Verity,** Complicial sets characterising the simplicial nerves of strict ω-categories, 2008

904 **William M. Goldman and Eugene Z. Xia,** Rank one Higgs bundles and representations of fundamental groups of Riemann surfaces, 2008

903 **Gail Letzter,** Invariant differential operators for quantum symmetric spaces, 2008

902 **Bertrand Toën and Gabriele Vezzosi,** Homotopical algebraic geometry II: Geometric stacks and applications, 2008

901 **Ron Donagi and Tony Pantev (with an appendix by Dmitry Arinkin),** Torus fibrations, gerbes, and duality, 2008

900 **Wolfgang Bertram,** Differential geometry, Lie groups and symmetric spaces over general base fields and rings, 2008

899 **Piotr Hajłasz, Tadeusz Iwaniec, Jan Malý, and Jani Onninen,** Weakly differentiable mappings between manifolds, 2008

898 **John Rognes,** Galois extensions of structured ring spectra/Stably dualizable groups, 2008

897 **Michael I. Ganzburg,** Limit theorems of polynomial approximation with exponential weights, 2008

896 **Michael Kapovich, Bernhard Leeb, and John J. Millson,** The generalized triangle inequalities in symmetric spaces and buildings with applications to algebra, 2008

895 **Steffen Roch,** Finite sections of band-dominated operators, 2008

894 **Martin Dindoš,** Hardy spaces and potential theory on C^1 domains in Riemannian manifolds, 2008

893 **Tadeusz Iwaniec and Gaven Martin,** The Beltrami Equation, 2008

892 **Jim Agler, John Harland, and Benjamin J. Raphael,** Classical function theory, operator dilation theory, and machine computation on multiply-connected domains, 2008

891 **John H. Hubbard and Peter Papadopol,** Newton's method applied to two quadratic equations in $\mathbb{C}^2$ viewed as a global dynamical system, 2008

890 **Steven Dale Cutkosky,** Toroidalization of dominant morphisms of 3-folds, 2007

889 **Michael Sever,** Distribution solutions of nonlinear systems of conservation laws, 2007

888 **Roger Chalkley,** Basic global relative invariants for nonlinear differential equations, 2007

887 **Charlotte Wahl,** Noncommutative Maslov index and eta-forms, 2007

886 **Robert M. Guralnick and John Shareshian,** Symmetric and alternating groups as monodromy groups of Riemann surfaces I: Generic covers and covers with many branch points, 2007

885 **Jae Choon Cha,** The structure of the rational concordance group of knots, 2007

TITLES IN THIS SERIES

884 **Dan Haran, Moshe Jarden, and Florian Pop,** Projective group structures as absolute Galois structures with block approximation, 2007

883 **Apostolos Beligiannis and Idun Reiten,** Homological and homotopical aspects of torsion theories, 2007

882 **Lars Inge Hedberg and Yuri Netrusov,** An axiomatic approach to function spaces, spec tral synthesis and Luzin approximation, 2007

881 **Tao Mei,** Operator valued Hardy spaces, 2007

880 **Bruce C. Berndt, Geumlan Choi, Youn-Seo Choi, Heekyoung Hahn, Boon Pin Yeap, Ae Ja Yee, Hamza Yesilyurt, and Jinhee Yi,** Ramanujan's forty identities for Rogers-Ramanujan functions, 2007

879 **O. García-Prada, P. B. Gothen, and V. Muñoz,** Betti numbers of the moduli space of rank 3 parabolic Higgs bundles, 2007

878 **Alessandra Celletti and Luigi Chierchia,** KAM stability and celestial mechanics, 2007

877 **María J. Carro, José A. Raposo, and Javier Soria,** Recent developments in the theory of Lorentz spaces and weighted inequalities, 2007

876 **Gabriel Debs and Jean Saint Raymond,** Borel liftings of Borel sets: Some decidable and undecidable statements, 2007

875 **C. Krattenthaler and T. Rivoal,** Hypergéométrie et fonction zêta de Riemann, 2007

874 **Sonia Natale,** Semisolvability of semisimple Hopf algebras of low dimension, 2007

873 **A. J. Duncan,** Exponential genus problems in one-relator products of groups, 2007

872 **Anthony V. Geramita, Tadahito Harima, Juan C. Migliore, and Yong Su Shin,** The Hilbert function of a level algebra, 2007

871 **Pascal Auscher,** On necessary and sufficient conditions for L^p-estimates of Riesz transforms associated to elliptic operators on $\mathbb{R}^n$ and related estimates, 2007

870 **Takuro Mochizuki,** Asymptotic behaviour of tame harmonic bundles and an application to pure twistor D-modules, Part 2, 2007

869 **Takuro Mochizuki,** Asymptotic behaviour of tame harmonic bundles and an application to pure twistor D-modules, Part 1, 2007

868 **Gelu Popescu,** Entropy and multivariable interpolation, 2006

867 **Vilmos Totik,** Metric properties of harmonic measures, 2006

866 **William Craig,** Semigroups underlying first-order logic, 2006

865 **Nathanial P. Brown,** Invariant means and finite representation theory of $C*$-algebras, 2006

864 **John M. Lee,** Fredholm operators and Einstein metrics on conformally compact manifolds, 2006

863 **M. Lübke and A. Teleman,** The Universal Kobayashi-Hitchin correspondence on Hermitian manifolds, 2006

862 **Alberto Canonaco,** The Beilinson complex and canonical rings of irregular surfaces, 2006

861 **Leon A. Takhtajan and Lee-Peng Teo,** Weil-Petersson metric on the universal Teichmüller space, 2006

860 **Thomas M. Fiore,** Pseudo limits, biadjoints and pseudo algebras: Categorical foundations of conformal field theory, 2006

859 **N. Arcozzi, R. Rochberg, and E. Sawyer,** Carleson measures and interpolating sequences for Besov spaces on complex balls, 2006

858 **Enrico Valdinoci, Berardino Sciunzi, and Vasile Ovidiu Savin,** Flat level set regularity of p-Laplace phase transitions, 2006

For a complete list of titles in this series, visit the
AMS Bookstore at **www.ams.org/bookstore/**.